The Safety Officer's CONCISE DESK REFERENCE

Daniel Patrick O'Brien

Manager of Environmental Health and Safety for
Carbon Black Operations

Sid Richardson Carbon Company
Borger, Texas

CRC Press
Taylor & Francis Group
Boca Raton London New York

CRC Press is an imprint of the
Taylor & Francis Group, an **informa** business

CRC Press
Taylor & Francis Group
6000 Broken Sound Parkway NW, Suite 300
Boca Raton, FL 33487-2742

First issued in paperback 2019

© 2002 by Taylor & Francis Group, LLC
CRC Press is an imprint of Taylor & Francis Group, an Informa business

No claim to original U.S. Government works

ISBN-13: 978-1-56670-407-6 (hbk)
ISBN-13: 978-0-367-39704-3 (pbk)

Library of Congress Cataloging-in-Publication Data

O'Brien, Daniel Patrick, 1955–
    The safety officer's concise desk reference / by Daniel Patrick O'Brien
        p. cm.
    ISBN 1-56670-407-3 (alk. paper)
    1. Industrial safety — Handbooks, manuals, etc. I. Title.

T55 .O24 2001
658.3'82—dc21
                                     2001029454
                                       CIP

Library of Congress Card Number 2001029454

Visit the Taylor & Francis Web site at
http://www.taylorandfrancis.com

and the CRC Press Web site at
http://www.crcpress.com

Dedication

This book is dedicated to all the practitioners who take on the noble responsibility of protecting lives. Some have "safety" in their title, some have "industrial hygiene" in theirs, and some don't even have a title, just the responsibility to protect their fellow workers. Regardless of what your title may say, if your livelihood revolves around protecting others from needless injury, I hope this book helps you do your job a lot better. There are other worthy professions; many are very important, but none so noble as that of protecting lives.

Preface

This book is not intended to serve as the *only* Safety, Health, and Industrial Hygiene reference a safety office requires. The intention of this book is to serve as a quick reference for valuable data. As each chapter was written, it became obvious that an entire book could have been written on any one of the chapters and still only scratched the surface of any given topic. Readers will likely note obvious omissions. That's good. The hope is that this book will have multiple editions in the future, each time becoming more and more of an essential desk reference.

Many fundamental pieces of information that the average safety professional should have on hand are included. Most every piece of data included in this book is probably already in the reader's library, note pads, or seminar manuals. The intention is to pull those most important points together in one easy reference source. Please let the publisher know what specific information would make future editions of this book even more valuable.

In many cases, only an example or a limited section of a topic is provided because inclusion of a thorough discussion on any one of the many subjects addressed here would immediately take over the entire book. The intention is to provide a template, example, or most important sections of a wider array of topics. Thus, customization is left to the reader, with the hope that he or she has been given tools to build systems and programs that include the critical elements and examples to transform easily into exactly what individual readers need.

This is not a continual-read book either. While it was not intended for that, it should be a text that a safety professional keeps close at hand to supply enough information to point him or her in the right direction.

The Author

Daniel Patrick O'Brien, CSP, is a Certified Safety Professional, and is currently Manager of Environmental, Health and Safety for Carbon Black Operations with Sid Richardson Carbon Company in Borger, Texas. He is a professional member of the ASSE and is past president of the Panhandle chapter. He holds a bachelor of science in industrial education and a master of science in industrial technology, both from West Texas State University.

Mr. O'Brien has served more than 8 years as Secretary of the North American Product Safety and Regulatory Committee for the International Carbon Black Association. He serves on the Industrial Hygiene, TLV, Environmental, and Hazcom subcommittees.

He is an adjunct professor at West Texas A&M University. Mr. O'Brien has authored numerous safety- and nonsafety-oriented articles in industry journals and trade publications and has presented training seminars internationally. His previous book was titled *Business Measurements for Safety Performance,* published by CRC Press in 1999.

Contents

1 Introduction

1.1 THE SAFETY CONTINUUM

A brief look at the big picture may be of some value. It is important to understand where an organization is in regard to safety. Likewise, it is important to understand why it reacts and handles safety-oriented issues as it does. The following paragraphs provide a better feel for where an organization is and why the people in it respond as they do to safety and health needs and concerns.

There are safety systems in every stage of development and implementation — some in the elementary stages of the safety culture and some on the cutting edge of safety thinking. Where an individual or an organization is on this continuum between elementary thinking and advanced thinking determines why and to what extent it participates in the safety system.

The most rudimentary position on the safety system continuum is that the individual or organization is totally *unaware* that a safety culture even exists. In this state, evaluation of the safety aspects of an operation is nonexistent. Tasks are performed with no thought or consideration given to how safety fits into the operation. In this stage, employee injuries are considered as part of the operation. No consideration is given to proactive approaches in preventing accidents. What little consideration has been given to safety quickly returns the verdict of too costly, too much trouble, or simply not worth it.

Advancing on the safety continuum, but still in the elementary stages, is an individual or organization that functions out of *responsibility*. That is, the individual or organization responds to safety needs and problems out of a responsibility that has been given or dictated to it. This group would just as soon not have the responsibility of safety. It is burdensome, probably one of the most hated responsibilities it must deal with in the work environment. Such individuals and organizations see little or no value for safety activities. Their participation is forced from a higher level of authority. Upper management is aware there is a need for safety but does not really know how to deal with safety issues other than to present a facade of importance for safety. Upper management may also feel somewhat burdened by safety issues and responsibilities. At this stage the safety manager would be responsible for all training, inspections, and permit writing as well as all other safety activities. Basically, if the task involves safety, the safety manager would be solely responsible for the planning, implementation, and success or failure of the safety system.

Proceeding on this continuum of safety development, the next stage is the *empowerment* stage. At this level the group or individual has the responsibility for safety and is empowered to make decisions that affect safety without a tremendous

1

amount of input or impedance from superiors. The safety manager begins to have involvement from others in the organization to accomplish safety-related tasks. Some would participate out of empowerment and some out of responsibility. This stage is often rife with hypocrisy. Management has moved its lips with empowerment but continues to wave the stick of micromanagement — in essence, trapping some of the newly empowered players into the mold of forced participation and unwanted responsibility.

In the empowerment stage, supervisors begin to plan and conduct safety meetings, perform area inspection, and watch for hazards. This first glimmer of light begins to breed *ownership*, the next stage on the safety system continuum. Ownership even begins to trickle down to the employees, often without input or mandates from management. In the ownership stage all levels of the organization are involved. Employees may facilitate safety meetings, conduct accident investigations, track program statistics, and perform job hazard evaluations, all with little or no participation from management. Employees at all levels are involved and proud of it. They sense the program is beneficial to them and their families.

Next is the stage of *motivation*. This stage, once obtained, reaps benefits beyond comprehension by previous levels. At this level, systems will often begin to run themselves and employees participate because they are driven to participate by an inner need. They want to participate. Employees recognize the benefits and want them for themselves. Participation becomes fun and rewarding. Even family members add additional incentive to participate and perform safely.

An organization probably does not fit snugly into any single stage of the safety continuum. It may be improving gradually from one stage to the next. It may even have characteristics of several stages. Regardless of where a specific organization is on the continuum, an evaluation of why it participates in safety may provide some insight into where the company is on the safety continuum.

With all said about the safety continuum, what value is it to safety professionals? The hope is that it will allow them to deal better with those in their organization and to achieve greater things within the organization in which they are currently working. Not all companies are Exxon-Mobil, just as not all are mom-and-pop operations.

1.2 HISTORICAL EVENTS RELEVANT TO SAFETY AND INDUSTRIAL HYGIENE

There are entire texts devoted to the evolution of the modern safety movement, and most books related to safety and industrial hygiene in any way have some specific historical episodes that relate closely with their focus of attention. For that reason, this book will address the history only in a wide, "historical," perspective. That is, it will not try to focus on any specific point or theory of development. The mission here is to provide a brief, thumbnail presentation of the most significant historical events related to safety and industrial hygiene.

Some events are historically important from a safety perspective, i.e., process safety management development, whereas others are historically valuable from an industrial hygiene, safety, or worker's compensation perspective. The following is not necessarily a chronological historical perspective, rather a significant historical implication perspective.

In 1910, a devastating fire in New York City's Triangle Clothing Factory killed 146 employees. This disaster, called the Triangle Fire, outraged the public and spurred demand for factory legislation and health and safety reform. After an amendment to the state constitution was approved in 1913 at the general election, a compulsory Workmen's Compensation Act finally became effective in mid-1914. This was among the first in the nation.

In Louisville, Kentucky in 1965, a unit that processed vinyl acetylene exploded, killing 12 and injuring 61. Estimated cost was $51 million in damages. The accident was attributed to mechanical failure in a process.

In 1974, 28 deaths and $167 million in damages occurred in Flixborough, England in a chemical explosion. The chemical involved was cyclohexane vapor. When the leaking cyclohexane reached a nearby hydrogen plant, an explosion occurred, causing one of the most highly publicized industrial accidents in history. As a result, knowledge of unconfined vapor cloud explosions (UVCE) has increased tremendously.

The Texas City, Texas explosion on May 30, 1978 was a significant event in the development of process safety management regulations. The explosion happened as a result of an overfilling of one of the storage tanks of isobutane. The tank was overpressured through an offsite pipeline pumping station. The initial explosion was followed by approximately 20 min of additional or secondary explosions caused by the earlier blast. The toll, 7 fatalities, 10 injuries on site, and about $100 million in damage, left the plant nearly destroyed.

In 1982, in Tacoa, Venezuela a power station with a half-filled storage tank containing 20,000 tons of fuel oil exploded and caught fire. A secondary explosion tore the roof from one of the storage tanks and spewed burning oil on emergency workers. Massive environmental damage was done to the nearby Caracas Sea. The outcome was 145 fatalities, 500 injured, and $70 million in damages. Local fire brigades lacked equipment to fight the blaze.

The San Juan Ixuatepec, Mexico explosion on November 19, 1984 occurred when approximately 11,000 m^3 of liquid propane gas ignited from a low-lying vapor cloud coming from a pipeline or tank. The exact source is not known. The flames and fireball from the initial blast triggered nine additional explosions. Of 48 cylinders in the complex, 44 were destroyed, some to the point that they were not able to be identified. Over 500 fatalities resulted and an additional 7000 people were injured. The cost was estimated at $25 million.

In Bhopal, India on December 2, 1984 a methyl isocyanate storage tank exploded at a Union Carbide plant. Several key pieces of equipment were not operational at the time of the explosion. Poor community warning sirens were in place at the time of the explosion, which contributed to the high fatality rate. Over 3400 were killed and over 10,000 people were injured. This situation was exacerbated by the lack of clear medical procedures for treating cyanide poisoning.

Chernobyl, the Ukraine, certainly among the most famous industrial accidents, caused 31 deaths (reported); 50,000 people were permanently evacuated and over

300 square miles remains uninhabitable. Safety systems were bypassed and multiple safety shutdown devices were not working. A runaway nuclear reaction developed and caused release of large amounts of radioactive materials.

Yet another famous disaster that changed American industry is the Pasadena, Texas explosion in 1989. A polyethylene plant exploded when a contract employee placed a valve in the wrong position. About 85,000 lb of ethylene and isobutane escaped in about 2 min; 23 people were killed and 130 injured; and there were about $744 million in damages.

These incidents played major roles in promulgating what is known today as Process Safety Management (PSM). They are stark reminders to everyone that safety is not just a good thing to do. It is a life-and-death choice made with every decision.

1.3   BENCHMARKING

Once it is understood where an organization falls in the spectrum of other industries and other businesses, chances are that the organization will want to do some benchmarking. Benchmarking focuses on specific items, rather than looking at whole cultures or philosophies. Benchmarking can provide hard data for parameters like "cost of accidents" or "percent of hand injuries." It can be steered in any direction, from very specific to much broader. For example: cost of back injuries in women employees vs. number of accidents per 100 employees.

Benchmarking can easily be thrown into the hat with other industry buzzwords such as *behavior modification*, *paradigm shifts*, and *globalization*. In a time when people are intensely concerned about whether their cars measure up to the neighbor's, it should be no surprise that industries have become very interested in comparing themselves with other industries. Thus, enter the newest buzzword in industry: *benchmarking*. Simply put, benchmarking is the comparison between one company's performance in a particular area and another company's performance in that same area. Benchmarking sounds simple enough, but all too often comparisons are offered with the Monsantos, Procter & Gambles, Occidental Chemicals, and Dow Chemicals of the world. This is a commendable aspiration, but most likely not very accurate or helpful. In fact, it is similar to comparing a Little League pitcher with Nolan Ryan: flattering, but useless.

Attempts to benchmark with another company or industry should be carefully thought out and organized before any step toward comparison is taken. For example, it would serve no useful purpose for "ACME Maintenance Company" to benchmark Procter & Gamble. The differences are obvious (i.e., company size, products, and market share), and there are no real connecting points between the two companies. This does not mean that Procter & Gamble does not have anything to teach ACME. It is just not the most productive tool for improving ACME.

Selection in the benchmarking process should include several criteria to obtain successful comparisons. The first consideration should be to determine exactly what is desired from the benchmarking. In most cases, comparisons in a very narrow field of evaluation will be most helpful and probably all that can be utilized for benefit at one time. For example, collecting benchmarking data for incentive programs may be beneficial, whereas data on OSHA recordables may not be truly comparative or useful. The point is that to benchmark "the safety program" will most likely only confuse and frustrate the recipient of the data. To benchmark a "safety program" would be similar to asking an auto parts person if there were any Fords in stock. Benchmarking should be used to initiate or improve a specific sector of a safety program. Some examples of benchmark sectors that might be of a helpful nature would be job hazard analysis utilization, incentive programs, employee observation programs, accident investigation techniques, return-to-work policies, employee rotation, ergonomic program implementation, and many others.

1.3.1 WHY DOES ONE BENCHMARK?

Benchmarking is a critical link in today's business environment because of the incredibly competitive nature of the marketplace as a whole. Companies are forced to produce more product with fewer people, at a faster and more consistent pace. Competition occurs on a global scale, which only increases the need for "best of the best" practices. Constant improvement in every area is essential to remain competitive on a global basis. Benchmarking allows program development in a specific area to take place in less time, with less expenditure, and all on a shorter learning curve. Benchmarking allows companies to heighten their market awareness with real-time, actual situations. It lessens the need to reinvent the wheel. Benchmarking allows learning across industry, company, and geographical boundaries.

1.3.2 WHO SHOULD BE BENCHMARKED?

Leaders in industry would be the easy answer, but not necessarily the correct answer. It is best to benchmark the specific leaders in industry that excel in the specific area a company wants to improve or in which it seeks to implement new programs. These leaders might be called "Niche Heroes": programs such as the Rohm and Haas home safety programs, Lockhead-Martin procedures awareness and employee interaction program, General Electric and American Ref-Fuel VPP programs. These are companies that have moved to the top of a particular part of the safety culture. They possess specific areas of excellence. In many cases these will be industry leaders, but most often they will be companies that have had all the elements of success come together in just the right mix. In some cases, the companies to benchmark have come to excellence through setbacks or even catastrophes. An example is the Phillips Petroleum Sweeny, Texas plant. After the 1989 explosion, the plant developed one of the best contractor interfaces in the industry.

1.3.3 WHERE DOES ONE START?

Most benchmarking is done over the phone. If benchmarking major programs, there may be a need for a site visit, but most of the time a simple chain of phone calls can get the needed information. Contacts through professional organizations make this step much easier. Involvement in the local chapter of the American Society of Safety Engineers (ASSE), American Industrial Hygiene Association, and other local, state, and national safety and health organizations proves to be an excellent opportunity to find and develop contacts worthy of benchmarking. Attendance at annual professional development conferences, such ASSE, the American Conference on Government Industrial Hygienists (ACGIH), and the Voluntary Protection Program Association (VPPA), all open the window of opportunity to benchmark the best.

A common place to start the benchmarking process is with competitors. This can be a good place to begin gathering information because there will probably be some natural similarities in size, equipment, cultures, etc. Starting with a competitor may have some drawbacks. First, there may be more reluctance to share data because of fears of antitrust litigation. Although this is a genuine concern, proper dialogue will avoid any conflicts in this area. Remember that safety and health-related information is sought, not process design, capacities, or any proprietary information. The contacts will probably be centered around the safety and health manager and issues of trade secrets and formulas should never come up. If they do come up, avoid them at all cost and end the contact immediately. In most cases, safety managers are more than willing to share their success stories, whether they are a competitor or not. If the area to be benchmarked has been properly defined, concerns of proprietary information should not be an issue.

Consideration for the person helping with the benchmarking is essential. Typically, a phone call requesting information on benchmarking "return-to-work policies" or benchmarking "hoist inspection programs" is easy to explain, is not considered confidential, and it does not take long to collect the necessary data. Keep the scope of benchmarking narrow and specific to avoid taking large amounts of time from the person supplying the information. It is also important to maintain the same benchmark with each contact. It is easy for the scope of the benchmarking to grow wider as information is accumulated. In some cases, the request to acquire benchmarking data will be refused. Although this is very uncommon, if it happens, just go on to the next company.

It is customary for the company wishing to benchmark to share what it is currently doing in the benchmark area. This can be done upfront and often alleviates any concerns the company might have. In addition, most companies that share benchmarking information will expect to receive some kind of report or follow-up information when the benchmarking is finished. Follow-through in this area is critical because failure to fulfill this end of the bargain could make obtaining benchmarking information in the future impossible.

Pitfalls in benchmarking can be much like following the wrong road on a road map. It is always best to be sure it is the right highway before speeding off into the sunset. Often in some companies' quest for greatness, benchmarks are often set too high, setting the company up for eventual failure. In choosing benchmarking standards, a company must pick benchmarks that are similar to existing needs, meaningful, high enough to require effort to achieve, but within reach for employees and management to strive for with reasonable expectation of achieving.

Benchmarks that are too broad in scope can also be detrimental to a company. Seldom would one company seek to benchmark every program a mentor company might have. More productive would be selection of specific programs that closely fit the company's culture and abilities. When specific goals are set, employees and management can picture attainment of the predetermined benchmark, and achievement of the benchmark is much more realistic and tangible.

Another pitfall of benchmarking is pushing the benchmark harder than the foundations of the safety program. In other words, it is important to keep the "main thing" the "main thing." When a company loses sight of its safety programs in pursuit of a benchmark goal, it ends up losing ground on safety.

Given these general parameters, benchmarking can be the tool of choice for companies to choose the "Best of the Best" for constant improvements in specific areas of safety and health programs. Benchmarking is but one of many tools that can be used to change the direction of an overall safety program.

2 Employment Situations

2.1 THE ROLE OF THE SAFETY PROFESSIONAL

In today's industry and business sectors, few occupations can afford the luxury of wearing a single hat when it comes to work-related responsibilities. With the downsizing, rightsizing, and outsourcing movements in recent years, most employees are required to wear multiple hats of specialization. For example, it is not unusual for employees to have portions of a downsized employee's workload placed on their responsibility list. In fact, in some severe situations, employees can find themselves doing the entire workload of an another employee or former employee.

The whole work environment is fueled by efficiency and competitiveness. This trend of work responsibility consolidation will not disappear from work environments any time soon. If companies and organizations are to remain competitive in the global economy, they must get everything they can out of every employee.

When looking specifically in the area of the safety manager, this business method is especially true. The very nature of what is considered safety related feeds this process even more. For example, the line between "safety" and "industrial hygiene" is often too slight to measure. That is not to say that everything that deals with safety automatically deals with industrial hygiene. It does say that those job functions typically related to safety are often considered to be the very tasks that are related to industrial hygiene work. The same scenarios exist in other areas as well. Safety and environmental systems have many tasks that are closely related, if not the same. Safety and fire protection systems in many cases would be difficult to distinguish from each other. This list of safety-related or safety-oriented workplace responsibilities is quite lengthy, and thus the typical safety manager would be expected to wear many hats. While these "hats" are closely related, they are sufficiently diversified that today's safety professional must be an expert in areas outside the area historically designated strictly to safety.

This book addresses many of these areas in a cursory perspective. Be mindful that in any one of these many areas that filter over into the safety profession, there are multitudes of books and articles dealing specifically with that area. For example, the field of industrial hygiene is treated as comprehensively as any field in industry today. Similarly, Risk Management, Hazardous Chemicals, Fire Protection, Security, Epidemiology, Human Resources, and many others all have complete fields of study devoted specifically to that area of "Safety." For this reason, this book takes the most liberal and widespread concept of what "safety" applies to and covers. Today's safety professionals most likely will be involved in just about all of these areas of expertise in the course of fulfilling their roles as "safety" professionals.

The old adage "jack of all trades — master of none" does not apply to the safety professional. Granted, safety professionals cannot be experts on every one of these many fields. But they must be versed enough in the different areas to serve their companies well. Just as a general practitioner (GP) medical doctor would not typically perform surgery, the GP must still be well versed on the general needs and procedures of a surgeon to work more effectively with the surgeon in the total treatment of the patient. Similarly, the surgeon must be familiar with the general needs and procedures of the GP to work more effectively with the GP. This example holds true for the safety professional dealing with air-sampling protocols that would typically be handled by the expert industrial hygienist. In this case it may not be necessary for the safety professional to be an expert in all aspects of air sampling. However, it is important to have a thorough working knowledge of air sampling techniques and to be familiar enough with industrial hygiene work to know when to summon the help of the industrial hygiene expert.

This book will try to provide many of those fundamentals from various areas of expertise. This book is not intended to provide the information needed to become an expert in any one of these many areas discussed here, but, rather, to provide some basic tips, information, commonly used factors, charts, and other useful information that will allow access to this information quickly.

Current trends in the safety field have as many different twists as there are organizations and industries. There are, however, several "trends" that should be understood. It will be useful to talk about the reporting line of the safety manager in a typical organization. Keep in mind that in one organization this position may be called a manager, whereas in another it may be a coordinator, and in still another it may be called a specialist. In most cases, the exact title is not as important as the job responsibilities that have been assigned to the position. The author knows of one organization where the company air plane pilot has the official title of "Pilot and Safety Manager." Granted, this is an atypical situation, but it clearly shows how job responsibilities can vary drastically between organizations.

More on the variations of safety positions later. How a safety position falls into the organization depends largely on the organization and the specific responsibilities assigned. If the safety position has a "hands-on" twist, it may be very common for the position to answer to the operations manager. If the position has more of an "administrative" twist, it might be more common to report to the human resources manager. In organizations where there is a large staff of safety, environmental, or industrial hygiene personnel it may be more common to have a specific department head who oversees this "safety-oriented" department. In smaller organizations, it is common for the person assigned with safety responsibilities to have additional responsibilities and to report wherever those primary responsibilities report. For example, a design engineer who also has safety responsibilities may report to the manager responsible for engineering.

In all these cases, safety can work well if given the proper attention and effort. There is one critical question that should always be asked in regard to where the safety person reports in the organization: How close to the top is the safety responsibility assigned? Typically, the head of an organization wants as few "reports" as possible. If the general manager has everyone reporting directly to him or her, it makes for a micromanaged organization, as effective management is impossible. On the other hand, if the general manager has only three major department heads reporting to him or her, it might escalate safety to an entirely new level if the reporting of the safety representative is also to the general manager. This is not always possible, but should be kept in mind when considering to whom the safety representative should report.

Another common reporting scheme for the safety representative is to divide the safety responsibilities. This is common in small organizations or in organizations that do not wish to hire a full-time safety representative. Although this can be successful, it is much more difficult to hold the safety representative responsible for the safety responsibilities.

In larger organizations, it will be common for the safety manager to be assigned duties relatively close to the field of safety. For example, safety may include some worker compensation duties as well as the safety duties. Safety and environmental and safety and industrial hygiene are common assignment constructs.

The exact makeup in an organization should only be determined by what works in that organization. What an organization can afford or what the culture will accept can only be determined by that organization. What works well for one organization may be a total disaster for another. The specific talents and experience of individuals within an organization will also make a significant difference in which reporting order and responsibility assignments will work best.

2.2   THE ROLE OF INDUSTRIAL HYGIENE

In today's industrial settings, the role of the industrial hygienist is ever increasing. With the proliferation of chemicals in use and the ever-growing list of new products and processes, this increased need for industrial hygiene (IH) professionals will never diminish. Even within the IH profession, there are many specialized areas of expertise that make the field unbelievably broad. To understand any area of the IH profession, one must have a clear understanding of the human body, including anatomy, physiology, and pathology. Similar knowledge of all forms of life is essential. Toxicology, epidemiology, ergonomics all revolve around detailed knowledge of how life interacts with its environment.

Safety is no longer a single-focus profession. To provide an injury-free workplace is only a small step in the direction of overall employee and community health protection. "Safety" must now include not only the immediate health of workers, but also long-term effects of the overall well-being of society as a whole. Product safety may mean understanding health effects on end users decades after the product was first made.

IH is concerned with all aspects of exposure. All routes of entry to the body and environment must be considered. The IH professional may study a chemical exposure to the skin, but must also be familiar with the entry or the chemical and its effect on the ground. The IH field is so wide and deep that many specialized areas are needed to cover the entire gamut. The industrialized world has created entire fields of study that did not even exist 25 years ago.

The workplace of today can contain thermal, nuclear, biological, chemical, and carcinogenic hazards. The effects of noise, radiation, and vibration all are of paramount interest to the industrial hygienist. While safety professionals must be aware of all of these areas, they need not be experts in every field. In fact, the professional must know enough about these different fields to know when an expert in the field is needed. The practitioner with no knowledge of a particular hazard may bring a false sense of security to the workplace. Even Certified Industrial Hygienists (CIHs) and Certified Safety Professionals (CSPs) should not try to cover every avenue of employee safety. The Code of Ethics for both the CIH and CSP states that they "should only perform services in the areas of their expertise."

The workplace of today has intertwined issues that involve the environment, health, safety, industrial hygiene, legal issues, and so on. It is becoming increasingly important that the safety practitioner be competent in more and more of these areas. That is not to say an expert in every field, but aware of the issues in a multitude of areas. The days of placing the safety and health of workers in the hands of an unskilled and uninterested person are long past.

2.3 SUPERVISOR SAFETY TRAINING

"Mike" was the best electrician in the shop. He has been with the company for 8 years and has not been involved in any problem situations. The department supervisor is retiring, and Mike has been selected as his replacement. Unfortunately, the company has just created another situation commonly referred to as "The Peter Principle." Mike has been promoted to the point of incompetence. This has become more the rule than the exception in corporate America. Mike's incompetence does not come from his unwillingness to perform; rather Mike has not been given any of the tools to succeed as a supervisor. Although Mike made an outstanding electrician, he has little or no supervisory skills to utilize in his newfound promotion.

New supervisors as well as those who have been in a supervisory role for years need tools to perform their function successfully. Just as an electrician might have a "tool pouch" of various tools for the trade, supervisors need a similar "tool pouch" to meet the challenges of day-to-day employee management. What are those tools and how are supervisors equipped with them? This section provides some general guidelines for making sure that supervisors are armed with the information, skills, and techniques to succeed.

It is important to remember that the supervisor's tool pouch of management skills and techniques is not necessarily filled to overflowing. Some supervisors may indeed have many management tools to use in their supervisory arsenal. The reality is much different, however. Most supervisors, even long-time supervisors, and especially new supervisors, have few if any of these skills necessary to be an efficient supervisor.

Although there are some essential elements in a supervisor's training, there is no secret list that will guarantee a supervisor is ready for any and all obstacles that will bar his or her path. Inclusion of the following blocks of training should go a long way in preparing supervisors for success. Absence of these basic building blocks will almost surely set them up for failure.

2.3.1 THE BASICS

Before supervisors can successfully implement safety policies and procedures, they must have a firm grasp of those policies and procedures. This goes farther than just knowing the policies, as was required as an employee. Now the supervisor will be responsible for being the "go-to person" for how things are handled. The supervisor will be representing the company. Safety policies and procedures must be a high priority for any supervisor before safety will have a high priority for employees. Some of these "basics" are included here:

- Lockout / Tagout
- Machine Guarding
- Personal Protective Equipment
- Housekeeping
- Walking and Working Surfaces
- Confined Spaces
- Hazardous Communications
- Hazardous Material
- Record Keeping
- Ergonomics
- Hazard Identification
- Accident Investigation

2.3.2 PERSONAL INTERACTION SKILLS

As a supervisor it will be more important than ever to communicate effectively with management as well as with employees under direction. Some people have natural gifts that allow them to interact in a constructive–positive way. But most supervisors will need some encouragement and development along the way to become effective in interacting with others in the supervisory role. Most often this is achieved by specific courses or seminars designed to equip the participant with skills and techniques that are not readily learned without that specific focus. These seminars can be taught by in-house personnel or outside consultants; they can be offered by vendors in stand-alone courses or as part of large conferences offered by hundreds of organizations across the country. Skills that should be acquired in this category include:

- Personality Trait Assessment
- Basic Speaking
- Conflict Resolution
- Discipline Skills and Techniques
- Stress Management
- Methods to Build Loyalty
- Listening Skills
- Cultural Diversity
- Delegation Techniques
- Motivation and Encouragement Principles

2.3.3 Management Skills

The supervisor must develop skills that were not needed or seldom used as an employee. Supervisors may find themselves feeling like the third-string quarterback who has been suddenly called off the bench and now is guiding the team. The game instantly takes on a whole new perspective from that position. Even long-time supervisors may find times when their skill in management is dwarfed by the enormity of the supervision situation that is confronting them waiting to be addressed. Some core management skills include:

- Presentation Skills
- Computer Skills
- Communication Methods
- Budgeting
- Coaching and Team Building
- Project Management
- Decision Making
- Performing Performance Evaluations
- Dealing with Difficult Employees
- Observation Techniques
- Documentation

2.3.4 Clear Understanding of Directions and Goals

Part of arming supervisors with the proper tools and skills to succeed is to bring them into the loop of information that runs the organization. In some organizations this is more difficult than in others. Many things may affect how difficult this is, such as the degree of micromanagement present in the organization, how clear the business strategy is to management, union–management relationships, size of the organization, and others. The goal should be to have supervisors working in the same direction, toward the same goals, and for the same reasons that management is doing what it is doing. Ultimately, this will lead to the ability of employees to be on the same channel as well, positioning the supervisor for easier management and greater chance of success. Supervisors who appear "out of the loop" or "out of touch" lose credibility with their employees. Issues and skills that must be addressed in this area include:

- Bringing the supervisor "up to speed" on company goals
- Strategies and directions shared and clearly communicated
- Clear communications of staff changes, new programs, and other announcements
- "Heads-up" notice on issues that will support the supervisory role
- Inclusion of the supervisor in as much "management" decision making as possible
- Financial compensation that clearly reflects the difference in responsibility and authority

2.3.5 CONCLUSION

One can easily see that great supervisors need a lot of tools, skills, knowledge, and training to be successful. There is no single course or training segment that will prepare a supervisor in all of the areas mentioned. This equipping does not come in 1 week or even 1 month; rather, it is a long-term commitment, strategy, and plan. There is constant talk of improvement in the business sectors but the most precious commodity, the employees, are often neglected. There must be a commitment to continuous improvement of the skills and tools of supervisors. The following list provides an outline of a basic program that, when put to use with the information in this section, will reap huge rewards for supervisors, but more importantly, everyone in the company, from the employees to top management.

- Create a checklist of essential supervisory skills. The lists above can serve as a starting point. Divide the needed skill into priorities. Maintain an ongoing status of supervisor development.
- Start with the basics. A supervisor who is not well grounded in the basics will have a rough and rocky road to travel.
- Create a plan for achieving completion of all training on the checklist. The plan should be realistic but also aggressive. Supervisors should have most of the elements mentioned in this section completed in the first 18 months or so of their supervisory career.
- Continuous improvement should be the ongoing theme for a supervisor's development.
- Each year's performance goals should include additional supervisor training objectives.
- Performance reviews should reflect whether or not the training objectives were accomplished.
- Do not use all in-house trainers to accomplish training needs. Familiar faces are often less effective and lack the specific knowledge to provide beneficial improvement in supervisory skills.
- Utilize outside consultants to provide much of the needed training. Use of outside people can add expertise as well as fun and excitement to sometimes dry subjects.
- Often "corporate" personnel can provide training that is useful and beneficial in these areas.
- Continually review training to ensure that it is viewed as beneficial and worthwhile.

Following these suggestions will not guarantee excellent supervisors, but they will ensure that the best people possible are leading the employees. These suggestions will give supervisors real tools and skills necessary to perform their jobs. The company will profit from a more efficient, educated, and motivated staff who in turn can convey those attributes to employees.

Although supervisor training is not specifically safety related, it is intricately involved, just as safety is involved with the bottom line of an organization and with every business sector within the organization.

3 Regulations

In this section, some of the major OSHA standards are included for easy reference and quick access. In some cases, unnecessary jargon or information of no consequence has been omitted to make the reading easier and less cumbersome. The intent is to provide a single source for the more commonly used standards referencing.

3.1 PERSONAL PROTECTIVE EQUIPMENT

29 CFR 1910.13256

(a)
Application. Protective equipment, including personal protective equipment for eyes, face, head, and extremities, protective clothing, respiratory devices, and protective shields and barriers, shall be provided, used, and maintained in a sanitary and reliable condition wherever it is necessary by reason of hazards of processes or environment, chemical hazards, radiological hazards, or mechanical irritants encountered in a manner capable of causing injury or impairment in the function of any part of the body through absorption, inhalation, or physical contact.

(b)
Employee-owned equipment. Where employees provide their own protective equipment, the employer shall be responsible to assure its adequacy, including proper maintenance, and sanitation of such equipment.

(c)
Design. All personal protective equipment shall be of safe design and construction for the work to be performed.

(d)
Hazard assessment and equipment selection.

(d)(1)
The employer shall assess the workplace to determine if hazards are present, or are likely to be present, which necessitate the use of personal protective equipment (PPE). If such hazards are present, or likely to be present, the employer shall:

(d)(1)(i)
Select and have each affected employee use the types of PPE that will protect the affected employee from the hazards identified in the hazard assessment;

(d)(1)(ii)
Communicate selection decisions to each affected employee; and

(d)(1)(iii)

Select PPE that properly fits each affected employee. *Note:* Nonmandatory Appendix B contains an example of procedures that would comply with the requirement for a hazard assessment.

(d)(2)

The employer shall verify that the required workplace hazard assessment has been performed through a written certification that identifies the workplace evaluated; the person certifying that the evaluation has been performed; the date(s) of the hazard assessment; and that identifies the document as a certification of hazard assessment.

(e)

Defective and damaged equipment. Defective or damaged PPE shall not be used.

(f)

Training.

(f)(1)

The employer shall provide training to each employee who is required by this section to use PPE. Each such employee shall be trained to know at least the following:

(f)(1)(i)

When PPE is necessary;

(f)(1)(ii)

What PPE is necessary;

(f)(1)(iii)

How to properly don, doff, adjust, and wear PPE;

(f)(1)(iv)

The limitations of the PPE; and

(f)(1)(v)

The proper care, maintenance, useful life, and disposal of the PPE.

(f)(2)

Each affected employee shall demonstrate an understanding of the training specified in paragraph (f)(1) of this section, and the ability to use PPE properly, before being allowed to perform work requiring the use of PPE.

(f)(3)

When the employer has reason to believe that any affected employee who has already been trained does not have the understanding and skill required by paragraph (f)(2) of this section, the employer shall retrain each such employee. Circumstances where retraining is required include, but are not limited to, situations where:

(f)(3)(i)

Changes in the workplace render previous training obsolete; or

(f)(3)(ii)
Changes in the types of PPE to be used render previous training obsolete; or

(f)(3)(iii)
Inadequacies in an affected employee's knowledge or use of assigned PPE indicate that the employee has not retained the requisite understanding or skill.

(f)(4)
The employer shall verify that each affected employee has received and understood the required training through a written certification that contains the name of each employee trained, the date(s) of training, and that identifies the subject of the certification.

3.2 EYE AND FACE PROTECTION

29 CFR 1910.133

(a)
General requirements.

(a)(1)
The employer shall ensure that each affected employee uses appropriate eye or face protection when exposed to eye or face hazards from flying particles, molten metal, liquid chemicals, acids or caustic liquids, chemical gases or vapors, or potentially injurious light radiation.

(a)(2)
The employer shall ensure that each affected employee uses eye protection that provides side protection when there is a hazard from flying objects. Detachable side protectors (e.g., clip-on or slide-on side shields) meeting the pertinent requirements of this section are acceptable.

(a)(3)
The employer shall ensure that each affected employee who wears prescription lenses while engaged in operations that involve eye hazards wears eye protection that incorporates the prescription in its design, or wears eye protection that can be worn over the prescription lenses without disturbing the proper position of the prescription lenses or the protective lenses.

(a)(4)
Eye and face PPE shall be distinctly marked to facilitate identification of the manufacturer.

(a)(5)
The employer shall ensure that each affected employee uses equipment with filter lenses that have a shade number appropriate for the work being performed for protection from injurious light radiation. The following is a listing of appropriate shade numbers for various operations.

Filter Lenses for Protection against Radiant Energy

Operations Arc	Electrode Size 1/32 in.	Current	Minimum[a] Protective Shade
Shielded metal arc welding			
	<3	<60	7
	3–5	60–160	8
	5–8	160–250	10
	>8	250–550	11
Gas metal arc welding and flux cored arc welding			
		<60	7
		60–160	10
		160–250	10
		250–500	10
Gas tungsten arc welding			
		<50	8
		50–150	8
		150–500	10
Air carbon	Light	<500	10
Arc cutting	Heavy	500–1000	11
Plasma arc welding		<20	6
		20–100	8
		100–400	10
		400–800	11
Plasma arc cutting	Light[b]	<300	8
	Medium[b]	300–400	9
	Heavy[b]	400–800	10
Torch brazing			3
Torch soldering			2
Carbon arc welding			14

Filter Lenses for Protection against Radiant Energy

Operations	Plate Thickness in.	mm	Minimum[a] Protective Shade
Gas Welding			
Light	<1/8	<3.2	4
Medium	1/8–1/2	3.2–12.7	5
Heavy	>1/2	>12.7	6
Oxygen cutting			
Light	<1	<25	3
Medium	1–6	25–150	4
Heavy	>6	>150	5

[a] As a rule of thumb, start with a shade that is too dark to see the weld zone. Then go to a lighter shade that gives sufficient view of the weld zone without going below the minimum. In oxyfuel gas welding or cutting where the torch produces a high yellow light, it is desirable to use a filter lens that absorbs the yellow or sodium line in the visible light of the (spectrum) operation.

[b] These values apply where the actual arc is clearly seen. Experience has shown that lighter filters may be used when the arc is hidden by the workpiece.

(b)
Criteria for protective eye and face devices.

(b)(1)
Protective eye and face devices purchased after July 5, 1994 shall comply with ANSI Z87.1-1989, "American National Standard Practice for Occupational and Educational Eye and Face Protection," which is incorporated by reference as specified in Sec. 1910.6.

(b)(2)
Eye and face protective devices purchased before July 5, 1994 shall comply with the ANSI "USA Standard for Occupational and Educational Eye and Face Protection," Z87.1-1968, which is incorporated by reference as specified in Sec. 1910.6, or shall be demonstrated by the employer to be equally effective.

3.3 RESPIRATORY PROTECTION

29 CFR 1910.134

This section applies to General Industry (part 1910), Shipyards (part 1915), Marine Terminals (part 1917), Longshoring (part 1918), and Construction (part 1926).

(a)
Permissible practice.

(a)(1)
In the control of those occupational diseases caused by breathing air contaminated with harmful dusts, fogs, fumes, mists, gases, smokes, sprays, or vapors, the primary objective shall be to prevent atmospheric contamination. This shall be accomplished as far as feasible by accepted engineering control measures (for example, enclosure or confinement of the operation, general and local ventilation, and substitution of less toxic materials). When effective engineering controls are not feasible, or while they are being instituted, appropriate respirators shall be used pursuant to this section.

(a)(2)
Respirators shall be provided by the employer when such equipment is necessary to protect the health of the employee. The employer shall provide the respirators that are applicable and suitable for the purpose intended. The employer shall be responsible for the establishment and maintenance of a respiratory protection program, which shall include the requirements outlined in paragraph **(c)** of this section.

(b)
Definitions. The following definitions are important terms used in the respiratory protection standard in this section:

"**Air-purifying respirator**" means a respirator with an air-purifying filter, cartridge, or canister that removes specific air contaminants by passing ambient air through the air-purifying element.

"**Assigned protection factor**" (APF) [Reserved]

"**Atmosphere-supplying respirator**" means a respirator that supplies the respirator user with breathing air from a source independent of the ambient atmosphere, and includes supplied-air respirators (SARs) and self-contained breathing apparatus (SCBA) units.

"**Canister**" or "**cartridge**" means a container with a filter, sorbent, or catalyst, or a combination of these items, which removes specific contaminants from the air passed through the container.

"**Demand respirator**" means an atmosphere-supplying respirator that admits breathing air to the facepiece only when a negative pressure is created inside the facepiece by inhalation.

"**Emergency situation**" means any occurrence such as, but not limited to, equipment failure, rupture of containers, or failure of control equipment that may or does result in an uncontrolled significant release of an airborne contaminant.

"**Employee exposure**" means exposure to a concentration of an airborne contaminant that would occur if the employees were not using respiratory protection.

"**End-of-service-life indicator**" (ESLI) means a system that warns the respirator user of the approach of the end of adequate respiratory protection, for example, that the sorbent is approaching saturation or is no longer effective.

"**Escape-only respirator**" means a respirator intended to be used only for emergency exit.

"**Filter**" or air-purifying element means a component used in respirators to remove solid or liquid aerosols from the inspired air.

"**Filtering facepiece**" (dust mask) means a negative-pressure particulate respirator with a filter as an integral part of the facepiece or with the entire facepiece composed of the filtering medium.

"**Fit factor**" means a quantitative estimate of the fit of a particular respirator to a specific individual, and typically estimates the ratio of the concentration of a substance in ambient air to its concentration inside the respirator when worn.

"**Fit test**" means the use of a protocol to qualitatively or quantitatively evaluate the fit of a respirator on an individual. (*See also* Qualitative fit test QLFT and Quantitative fit test QNFT.)

"**Helmet**" means a rigid respiratory inlet covering that also provides head protection against impact and penetration.

"**High-efficiency particulate air (HEPA) filter**" means a filter that is at least 99.97% efficient in removing monodisperse particles of 0.3 mm in diameter. The equivalent NIOSH 42 CFR 84 particulate filters are the N100, R100, and P100 filters.

"**Hood**" means a respiratory inlet covering that completely covers the head and neck and may also cover portions of the shoulders and torso.

"Immediately dangerous to life or health" (IDLH) means an atmosphere that poses an immediate threat to life, would cause irreversible adverse health effects, or would impair an individual's ability to escape from a dangerous atmosphere.

"Interior structural firefighting" means the physical activity of fire suppression, rescue or both, inside buildings or enclosed structures that are involved in a fire situation beyond the incipient stage. (*See* 29 1910.155.)

"Loose-fitting facepiece" means a respiratory inlet covering that is designed to form a partial seal with the face.

"Maximum use concentration" (MUC) [Reserved].

"Negative pressure respirator" (tight fitting) means a respirator in which the air pressure inside the facepiece is negative during inhalation with respect to the ambient air pressure outside the respirator.

"Oxygen deficient atmosphere" means an atmosphere with an oxygen content below 19.5% by volume.

"Physician or other licensed health care professional" (PLHCP) means an individual whose legally permitted scope of practice (i.e., license, registration, or certification) allows him or her to independently provide, or be delegated the responsibility to provide, some or all of the health-care services required by paragraph **(e)** of this section.

"Positive-pressure respirator" means a respirator in which the pressure inside the respiratory inlet covering exceeds the ambient air pressure outside the respirator.

"Powered air-purifying respirator" (PAPR) means an air-purifying respirator that uses a blower to force the ambient air through air-purifying elements to the inlet covering.

"Pressure demand respirator" means a positive-pressure atmosphere-supplying respirator that admits breathing air to the facepiece when the positive pressure is reduced inside the facepiece by inhalation.

"Qualitative fit test" (QLFT) means a pass/fail fit test to assess the adequacy of respirator fit that relies on the individual's response to the test agent.

"Quantitative fit test" (QNFT) means an assessment of the adequacy of respirator fit by numerically measuring the amount of leakage into the respirator.

"Respiratory inlet covering" means that portion of a respirator that forms the protective barrier between the user's respiratory tract and an air-purifying device or breathing air source, or both. It may be a facepiece, helmet, hood, suit, or a mouthpiece respirator with nose clamp.

"Self-contained breathing apparatus" (SCBA) means an atmosphere-supplying respirator for which the breathing air source is designed to be carried by the user.

"Service life" means the period of time that a respirator, filter, or sorbent, or other respiratory equipment provides adequate protection to the wearer.

"Supplied-air respirator" (SAR) or airline respirator means an atmosphere-supplying respirator for which the source of breathing air is not designed to be carried by the user.

"This section" means this respiratory protection standard.

"Tight-fitting facepiece" means a respiratory inlet covering that forms a complete seal with the face.

"User seal check" means an action conducted by the respirator user to determine if the respirator is properly seated to the face.

(c)
Respiratory protection program. This paragraph requires the employer to develop and implement a written respiratory protection program with required worksite-specific procedures and elements for required respirator use. The program must be administered by a suitably trained program administrator. In addition, certain program elements may be required for voluntary use to prevent potential hazards associated with the use of the respirator. The "Small Entity Compliance Guide" contains criteria for the selection of a program administrator and a sample program that meets the requirements of this paragraph. Copies of the "Small Entity Compliance Guide" will be available on or about April 8, 1998 from the Occupational Safety and Health Administration's Office of Publications, Room N 3101, 200 Constitution Avenue, NW, Washington, D.C., 20210 (202-219-4667).

(c)(1)
In any workplace where respirators are necessary to protect the health of the employee or whenever respirators are required by the employer, the employer shall establish and implement a written respiratory protection program with worksite-specific procedures. The program shall be updated as necessary to reflect those changes in workplace conditions that affect respirator use. The employer shall include in the program the following provisions of this section, as applicable:

(c)(1)(i)
Procedures for selecting respirators for use in the workplace;

(c)(1)(ii)
Medical evaluations of employees required to use respirators;

(c)(1)(iii)
Fit testing procedures for tight-fitting respirators;

(c)(1)(iv)
Procedures for proper use of respirators in routine and reasonably foreseeable emergency situations;

(c)(1)(v)
Procedures and schedules for cleaning, disinfecting, storing, inspecting, repairing, discarding, and otherwise maintaining respirators;

(c)(1)(vi)
Procedures to ensure adequate air quality, quantity, and flow of breathing air for atmosphere-supplying respirators;

(c)(1)(vii)
Training of employees in the respiratory hazards to which they are potentially exposed during routine and emergency situations;

(c)(1)(viii)
Training of employees in the proper use of respirators, including putting them on and removing them, any limitations on their use, and their maintenance; and

(c)(1)(ix)
Procedures for regularly evaluating the effectiveness of the program.

(c)(2)
Where respirator use is not required:

(c)(2)(i)
An employer may provide respirators at the request of employees or permit employees to use their own respirators, if the employer determines that such respirator use will not in itself create a hazard. If the employer determines that any voluntary respirator use is permissible, the employer shall provide the respirator users with the information contained in Appendix D to this section ("Information for Employees Using Respirators When Not Required under the Standard"); and

(c)(2)(ii)
In addition, the employer must establish and implement those elements of a written respiratory protection program necessary to ensure that any employee using a respirator voluntarily is medically able to use that respirator, and that the respirator is cleaned, stored, and maintained so that its use does not present a health hazard to the user. *Exception:* Employers are not required to include in a written respiratory protection program those employees whose only use of respirators involves the voluntary use of filtering facepieces (dust masks).

(c)(3)
The employer shall designate a program administrator who is qualified by appropriate training or experience that is commensurate with the complexity of the program to administer or oversee the respiratory protection program and conduct the required evaluations of program effectiveness.

(c)(4)
The employer shall provide respirators, training, and medical evaluations at no cost to the employee.

(d)
Selection of respirators. This paragraph requires the employer to evaluate respiratory hazard(s) in the workplace, to identify relevant workplace and user factors, and to base respirator selection on these factors. The paragraph also specifies appropriately

protective respirators for use in IDLH atmospheres, and limits the selection and use of air-purifying respirators.

(d)(1)
General requirements.

(d)(1)(i)
The employer shall select and provide an appropriate respirator based on the respiratory hazard(s) to which the worker is exposed and workplace and user factors that affect respirator performance and reliability.

(d)(1)(ii)
The employer shall select a NIOSH-certified respirator. The respirator shall be used in compliance with the conditions of its certification.

(d)(1)(iii)
The employer shall identify and evaluate the respiratory hazard(s) in the workplace; this evaluation shall include a reasonable estimate of employee exposures to respiratory hazard(s) and an identification of the contaminant's chemical state and physical form. Where the employer cannot identify or reasonably estimate the employee exposure, the employer shall consider the atmosphere to be IDLH.

(d)(1)(iv)
The employer shall select respirators from a sufficient number of respirator models and sizes so that the respirator is acceptable to, and correctly fits, the user.

(d)(2)
Respirators for IDLH atmospheres.

(d)(2)(i)
The employer shall provide the following respirators for employee use in IDLH atmospheres:

(d)(2)(i)(A)
A full facepiece pressure-demand SCBA certified by NIOSH for a minimum service life of 30 min, or

(d)(2)(i)(B)
A combination full facepiece pressure-demand supplied-air respirator (SAR) with auxiliary self-contained air supply.

(d)(2)(ii)
Respirators provided only for escape from IDLH atmospheres shall be NIOSH-certified for escape from the atmosphere in which they will be used.

(d)(2)(iii)
All oxygen-deficient atmospheres shall be considered IDLH. Exception: If the employer demonstrates that, under all foreseeable conditions, the oxygen concentration can be maintained within the ranges . . . (of acceptable oxygen concentrations, i.e., 19.5–21.0) . . . then any atmosphere-supplying respirator may be used.

(d)(3)
Respirators for atmospheres that are not IDLH.

(d)(3)(i)
The employer shall provide a respirator that is adequate to protect the health of the employee and ensure compliance with all other OSHA statutory and regulatory requirements, under routine and reasonably foreseeable emergency situations.

(d)(3)(i)(A)
Assigned Protection Factors (APFs) [Reserved]

(d)(3)(i)(B)
Maximum Use Concentration (MUC) [Reserved]

(d)(3)(ii)
The respirator selected shall be appropriate for the chemical state and physical form of the contaminant.

(d)(3)(iii)
For protection against gases and vapors, the employer shall provide:

(d)(3)(iii)(A)
An atmosphere-supplying respirator, or

(d)(3)(iii)(B)
An air-purifying respirator, provided that:

(d)(3)(iii)(B)(1)
The respirator is equipped with an end-of-service-life indicator (ESLI) certified by NIOSH for the contaminant; or

(d)(3)(iii)(B)(2)
If there is no ESLI appropriate for conditions in the employer's workplace, the employer implements a change schedule for canisters and cartridges that is based on objective information or data that will ensure that canisters and cartridges are changed before the end of their service life. The employer shall describe in the respirator program the information and data relied upon, the basis for the canister and cartridge change schedule, and the basis for reliance on the data.

(d)(3)(iv)
For protection against particulates, the employer shall provide:

(d)(3)(iv)(A)
An atmosphere-supplying respirator; or

(d)(3)(iv)(B)
An air-purifying respirator equipped with a filter certified by NIOSH under 30 CFR part 11 as a high-efficiency particulate air (HEPA) filter, or an air-purifying respirator equipped with a filter certified for particulates by NIOSH under 42 CFR part 84; or

(d)(3)(iv)(C)
For contaminants consisting primarily of particles with mass median aerodynamic diameters (MMAD) of at least 2 μm, an air-purifying respirator equipped with any filter certified for particulates by NIOSH.

(e)
Medical evaluation. Using a respirator may place a physiological burden on employees that varies with the type of respirator worn, the job and workplace conditions in which the respirator is used, and the medical status of the employee. Accordingly, this paragraph specifies the minimum requirements for medical evaluation that employers must implement to determine the employee's ability to use a respirator.

(e)(1)
General. The employer shall provide a medical evaluation to determine the employee's ability to use a respirator, before the employee is fit-tested or required to use the respirator in the workplace. The employer may discontinue an employee's medical evaluations when the employee is no longer required to use a respirator.

(e)(2)
Medical evaluation procedures.

(e)(2)(i)
The employer shall identify a physician or other licensed health care professional (PLHCP) to perform medical evaluations using a medical questionnaire or an initial medical examination that obtains the same information as the medical questionnaire.

(e)(2)(ii)
The medical evaluation shall obtain the information requested by the questionnaire in Sections 1 and 2, Part A of Appendix C of this section.

(e)(3)
Follow-up medical examination.

(e)(3)(i)
The employer shall ensure that a follow-up medical examination is provided for an employee who gives a positive response to any question among questions 1 through 8 in Section 2, Part A of Appendix C or whose initial medical examination demonstrates the need for a follow-up medical examination.

(e)(3)(ii)
The follow-up medical examination shall include any medical tests, consultations, or diagnostic procedures that the PLHCP deems necessary to make a final determination.

(e)(4)
Administration of the medical questionnaire and examinations.

(e)(4)(i)
The medical questionnaire and examinations shall be administered confidentially during the employee's normal working hours or at a time and place convenient to

the employee. The medical questionnaire shall be administered in a manner that ensures that the employee understands its content.

(e)(4)(ii)
The employer shall provide the employee with an opportunity to discuss the questionnaire and examination results with the PLHCP.

(e)(5)
Supplemental information for the PLHCP.

(e)(5)(i)
The following information must be provided to the PLHCP before the PLHCP makes a recommendation concerning an employee's ability to use a respirator:

(e)(5)(i)(A)
The type and weight of the respirator to be used by the employee;

(e)(5)(i)(B)
The duration and frequency of respirator use (including use for rescue and escape);

(e)(5)(i)(C)
The expected physical work effort;

(e)(5)(i)(D)
Additional protective clothing and equipment to be worn; and

(e)(5)(i)(E)
Temperature and humidity extremes that may be encountered.

(e)(5)(ii)
Any supplemental information provided previously to the PLHCP regarding an employee need not be provided for a subsequent medical evaluation if the information and the PLHCP remain the same.

(e)(5)(iii)
The employer shall provide the PLHCP with a copy of the written respiratory protection program and a copy of this section.

Note: When the employer replaces a PLHCP, the employer must ensure that the new PLHCP obtains this information, either by providing the documents directly to the PLHCP or by having the documents transferred from the former PLHCP to the new PLHCP. However, OSHA does not expect employers to have employees medically reevaluated solely because a new PLHCP has been selected.

(e)(6)
Medical determination. In determining the employee's ability to use a respirator, the employer shall:

(e)(6)(i)
Obtain a written recommendation regarding the employee's ability to use the respirator from the PLHCP. The recommendation shall provide only the following information:

(e)(6)(i)(A)

Any limitations on respirator use related to the medical condition of the employee, or relating to the workplace conditions in which the respirator will be used, including whether or not the employee is medically able to use the respirator;

(e)(6)(i)(B)

The need, if any, for follow-up medical evaluations; and

(e)(6)(i)(C)

A statement that the PLHCP has provided the employee with a copy of the PLHCP's written recommendation.

(e)(6)(ii)

If the respirator is a negative pressure respirator and the PLHCP finds a medical condition that may place the employee's health at increased risk if the respirator is used, the employer shall provide a PAPR if the PLHCP's medical evaluation finds that the employee can use such a respirator; if a subsequent medical evaluation finds that the employee is medically able to use a negative pressure respirator, then the employer is no longer required to provide a PAPR.

(e)(7)

Additional medical evaluations. At a minimum, the employer shall provide additional medical evaluations that comply with the requirements of this section if:

(e)(7)(i)

An employee reports medical signs or symptoms that are related to the ability to use a respirator;

(e)(7)(ii)

A PLHCP, supervisor, or the respirator program administrator informs the employer that an employee needs to be reevaluated;

(e)(7)(iii)

Information from the respiratory protection program, including observations made during fit testing and program evaluation, indicates a need for employee reevaluation; or

(e)(7)(iv)

A change occurs in workplace conditions (e.g., physical work effort, protective clothing, temperature) that may result in a substantial increase in the physiological burden placed on an employee.

(f)

Fit testing. This paragraph requires that, before an employee may be required to use any respirator with a negative or positive pressure tight-fitting facepiece, the employee must be fit tested with the same make, model, style, and size of respirator that will be used. This paragraph specifies the kinds of fit tests allowed, the proce-dures for conducting them, and how the results of the fit tests must be used.

(f)(1)
The employer shall ensure that employees using a tight-fitting facepiece respirator pass an appropriate qualitative fit test (QLFT) or quantitative fit test (QNFT) as stated in this paragraph.

(f)(2)
The employer shall ensure that an employee using a tight-fitting facepiece respirator is fit tested prior to initial use of the respirator, whenever a different respirator facepiece (size, style, model, or make) is used and at least annually thereafter.

(f)(3)
The employer shall conduct an additional fit test whenever the employee reports, or the employer, PLHCP, supervisor, or program administrator makes visual observations of, changes in the employee's physical condition that could affect respirator fit. Such conditions include, but are not limited to, facial scarring, dental changes, cosmetic surgery, or an obvious change in body weight.

(f)(4)
If after passing a QLFT or QNFT the employee subsequently notifies the employer, program administrator, supervisor, or PLHCP that the fit of the respirator is unacceptable, the employee shall be given a reasonable opportunity to select a different respirator facepiece and to be retested.

(f)(5)
The fit test shall be administered using an OSHA-accepted QLFT or QNFT protocol. The OSHA-accepted QLFT and QNFT protocols and procedures are contained in Appendix A of this section.

(f)(6)
QLFT may only be used to fit test negative pressure air-purifying respirators that must achieve a fit factor of 100 or less.

(f)(7)
If the fit factor, as determined through an OSHA-accepted QNFT protocol, is equal to or greater than 100 for tight-fitting half facepieces, or equal to or greater than 500 for tight-fitting full facepieces, the QNFT has been passed with that respirator.

(f)(8)
Fit testing of tight-fitting atmosphere-supplying respirators and tight-fitting powered air-purifying respirators shall be accomplished by performing quantitative or qualitative fit testing in the negative pressure mode, regardless of the mode of operation (negative or positive pressure) that is used for respiratory protection.

(f)(8)(i)
Qualitative fit testing of these respirators shall be accomplished by temporarily converting the respirator user's actual facepiece into a negative-pressure respirator with appropriate filters, or by using an identical negative pressure air-purifying respirator facepiece with the same sealing surfaces as a surrogate for the atmosphere-supplying or powered air-purifying respirator facepiece.

(f)(8)(ii)

Quantitative fit testing of these respirators shall be accomplished by modifying the facepiece to allow sampling inside the facepiece in the breathing zone of the user, midway between the nose and mouth. This requirement shall be accomplished by installing a permanent sampling probe onto a surrogate facepiece, or by using a sampling adapter designed to provide temporarily a means of sampling air from inside the facepiece.

(f)(8)(iii)

Any modifications to the respirator facepiece for fit testing shall be completely removed, and the facepiece restored to NIOSH-approved configuration, before that facepiece can be used in the workplace.

(g)

Use of respirators. This paragraph requires employers to establish and implement procedures for the proper use of respirators. These requirements include prohibiting conditions that may result in facepiece seal leakage, preventing employees from removing respirators in hazardous environments, taking actions to ensure continued effective respirator operation throughout the work shift, and establishing procedures for the use of respirators in IDLH atmospheres or in interior structural firefighting situations.

(g)(1)

Facepiece seal protection.

(g)(1)(i)

The employer shall not permit respirators with tight-fitting facepieces to be worn by employees who have:

(g)(1)(i)(A)

Facial hair that comes between the sealing surface of the facepiece and the face or that interferes with valve function; or

(g)(1)(i)(B)

Any condition that interferes with the face-to-facepiece seal or valve function.

(g)(1)(ii)

If an employee wears corrective glasses or goggles or other personal protective equipment, the employer shall ensure that such equipment is worn in a manner that does not interfere with the seal of the facepiece to the face of the user.

(g)(1)(iii)

For all tight-fitting respirators, the employer shall ensure that employees perform a user seal check each time they put on the respirator using the procedures in Appendix B-1 or procedures recommended by the respirator manufacturer that the employer demonstrates are as effective as those in Appendix B-1 of this section.

(g)(2)

Continuing respirator effectiveness.

(g)(2)(i)

Appropriate surveillance shall be maintained of work area conditions and degree of employee exposure or stress. When there is a change in work area conditions or degree of employee exposure or stress that may affect respirator effectiveness, the employer shall reevaluate the continued effectiveness of the respirator.

(g)(2)(ii)

The employer shall ensure that employees leave the respirator use area:

(g)(2)(ii)(A)

To wash their faces and respirator facepieces as necessary to prevent eye or skin irritation associated with respirator use; or

(g)(2)(ii)(B)

If they detect vapor or gas breakthrough, changes in breathing resistance, or leakage of the facepiece; or

(g)(2)(ii)(C)

To replace the respirator or the filter, cartridge, or canister elements.

(g)(2)(iii)

If the employee detects vapor or gas breakthrough, changes in breathing resistance, or leakage of the facepiece, the employer must replace or repair the respirator before allowing the employee to return to the work area.

(g)(3)

Procedures for IDLH atmospheres. For all IDLH atmospheres, the employer shall ensure that:

(g)(3)(i)

One employee or, when needed, more than one employee is located outside the IDLH atmosphere;

(g)(3)(ii)

Visual, voice, or signal line communication is maintained between the employee(s) in the IDLH atmosphere and the employee(s) located outside the IDLH atmosphere;

(g)(3)(iii)

The employee(s) located outside the IDLH atmosphere are trained and equipped to provide effective emergency rescue;

(g)(3)(iv)

The employer or designee is notified before the employee(s) located outside the IDLH atmosphere enter the IDLH atmosphere to provide emergency rescue;

(g)(3)(v)

The employer or designee authorized to do so by the employer, once notified, provides necessary assistance appropriate to the situation;

(g)(3)(vi)

Employee(s) located outside the IDLH atmospheres are equipped with:

(g)(3)(vi)(A)
Pressure-demand or other positive-pressure SCBAs, or a pressure-demand or other positive-pressure supplied-air respirator with auxiliary SCBA; and either

(g)(3)(vi)(B)
Appropriate retrieval equipment for removing the employee(s) who enter(s) these hazardous atmospheres where retrieval equipment would contribute to the rescue of the employee(s) and would not increase the overall risk resulting from entry; or

(g)(3)(vi)(C)
Equivalent means for rescue where retrieval equipment is not required under paragraph (g)(3)(vi)(B).

(g)(4)
Procedures for interior structural firefighting. In addition to the requirements set forth under paragraph (g)(3), in interior structural fires, the employer shall ensure that:

(g)(4)(i)
At least two employees enter the IDLH atmosphere and remain in visual or voice contact with one another at all times;

(g)(4)(ii)
At least two employees are located outside the IDLH atmosphere; and

(g)(4)(iii)
All employees engaged in interior structural firefighting use SCBAs.

Note 1: One of the two individuals located outside the IDLH atmosphere may be assigned to an additional role, such as incident commander in charge of the emergency or safety officer, so long as this individual is able to perform assistance or rescue activities without jeopardizing the safety or health of any firefighter working at the incident.

Note 2: Nothing in this section is meant to preclude firefighters from performing emergency rescue activities before an entire team has assembled.

(h)
Maintenance and care of respirators. This paragraph requires the employer to provide for the cleaning and disinfecting, storage, inspection, and repair of respirators used by employees.

(h)(1)
Cleaning and disinfecting. The employer shall provide each respirator user with a respirator that is clean, sanitary, and in good working order. The employer shall ensure that respirators are cleaned and disinfected using the procedures in Appendix B-2 of this section, or procedures recommended by the respirator manufacturer, provided that such procedures are of equivalent effectiveness. The respirators shall be cleaned and disinfected at the following intervals:

(h)(1)(i)
Respirators issued for the exclusive use of an employee shall be cleaned and disinfected as often as necessary to be maintained in a sanitary condition;

(h)(1)(ii)
Respirators issued to more than one employee shall be cleaned and disinfected before being worn by different individuals;

(h)(1)(iii)
Respirators maintained for emergency use shall be cleaned and disinfected after each use; and

(h)(1)(iv)
Respirators used in fit-testing and training shall be cleaned and disinfected after each use.

(h)(2)
Storage. The employer shall ensure that respirators are stored as follows:

(h)(2)(i)
All respirators shall be stored to protect them from damage, contamination, dust, sunlight, extreme temperatures, excessive moisture, and damaging chemicals, and they shall be packed or stored to prevent deformation of the facepiece and exhalation valve.

(h)(2)(ii)
In addition to the requirements of paragraph (h)(2)(i) of this section, emergency respirators shall be:

(h)(2)(ii)(A)
Kept accessible to the work area;

(h)(2)(ii)(B)
Stored in compartments or in covers that are clearly marked as containing emergency respirators; and

(h)(2)(ii)(C)
Stored in accordance with any applicable manufacturer instructions.

(h)(3)
Inspection.

(h)(3)(i)
The employer shall ensure that respirators are inspected as follows:

(h)(3)(i)(A)
All respirators used in routine situations shall be inspected before each use and during cleaning;

(h)(3)(i)(B)
All respirators maintained for use in emergency situations shall be inspected at least monthly and in accordance with the manufacturer's recommendations, and shall be checked for proper function before and after each use; and

(h)(3)(i)(C)
Emergency escape-only respirators shall be inspected before being carried into the workplace for use.

(h)(3)(ii)
The employer shall ensure that respirator inspections include the following:

(h)(3)(ii)(A)
A check of respirator function, tightness of connections, and the condition of the various parts including, but not limited to, the facepiece, head straps, valves, connecting tube, and cartridges, canisters or filters; and

(h)(3)(ii)(B)
A check of elastomeric parts for pliability and signs of deterioration.

(h)(3)(iii)
In addition to the requirements of paragraphs (h)(3)(i) and (ii) of this section, self-contained breathing apparatus shall be inspected monthly. Air and oxygen cylinders shall be maintained in a fully charged state and shall be recharged when the pressure falls to 90% of the manufacturer's recommended pressure level. The employer shall determine that the regulator and warning devices function properly.

(h)(3)(iv)
For respirators maintained for emergency use, the employer shall:

(h)(3)(iv)(A)
Certify the respirator by documenting the date the inspection was performed, the name (or signature) of the person who made the inspection, the findings, required remedial action, and a serial number or other means of identifying the inspected respirator; and

(h)(3)(iv)(B)
Provide this information on a tag or label that is attached to the storage compartment for the respirator, is kept with the respirator, or is included in inspection reports stored as paper or electronic files. This information shall be maintained until replaced following a subsequent certification.

(h)(4)
Repairs. The employer shall ensure that respirators that fail an inspection or are otherwise found to be defective are removed from service, and are discarded or repaired or adjusted in accordance with the following procedures:

(h)(4)(i)
Repairs or adjustments to respirators are to be made only by persons appropriately trained to perform such operations and shall use only the respirator manufacturer's NIOSH-approved parts designed for the respirator;

(h)(4)(ii)
Repairs shall be made according to the manufacturer's recommendations and specifications for the type and extent of repairs to be performed; and

(h)(4)(iii)
Reducing and admission valves, regulators, and alarms shall be adjusted or repaired only by the manufacturer or a technician trained by the manufacturer.

(i)
Breathing air quality and use. This paragraph requires the employer to provide employees using atmosphere-supplying respirators (supplied-air and SCBA) with breathing gases of high purity.

(i)(1)
The employer shall ensure that compressed air, compressed oxygen, liquid air, and liquid oxygen used for respiration accords with the following specifications:

(i)(1)(i)
Compressed and liquid oxygen shall meet the U.S Pharmacopoeia requirements for medical or breathing oxygen; and

(i)(1)(ii)
Compressed breathing air shall meet at least the requirements for Grade D breathing air described in ANSI/Compressed Gas Association Commodity Specification for Air, G-7.1-1989, to include:

(i)(1)(ii)(A)
Oxygen content (v/v) of 19.5 to 23.5%;

(i)(1)(ii)(B)
Hydrocarbon (condensed) content of 5 mg/m^3 of air or less;

(i)(1)(ii)(C)
Carbon monoxide (CO) content of 10 ppm or less;

(i)(1)(ii)(D)
Carbon dioxide content of 1000 ppm or less; and

(i)(1)(ii)(E)
Lack of noticeable odor.

(i)(2)
The employer shall ensure that compressed oxygen is not used in atmosphere-supplying respirators that have previously used compressed air.

(i)(3)
The employer shall ensure that oxygen concentrations greater than 23.5% are used only in equipment designed for oxygen service or distribution.

(i)(4)
The employer shall ensure that cylinders used to supply breathing air to respirators meet the following requirements:

(i)(4)(i)
Cylinders are tested and maintained as prescribed in the Shipping Container Specification Regulations of the Department of Transportation (49 CFR part 173 and part 178);

(i)(4)(ii)
Cylinders of purchased breathing air have a certificate of analysis from the supplier that the breathing air meets the requirements for Grade D breathing air; and

(i)(4)(iii)
The moisture content in the cylinder does not exceed a dew point of $-50°F$ ($-45.6°C$) at 1 atm pressure.

(i)(5)
The employer shall ensure that compressors used to supply breathing air to respirators are constructed and situated so as to:

(i)(5)(i)
Prevent entry of contaminated air into the air-supply system;

(i)(5)(ii)
Minimize moisture content so that the dew point at 1 atm pressure is $10°F$ ($5.56°C$) below the ambient temperature;

(i)(5)(iii)
Have suitable in-line air-purifying sorbent beds and filters to further ensure breathing air quality. Sorbent beds and filters shall be maintained and replaced or refurbished periodically following the manufacturer's instructions.

(i)(5)(iv)
Have a tag containing the most recent change date and the signature of the person authorized by the employer to perform the change. The tag shall be maintained at the compressor.

(i)(6)
For compressors that are not oil-lubricated, the employer shall ensure that carbon monoxide levels in the breathing air do not exceed 10 ppm.

(i)(7)
For oil-lubricated compressors, the employer shall use a high-temperature or carbon monoxide alarm, or both, to monitor carbon monoxide levels. If only high-temperature alarms are used, the air supply shall be monitored at intervals sufficient to prevent carbon monoxide in the breathing air from exceeding 10 ppm.

(i)(8)
The employer shall ensure that breathing air couplings are incompatible with outlets for nonrespirable worksite air or other gas systems. No asphyxiating substance shall be introduced into breathing air lines.

(i)(9)
The employer shall use breathing gas containers marked in accordance with the NIOSH respirator certification standard, 42 CFR part 84.

(j)
Identification of filters, cartridges, and canisters. The employer shall ensure that all filters, cartridges, and canisters used in the workplace are labeled and color-coded with the NIOSH approval label and that the label is not removed and remains legible.

(k)
Training and information. This paragraph requires the employer to provide effective training to employees who are required to use respirators. The training must be comprehensive, understandable, and recur annually, and more often if necessary. This paragraph also requires the employer to provide the basic information on respirators in Appendix D of this section to employees who wear respirators when not required by this section or by the employer to do so.

(k)(1)
The employer shall ensure that each employee can demonstrate knowledge of at least the following:

(k)(1)(i)
Why the respirator is necessary and how improper fit, usage, or maintenance can compromise the protective effect of the respirator;

(k)(1)(ii)
What the limitations and capabilities of the respirator are;

(k)(1)(iii)
How to use the respirator effectively in emergency situations, including situations in which the respirator malfunctions;

(k)(1)(iv)
How to inspect, put on and remove, use, and check the seals of the respirator;

(k)(1)(v)
What the procedures are for maintenance and storage of the respirator;

(k)(1)(vi)
How to recognize medical signs and symptoms that may limit or prevent the effective use of respirators; and

(k)(1)(vii)
The general requirements of this section.

(k)(2)
The training shall be conducted in a manner that is understandable to the employee.

(k)(3)
The employer shall provide the training prior to requiring the employee to use a respirator in the workplace.

(k)(4)

An employer who is able to demonstrate that a new employee has received training within the last 12 months that addresses the elements specified in paragraph (k)(1)(i) through (vii) is not required to repeat such training provided that, as required by paragraph (k)(1), the employee can demonstrate knowledge of those element(s). Previous training not repeated initially by the employer must be provided no later than 12 months from the date of the previous training.

(k)(5)

Retraining shall be administered annually, and when the following situations occur:

(k)(5)(i)

Changes in the workplace or the type of respirator render previous training obsolete;

(k)(5)(ii)

Inadequacies in the employee's knowledge or use of the respirator indicate that the employee has not retained the requisite understanding or skill; or

(k)(5)(iii)

Any other situation arises in which retraining appears necessary to ensure safe respirator use.

(k)(6)

The basic advisory information on respirators, as presented in Appendix D of this section, shall be provided by the employer in any written or oral format, to employees who wear respirators when such use is not required by this section or by the employer.

(l)

Program evaluation. This section requires the employer to conduct evaluations of the workplace to ensure that the written respiratory protection program is being properly implemented, and to consult employees to ensure that they are using the respirators properly.

(l)(1)

The employer shall conduct evaluations of the workplace as necessary to ensure that the provisions of the current written program are being effectively implemented and that it continues to be effective.

(l)(2)

The employer shall regularly consult employees required to use respirators to assess the employees' views on program effectiveness and to identify any problems. Any problems that are identified during this assessment shall be corrected. Factors to be assessed include, but are not limited to:

(l)(2)(i)

Respirator fit (including the ability to use the respirator without interfering with effective workplace performance);

(l)(2)(ii)

Appropriate respirator selection for the hazards to which the employee is exposed;

(l)(2)(iii)
Proper respirator use under the workplace conditions the employee encounters; and

(l)(2)(iv)
Proper respirator maintenance.

(m)
Record keeping. This section requires the employer to establish and retain written information regarding medical evaluations, fit testing, and the respirator program. This information will facilitate employee involvement in the respirator program, assist the employer in auditing the adequacy of the program, and provide a record for compliance determinations by OSHA.

(m)(1)
Medical evaluation. Records of medical evaluations required by this section must be retained and made available in accordance with 29 CFR 1910.1020.

(m)(2)
Fit testing.

(m)(2)(i)
The employer shall establish a record of the qualitative and quantitative fit tests administered to an employee including:

(m)(2)(i)(A)
The name or identification of the employee tested;

(m)(2)(i)(B)
Type of fit test performed;

(m)(2)(i)(C)
Specific make, model, style, and size of respirator tested;

(m)(2)(i)(D)
Date of test; and

(m)(2)(i)(E)
The pass/fail results for QLFTs or the fit factor and strip chart recording or other recording of the test results for QNFTs.

(m)(2)(ii)
Fit test records shall be retained for respirator users until the next fit test is administered.

(m)(3)
A written copy of the current respirator program shall be retained by the employer.

(m)(4)
Written materials required to be retained under this paragraph shall be made available upon request to affected employees and to the Assistant Secretary or designee for examination and copying.

(n)
Dates.

(n)(1)
Effective date. This section is effective April 8, 1998. The obligations imposed by this section commence on the effective date unless otherwise noted in this paragraph. Compliance with obligations that do not commence on the effective date shall occur no later than the applicable start-up date.

(n)(2)
Compliance dates. All obligations of this section commence on the effective date except as follows:

(n)(2)(i)
The determination that respirator use is required (paragraph **(a)**) shall be completed no later than September 8, 1998.

(o)
Appendices.

(o)(1)
Compliance with Appendix A, Appendix B-1, Appendix B-2, and Appendix C of this section is mandatory.

(o)(2)
Appendix D of this section is nonmandatory and is not intended to create any additional obligations not otherwise imposed or to detract from any existing obligations.

APPENDIX A TO §1910.134: FIT-TESTING PROCEDURES (MANDATORY)

Part I. OSHA-Accepted Fit-Test Protocols

A. Fit-Testing Procedures — General Requirements

The employer shall conduct fit testing using the following procedures. The requirements in this appendix apply to all OSHA-accepted fit-test methods, both QLFT and QNFT.

1. The test subject shall be allowed to pick the most acceptable respirator from a sufficient number of respirator models and sizes so that the respirator is acceptable to, and correctly fits, the user.

2. Prior to the selection process, the test subject shall be shown how to put on a respirator, how it should be positioned on the face, how to set strap tension, and how to determine an acceptable fit. A mirror shall be available to assist the subject in evaluating the fit and positioning of the respirator. This instruction may not constitute the subject's formal training on respirator use, because it is only a review.

3. The test subject shall be informed that he or she is being asked to select the respirator that provides the most acceptable fit. Each respirator represents a different size and shape, and if fitted and used properly, will provide adequate protection.

4. The test subject shall be instructed to hold each chosen facepiece up to the face and eliminate those that obviously do not give an acceptable fit.

5. The more acceptable facepieces are noted in case the one selected proves unacceptable; the most comfortable mask is donned and worn for at least 5 min to assess comfort. Assistance in assessing comfort can be given by discussing the points in the following item A.6. If the test subject is not familiar with using a particular respirator, the test subject shall be directed to don the mask several times and to adjust the straps each time to become adept at setting proper tension on the straps.

6. Assessment of comfort shall include a review of the following points with the test subject and allowing the test subject adequate time to determine the comfort of the respirator:

(a) Position of the mask on the nose

(b) Room for eye protection

(c) Room to talk

(d) Position of mask on face and cheeks

7. The following criteria shall be used to help determine the adequacy of the respirator fit:

(a) Chin properly placed;

(b) Adequate strap tension, not overly tightened;

(c) Fit across nose bridge;

(d) Respirator of proper size to span distance from nose to chin;

(e) Tendency of respirator to slip;

(f) Self-observation in mirror to evaluate fit and respirator position.

8. The test subject shall conduct a user seal check, either the negative and positive pressure seal checks described in Appendix B-1 of this section or those recommended by the respirator manufacturer which provide equivalent protection to the procedures in Appendix B-1. Before conducting the negative and positive pressure checks, the subject shall be told to seat the mask on the face by moving the head from side-to-side and up and down slowly while taking in a few slow deep breaths. Another facepiece shall be selected and retested if the test subject fails the user seal check tests.

9. The test shall not be conducted if there is any hair growth between the skin and the facepiece sealing surface, such as stubble beard growth, beard, mustache, or sideburns that cross the respirator sealing surface. Any type of apparel that interferes with a satisfactory fit shall be altered or removed.

10. If a test subject exhibits difficulty in breathing during the tests, he or she shall be referred to a physician or other licensed health-care professional, as appropriate, to determine whether the test subject can wear a respirator while performing his or her duties.

11. If the employee finds the fit of the respirator unacceptable, the test subject shall be given the opportunity to select a different respirator and to be retested.

12. Exercise regimen. Prior to the commencement of the fit test, the test subject shall be given a description of the fit test and the test subject's responsibilities during the test procedure. The description of the process shall include a description of the test exercises that the subject will be performing. The respirator to be tested shall be worn for at least 5 min before the start of the fit test.

13. The fit test shall be performed while the test subject is wearing any applicable safety equipment that may be worn during actual respirator use which could interfere with respirator fit.

14. Test exercises.

(a) The following test exercises are to be performed for all fit testing methods prescribed in this appendix, except for the CNP method. A separate fit testing exercise regimen is contained in the CNP protocol. The test subject shall perform exercises, in the test environment, in the following manner:

(1) Normal breathing. In a normal standing position, without talking, the subject shall breathe normally.

(2) Deep breathing. In a normal standing position, the subject shall breathe slowly and deeply, taking caution not to hyperventilate.

(3) Turning head side to side. Standing in place, the subject shall slowly turn his or her head from side to side between the extreme positions on each side. The head shall be held at each extreme momentarily so the subject can inhale at each side.

(4) Moving head up and down. Standing in place, the subject shall slowly move his or her head up and down. The subject shall be instructed to inhale in the up position (i.e., when looking toward the ceiling).

(5) Talking. The subject shall talk out loud slowly and loud enough to be heard clearly by the test conductor. The subject can read from a prepared text such as the Rainbow Passage, count backward from 100, or recite a memorized poem or song.

Rainbow Passage
When the sunlight strikes raindrops in the air, they act like a prism and form a rainbow. The rainbow is a division of white light into many beautiful colors. These take the shape of a long round arch, with its path high above, and its two ends apparently beyond the horizon. There is, according to legend, a boiling pot of gold at one end. People look, but no one ever finds it. When a man looks for something beyond reach, his friends say he is looking for the pot of gold at the end of the rainbow.

(6) Grimace. The test subject shall grimace by smiling or frowning. (This applies only to QNFT testing; it is not performed for QLFT.)

(7) Bending over. The test subject shall bend at the waist as if he or she were to touch his or her toes. Jogging in place shall be substituted for this exercise in those test environments such as shroud type QNFT or QLFT units that do not permit bending over at the waist.

(8) Normal breathing. Same as exercise (1).

(b) Each test exercise shall be performed for 1 min except for the grimace exercise, which shall be performed for 15 s. The test subject shall be questioned by the test conductor regarding the comfort of the respirator upon completion of the protocol. If it has become unacceptable, another model of respirator shall be tried. The respirator shall not be adjusted once the fit-test exercises begin. Any adjustment voids the test, and the fit test must be repeated.

B. Qualitative Fit-Test (QLFT) Protocols

1. General

(a) The employer shall ensure that persons administering QLFT are able to prepare test solutions, calibrate equipment, perform tests properly, recognize invalid tests, and ensure that test equipment is in proper working order.

(b) The employer shall ensure that QLFT equipment is kept clean and well maintained so that it operates within the parameters for which it was designed.

2. Isoamyl Acetate Protocol

Note: This protocol is not appropriate to use for the fit testing of particulate respirators. If used to fit test particulate respirators, the respirator must be equipped with an organic vapor filter.

(a) Odor Threshold Screening. Odor threshold screening, performed without wearing a respirator, is intended to determine if the individual tested can detect the odor of isoamyl acetate (IAA) at low levels.

(1) Three 1 liter glass jars with metal lids are required.

(2) Odor-free water (e.g., distilled or spring water) at approximately 25°C (77°F) shall be used for the solutions.

(3) The IAA (also known at isopentyl acetate) stock solution is prepared by adding 1 ml of pure IAA to 800 ml of odor-free water in a 1 liter jar, closing the lid and shaking for 30 s. A new solution shall be prepared at least weekly.

(4) The screening test shall be conducted in a room separate from the room used for actual fit testing. The two rooms shall be well ventilated to prevent the odor of IAA from becoming evident in the general room air where testing takes place.

(5) The odor test solution is prepared in a second jar by placing 0.4 ml of the stock solution into 500 ml of odor-free water using a clean dropper or pipette. The solution

shall be shaken for 30 s and allowed to stand for 2 to 3 min so that the IAA concentration above the liquid may reach equilibrium. This solution shall be used for only 1 day.

(6) A test blank shall be prepared in a third jar by adding 500 cc of odor-free water.

(7) The odor test and test blank jar lids shall be labeled (e.g., 1 and 2) for jar identification. Labels shall be placed on the lids so that they can be peeled off periodically and switched to maintain the integrity of the test.

(8) The following instruction shall be typed on a card and placed on the table in front of the two test jars (i.e., 1 and 2): "The purpose of this test is to determine if you can smell banana oil at a low concentration. The two bottles in front of you contain water. One of these bottles also contains a small amount of banana oil. Be sure the covers are on tight, then shake each bottle for 2 s. Unscrew the lid of each bottle, one at a time, and sniff at the mouth of the bottle. Indicate to the test conductor which bottle contains banana oil."

(9) The mixtures used in the IAA odor detection test shall be prepared in an area separate from where the test is performed, in order to prevent olfactory fatigue in the subject.

(10) If the test subject is unable to correctly identify the jar containing the odor test solution, the IAA qualitative fit test shall not be performed.

(11) If the test subject correctly identifies the jar containing the odor test solution, the test subject may proceed to respirator selection and fit testing.

(b) Isoamyl Acetate Fit Test

(1) The fit-test chamber shall be a clear 55-gal drum liner suspended inverted over a 2-ft-diameter frame so that the top of the chamber is about 6 in. above the test subject's head. If no drum liner is available, a similar chamber shall be constructed using plastic sheeting. The inside top center of the chamber shall have a small hook attached.

(2) Each respirator used for the fitting and fit testing shall be equipped with organic vapor cartridges or offer protection against organic vapors.

(3) After selecting, donning, and properly adjusting a respirator, the test subject shall wear it to the fit testing room. This room shall be separate from the room used for odor threshold screening and respirator selection, and shall be well ventilated, by an exhaust fan or laboratory hood, to prevent general room contamination.

(4) A copy of the test exercises and any prepared text from which the subject is to read shall be taped to the inside of the test chamber.

(5) Upon entering the test chamber, the test subject shall be given a 6- × 5-in. piece of paper towel, or other porous, absorbent, single-ply material, folded in half and wetted with 0.75 ml of pure IAA. The test subject shall hang the wet towel on the hook at the top of the chamber. An IAA test swab or ampule may be substituted for the IAA wetted paper towel provided it has been demonstrated that the alternative

IAA source will generate an IAA test atmosphere with a concentration equivalent to that generated by the paper towel method.

(6) Allow 2 min for the IAA test concentration to stabilize before starting the fit-test exercises. This would be an appropriate time to talk with the test subject; to explain the fit test, the importance of his or her cooperation, and the purpose for the test exercises; or to demonstrate some of the exercises.

(7) If at any time during the test, the subject detects the banana-like odor of IAA, the test is failed. The subject shall quickly exit from the test chamber and leave the test area to avoid olfactory fatigue.

(8) If the test is failed, the subject shall return to the selection room and remove the respirator. The test subject shall repeat the odor sensitivity test, select and put on another respirator, return to the test area, and again begin the fit-test procedure described in (b)(1) through (7) above. The process continues until a respirator that fits well has been found. Should the odor sensitivity test be failed, the subject shall wait at least 5 min before retesting. Odor sensitivity will usually have returned by this time.

(9) If the subject passes the test, the efficiency of the test procedure shall be demonstrated by having the subject break the respirator face seal and take a breath before exiting the chamber.

(10) When the test subject leaves the chamber, the subject shall remove the saturated towel and return it to the person conducting the test, so that there is no significant IAA concentration buildup in the chamber during subsequent tests. The used towels shall be kept in a self-sealing plastic bag to keep the test area from being contaminated.

3. Saccharin Solution Aerosol Protocol

The entire screening and testing procedure shall be explained to the test subject prior to the conduct of the screening test.

(a) Taste threshold screening. The saccharin taste threshold screening, performed without wearing a respirator, is intended to determine whether the individual being tested can detect the taste of saccharin.

(1) During threshold screening as well as during fit testing, subjects shall wear an enclosure about the head and shoulders that is approximately 12 in. in diameter by 14 in. tall with at least the front portion clear and that allows free movements of the head when a respirator is worn. An enclosure substantially similar to the 3M hood assembly, parts # FT 14 and # FT 15 combined, is adequate.

(2) The test enclosure shall have a 3/4-in. (1.9 cm) hole in front of the test subject's nose and mouth area to accommodate the nebulizer nozzle.

(3) The test subject shall don the test enclosure. Throughout the threshold screening test, the test subject shall breathe through his or her slightly open mouth with tongue extended. The subject is instructed to report when he or she detects a sweet taste.

(4) Using a DeVilbiss Model 40 Inhalation Medication Nebulizer or equivalent, the test conductor shall spray the threshold check solution into the enclosure. The nozzle is directed away from the nose and mouth of the person. This nebulizer shall be clearly marked to distinguish it from the fit-test solution nebulizer.

(5) The threshold check solution is prepared by dissolving 0.83 g of sodium saccharin USP in 100 ml of warm water. It can be prepared by putting 1 ml of the fit-test solution (see (b)(5) below) in 100 ml of distilled water.

(6) To produce the aerosol, the nebulizer bulb is firmly squeezed so that it collapses completely, then released and allowed to fully expand.

(7) Ten squeezes are repeated rapidly and then the test subject is asked whether the saccharin can be tasted. If the test subject reports tasting the sweet taste during the ten squeezes, the screening test is completed. The taste threshold is noted as ten regardless of the number of squeezes actually completed.

(8) If the first response is negative, ten more squeezes are repeated rapidly and the test subject is again asked whether the saccharin is tasted. If the test subject reports tasting the sweet taste during the second ten squeezes, the screening test is completed. The taste threshold is noted as 20 regardless of the number of squeezes actually completed.

(9) If the second response is negative, ten more squeezes are repeated rapidly and the test subject is again asked whether the saccharin is tasted. If the test subject reports tasting the sweet taste during the third set of ten squeezes, the screening test is completed. The taste threshold is noted as 30 regardless of the number of squeezes actually completed.

(10) The test conductor will take note of the number of squeezes required to solicit a taste response.

(11) If the saccharin is not tasted after 30 squeezes (step 10), the test subject is unable to taste saccharin and may not perform the saccharin fit test.

Note: If the test subject eats or drinks something sweet before the screening test, he/she may be unable to taste the weak saccharin solution.

(12) If a taste response is elicited, the test subject shall be asked to take note of the taste for reference in the fit test.

(13) Correct use of the nebulizer means that approximately 1 ml of liquid is used at a time in the nebulizer body.

(14) The nebulizer shall be thoroughly rinsed in water, shaken dry, and refilled at least each morning and afternoon or at least every 4 h.

(b) Saccharin solution aerosol fit-test procedure.

(1) The test subject may not eat, drink (except plain water), smoke, or chew gum for 15 min before the test.

(2) The fit test uses the same enclosure described in 3 (a) above.

(3) The test subject shall don the enclosure while wearing the respirator selected in Section I.A of this appendix. The respirator shall be properly adjusted and equipped with a particulate filter(s).

(4) A second DeVilbiss Model 40 Inhalation Medication Nebulizer or equivalent is used to spray the fit-test solution into the enclosure. This nebulizer shall be clearly marked to distinguish it from the screening test solution nebulizer.

(5) The fit-test solution is prepared by adding 83 g of sodium saccharin to 100 ml of warm water.

(6) As before, the test subject shall breathe through the slightly open mouth with tongue extended, and report if he or she tastes the sweet taste of saccharin.

(7) The nebulizer is inserted into the hole in the front of the enclosure and an initial concentration of saccharin fit-test solution is sprayed into the enclosure using the same number of squeezes (10, 20, or 30 squeezes) based on the number of squeezes required to elicit a taste response as noted during the screening test. A minimum of 10 squeezes is required.

(8) After generating the aerosol, the test subject shall be instructed to perform the exercises in Section I.A.14. of this appendix.

(9) Every 30 s the aerosol concentration shall be replenished using one half the original number of squeezes used initially (e.g., 5, 10, or 15).

(10) The test subject shall indicate to the test conductor if at any time during the fit test the taste of saccharin is detected. If the test subject does not report tasting the saccharin, the test is passed.

(11) If the taste of saccharin is detected, the fit is deemed unsatisfactory and the test is failed. A different respirator shall be tried and the entire test procedure is repeated (taste threshold screening and fit testing).

(12) Since the nebulizer has a tendency to clog during use, the test operator must make periodic checks of the nebulizer to ensure that it is not clogged. If clogging is found at the end of the test session, the test is invalid.

4. Bitrex™ (Denatonium Benzoate) Solution Aerosol Qualitative Fit-Test Protocol

The Bitrex™ (Denatonium benzoate) solution aerosol QLFT protocol uses the published saccharin test protocol because that protocol is widely accepted. Bitrex is routinely used as a taste-aversion agent in household liquids that children should not be drinking and is endorsed by the American Medical Association, the National Safety Council, and the American Association of Poison Control Centers. The entire screening and testing procedure shall be explained to the test subject prior to the conduct of the screening test.

(a) Taste Threshold Screening. The Bitrex taste threshold screening, performed without wearing a respirator, is intended to determine whether the individual being tested can detect the taste of Bitrex.

(1) During threshold screening as well as during fit-testing, subjects shall wear an enclosure about the head and shoulders that is approximately 12 in (30.5 cm) in diameter by 14 in (35.6 cm) tall. The front portion of the enclosure shall be clear from the respirator and allow free movement of the head when a respirator is worn. An enclosure substantially similar to the 3M hood assembly, parts # FT 14 and # FT 15 combined, is adequate.

(2) The test enclosure shall have a 3/4-in. (1.9-cm) hole in front of the test subject's nose and mouth area to accommodate the nebulizer nozzle.

(3) The test subject shall don the test enclosure. Throughout the threshold screening test, the test subject shall breathe through his or her slightly open mouth with tongue extended. The subject is instructed to report when he or she detects a bitter taste.

(4) Using a DeVilbiss Model 40 Inhalation Medication Nebulizer or equivalent, the test conductor shall spray the Threshold Check Solution into the enclosure. This nebulizer shall be clearly marked to distinguish it from the fit-test solution nebulizer.

(5) The Threshold Check Solution is prepared by adding 13.5 mg of Bitrex to 100 ml of 5% salt (NaCl) solution in distilled water.

(6) To produce the aerosol, the nebulizer bulb is firmly squeezed so that the bulb collapses completely, and is then released and allowed to expand fully.

(7) An initial ten squeezes are repeated rapidly and then the test subject is asked whether the Bitrex can be tasted. If the test subject reports tasting the bitter taste during the ten squeezes, the screening test is completed. The taste threshold is noted as ten regardless of the number of squeezes actually completed.

(8) If the first response is negative, ten more squeezes are repeated rapidly and the test subject is again asked whether the Bitrex is tasted. If the test subject reports tasting the bitter taste during the second ten squeezes, the screening test is completed. The taste threshold is noted as 20 regardless of the number of squeezes actually completed.

(9) If the second response is negative, ten more squeezes are repeated rapidly and the test subject is again asked whether the Bitrex is tasted. If the test subject reports tasting the bitter taste during the third set of ten squeezes, the screening test is completed. The taste threshold is noted as 30 regardless of the number of squeezes actually completed.

(10) The test conductor will take note of the number of squeezes required to solicit a taste response.

(11) If the Bitrex is not tasted after 30 squeezes (step 10), the test subject is unable to taste Bitrex and may not perform the Bitrex fit test.

(12) If a taste response is elicited, the test subject shall be asked to take note of the taste for reference in the fit test.

(13) Correct use of the nebulizer means that approximately 1 ml of liquid is used at a time in the nebulizer body.

(14) The nebulizer shall be thoroughly rinsed in water, shaken to dry, and refilled at least each morning and afternoon or at least every 4 hours.

(b) Bitrex Solution Aerosol Fit Test Procedure

(1) The test subject may not eat, drink (except plain water), smoke, or chew gum for 15 min before the test.

(2) The fit test uses the same enclosure as that described in 4(a) above.

(3) The test subject shall don the enclosure while wearing the respirator selected according to Section I.A. of this appendix. The respirator shall be properly adjusted and equipped with any type of particulate filter(s).

(4) A second DeVilbiss Model 40 Inhalation Medication Nebulizer or equivalent is used to spray the fit test solution into the enclosure. This nebulizer shall be clearly marked to distinguish it from the screening test solution nebulizer.

(5) The fit test solution is prepared by adding 337.5 mg of Bitrex to 200 ml of a 5% salt (NaCl) solution in warm water.

(6) As before, the test subject shall breathe through his or her slightly open mouth with tongue extended, and be instructed to report if he or she tastes the bitter taste of Bitrex.

(7) The nebulizer is inserted into the hole in the front of the enclosure and an initial concentration of the fit-test solution is sprayed into the enclosure using the same number of squeezes (either 10, 20, or 30 squeezes) based on the number of squeezes required to elicit a taste response as noted during the screening test.

(8) After generating the aerosol, the test subject shall be instructed to perform the exercises in Section I.A.14. of this appendix.

(9) Every 30 s the aerosol concentration shall be replenished using one half the number of squeezes used initially (e.g., 5, 10, or 15).

(10) The test subject shall indicate to the test conductor if at any time during the fit test the taste of Bitrex is detected. If the test subject does not report tasting the Bitrex, the test is passed.

(11) If the taste of Bitrex is detected, the fit is deemed unsatisfactory and the test is failed. A different respirator shall be tried and the entire test procedure is repeated (taste threshold screening and fit testing).

5. Irritant Smoke (Stannic Chloride) Protocol

This qualitative fit test uses a person's response to the irritating chemicals released in the "smoke" produced by a stannic chloride ventilation smoke tube to detect leakage into the respirator.

(a) General Requirements and Precautions

(1) The respirator to be tested shall be equipped with high-efficiency particulate air (HEPA) or P100 series filter(s).

(2) Only stannic chloride smoke tubes shall be used for this protocol.

(3) No form of test enclosure or hood for the test subject shall be used.

(4) The smoke can be irritating to the eyes, lungs, and nasal passages. The test conductor shall take precautions to minimize the test subject's exposure to irritant smoke. Sensitivity varies, and certain individuals may respond to a greater degree to irritant smoke. Care shall be taken when performing the sensitivity screening checks that determine whether the test subject can detect irritant smoke to use only the minimum amount of smoke necessary to elicit a response from the test subject.

(5) The fit test shall be performed in an area with adequate ventilation to prevent exposure of the person conducting the fit test or the buildup of irritant smoke in the general atmosphere.

(b) Sensitivity Screening Check. The person to be tested must demonstrate his or her ability to detect a weak concentration of the irritant smoke.

(1) The test operator shall break both ends of a ventilation smoke tube containing stannic chloride, and attach one end of the smoke tube to a low-flow air pump set to deliver 200 ml/min, or an aspirator squeeze bulb. The test operator shall cover the other end of the smoke tube with a short piece of tubing to prevent potential injury from the jagged end of the smoke tube.

(2) The test operator shall advise the test subject that the smoke can be irritating to the eyes, lungs, and nasal passages and instruct the subject to keep his or her eyes closed while the test is performed.

(3) The test subject shall be allowed to smell a weak concentration of the irritant smoke before the respirator is donned to become familiar with its irritating properties and to determine if he or she can detect the irritating properties of the smoke. The test operator shall carefully direct a small amount of the irritant smoke in the test subject's direction to determine that he or she can detect it.

(c) Irritant Smoke Fit-Test Procedure

(1) The person being fit-tested shall don the respirator without assistance, and perform the required user seal check(s).

(2) The test subject shall be instructed to keep his or her eyes closed.

(3) The test operator shall direct the stream of irritant smoke from the smoke tube toward the face-seal area of the test subject, using the low-flow pump or the squeeze bulb. The test operator shall begin at least 12 in. from the facepiece and move the smoke stream around the whole perimeter of the mask. The operator shall gradually make two more passes around the perimeter of the mask, moving to within 6 in. of the respirator.

(4) If the person being tested has not had an involuntary response and/or detected the irritant smoke, proceed with the test exercises.

(5) The exercises identified in Section I.A.14 of this appendix shall be performed by the test subject while the respirator seal is being continually challenged by the smoke, directed around the perimeter of the respirator at a distance of 6 in.

(6) If the person being fit-tested reports detecting the irritant smoke at any time, the test is failed. The person being retested must repeat the entire sensitivity check and fit-test procedure.

(7) Each test subject passing the irritant smoke test without evidence of a response (involuntary cough, irritation) shall be given a second sensitivity screening check, with the smoke from the same smoke tube used during the fit test, once the respirator has been removed, to determine whether he or she still reacts to the smoke. Failure to evoke a response shall void the fit test.

(8) If a response is produced during this second sensitivity check, then the fit test is passed.

C. Quantitative Fit-Testing (QNFT) Protocols

The following quantitative fit-testing procedures have been demonstrated to be acceptable: Quantitative fit testing using a nonhazardous test aerosol, such as corn oil, polyethylene glycol 400 (PEG 400), di-2-ethyl hexyl sebacate (DEHS), or sodium chloride, generated in a test chamber, and employing instrumentation to quantify the fit of the respirator; Quantitative fit testing using ambient aerosol as the test agent and appropriate instrumentation (condensation nuclei counter) to quantify the respirator fit; Quantitative fit testing using controlled negative pressure and appropriate instrumentation to measure the volumetric leak rate of a facepiece to quantify the respirator fit.

1. General

(a) The employer shall ensure that persons administering QNFT are able to calibrate equipment and perform tests properly, recognize invalid tests, calculate fit factors properly, and ensure that test equipment is in proper working order.

(b) The employer shall ensure that QNFT equipment is kept clean, and is maintained and calibrated according to the manufacturer's instructions so that it operates at the parameters for which it was designed.

2. Generated Aerosol Quantitative Fit-Testing Protocol

(a) Apparatus.

(1) Instrumentation. Aerosol generation, dilution, and measurement systems using particulates, such as corn oil, polyethylene glycol 400 (PEG 400), di-2-ethyl hexyl sebacate (DEHS), or sodium chloride, as test aerosols shall be used for quantitative fit testing.

(2) Test chamber. The test chamber shall be large enough to permit all test subjects to perform freely all required exercises without disturbing the test agent concentration or the measurement apparatus. The test chamber shall be equipped and con-

structed so that the test agent is effectively isolated from the ambient air, yet uniform in concentration throughout the chamber.

(3) When testing air-purifying respirators, the normal filter or cartridge element shall be replaced with a high-efficiency particulate air (HEPA) or P100 series filter supplied by the same manufacturer.

(4) The sampling instrument shall be selected so that a computer record or strip chart record may be made of the test showing the rise and fall of the test agent concentration with each inspiration and expiration at fit factors of at least 2000. Integrators or computers that integrate the amount of test agent penetration leakage into the respirator for each exercise may be used provided a record of the readings is made.

(5) The combination of substitute air-purifying elements, test agent and test agent concentration shall be such that the test subject is not exposed in excess of an established exposure limit for the test agent at any time during the testing process, based upon the length of the exposure and the exposure limit duration.

(6) The sampling port on the test specimen respirator shall be placed and constructed so that no leakage occurs around the port (e.g., where the respirator is probed), a free air flow is allowed into the sampling line at all times, and there is no interference with the fit or performance of the respirator. The in-mask sampling device (probe) shall be designed and used so that the air sample is drawn from the breathing zone of the test subject, midway between the nose and mouth and with the probe extending into the facepiece cavity at least 1/4 in.

(7) The test setup shall permit the person administering the test to observe the test subject inside the chamber during the test.

(8) The equipment generating the test atmosphere shall maintain the concentration of test agent constant to within a 10% variation for the duration of the test.

(9) The time lag (interval between an event and the recording of the event on the strip chart or computer or integrator) shall be kept to a minimum. There shall be a clear association between the occurrence of an event and its being recorded.

(10) The sampling line tubing for the test chamber atmosphere and for the respirator sampling port shall be of equal diameter and of the same material. The length of the two lines shall be equal.

(11) The exhaust flow from the test chamber shall pass through an appropriate filter (i.e., high-efficiency particulate filter) before release.

(12) When sodium chloride aerosol is used, the relative humidity inside the test chamber shall not exceed 50%.

(13) The limitations of instrument detection shall be taken into account when determining the fit factor.

(14) Test respirators shall be maintained in proper working order and be inspected regularly for deficiencies such as cracks or missing valves and gaskets.

(b) Procedural Requirements

(1) When performing the initial user seal check using a positive or negative pressure check, the sampling line shall be crimped closed to avoid air pressure leakage during either of these pressure checks.

(2) The use of an abbreviated screening QLFT test is optional. Such a test may be utilized to quickly identify poorly fitting respirators that passed the positive and/or negative pressure test and reduce the amount of QNFT time. The use of the CNC QNFT instrument in the count mode is another optional method to obtain a quick estimate of fit and eliminate poorly fitting respirators before going on to perform a full QNFT.

(3) A reasonably stable test agent concentration shall be measured in the test chamber prior to testing. For canopy or shower curtain types of test units, the determination of the stability of the test agent may be established after the test subject has entered the test environment.

(4) Immediately after the subject enters the test chamber, the test agent concentration inside the respirator shall be measured to ensure that the peak penetration does not exceed 5% for a half mask or 1% for a full-facepiece respirator.

(5) A stable test agent concentration shall be obtained prior to the actual start of testing.

(6) Respirator restraining straps shall not be overtightened for testing. The straps shall be adjusted by the wearer without assistance from other persons to give a reasonably comfortable fit typical of normal use. The respirator shall not be adjusted once the fit test exercises begin.

(7) The test shall be terminated whenever any single peak penetration exceeds 5% for half masks and 1% for full-facepiece respirators. The test subject shall be refitted and retested.

(8) Calculation of fit factors.

(i) The fit factor shall be determined for the quantitative fit test by taking the ratio of the average chamber concentration to the concentration measured inside the respirator for each test exercise except the grimace exercise.

(ii) The average test chamber concentration shall be calculated as the arithmetic average of the concentration measured before and after each test (i.e., seven exercises) or the arithmetic average of the concentration measured before and after each exercise or the true average measured continuously during the respirator sample.

(iii) The concentration of the challenge agent inside the respirator shall be determined by one of the following methods:

(A) Average peak penetration method means the method of determining test agent penetration into the respirator utilizing a strip chart recorder, integrator, or computer. The agent penetration is determined by an average of the peak heights on the graph or by computer integration, for each exercise except the grimace exercise. Integrators or computers that calculate the actual test agent penetration into the respirator for each exercise will also be considered to meet the requirements of the average peak penetration method.

(B) Maximum peak penetration method means the method of determining test agent penetration in the respirator as determined by strip chart recordings of the test. The highest peak penetration for a given exercise is taken to be representative of average penetration into the respirator for that exercise.

(C) Integration by calculation of the area under the individual peak for each exercise except the grimace exercise. This includes computerized integration.

(D) The calculation of the overall fit factor using individual exercise fit factors involves first converting the exercise fit factors to penetration values, determining the average, and then converting that result back to a fit factor. This procedure is described in the following equation:

$$\text{Overall Fit Factor} = \frac{\text{Number of Exercises}}{1/\text{ff}_1 + 1/\text{ff}_2 + 1/\text{ff}_3 + 1/\text{ff}_4 + 1/\text{ff}_5 + 1/\text{ff}_6 + 1/\text{ff}_7 + 1/\text{ff}_8}$$

where ff_1, ff_2, ff_3, etc. are the fit factors for exercises 1, 2, 3, etc.

(9) The test subject shall not be permitted to wear a half mask or quarter-facepiece respirator unless a minimum fit factor of 100 is obtained, or a full-facepiece respirator unless a minimum fit factor of 500 is obtained.

(10) Filters used for quantitative fit testing shall be replaced whenever increased breathing resistance is encountered, or when the test agent has altered the integrity of the filter media.

3. Ambient Aerosol Condensation Nuclei Counter (CNC) Quantitative Fit-Testing Protocol

The ambient aerosol CNC quantitative fit-testing (Portacount™) protocol quantitatively fit tests respirators with the use of a probe. The probed respirator is only used for quantitative fit tests. A probed respirator has a special sampling device, installed on the respirator, that allows the probe to sample the air from inside the mask. A probed respirator is required for each make, style, model, and size that the employer uses and can be obtained from the respirator manufacturer or distributor. The CNC instrument manufacturer, TSI Inc., also provides probe attachments (TSI sampling adapters) that permit fit testing in an employee's own respirator. A minimum fit factor pass level of at least 100 is necessary for a half-mask respirator and a minimum fit factor pass level of at least 500 is required for a full-facepiece negative-pressure respirator. The entire screening and testing procedure shall be explained to the test subject prior to the conduct of the screening test.

(a) Portacount Fit-Test Requirements

(1) Check the respirator to make sure the sampling probe and line are properly attached to the facepiece and that the respirator is fitted with a particulate filter capable of preventing significant penetration by the ambient particles used for the

fit test (e.g., NIOSH 42 CFR 84 series 100, series 99, or series 95 particulate filter) per manufacturer's instruction.

(2) Instruct the person to be tested to don the respirator for 5 min before the fit test starts. This purges the ambient particles trapped inside the respirator and permits the wearer to make certain the respirator is comfortable. This individual shall already have been trained on how to wear the respirator properly.

(3) Check the following conditions for the adequacy of the respirator fit: chin properly placed; adequate strap tension, not overly tightened; fit across nose bridge; respirator of proper size to span distance from nose to chin; tendency of the respirator to slip; self-observation in a mirror to evaluate fit and respirator position.

(4) Have the person wearing the respirator do a user seal check. If leakage is detected, determine the cause. If leakage is from a poorly fitting facepiece, try another size of the same model respirator, or another model of respirator.

(5) Follow the manufacturer's instructions for operating the Portacount and proceed with the test.

(6) The test subject shall be instructed to perform the exercises in Section I.A.14 of this appendix.

(7) After the test exercises, the test subject shall be questioned by the test conductor regarding the comfort of the respirator upon completion of the protocol. If it has become unacceptable, another model of respirator shall be tried.

(b) Portacount Test Instrument

(1) The Portacount will automatically stop and calculate the overall fit factor for the entire set of exercises. The overall fit factor is what counts. The Pass or Fail message will indicate whether or not the test was successful. If the test was a Pass, the fit test is over.

(2) Since the pass or fail criterion of the Portacount is user programmable, the test operator shall ensure that the pass or fail criterion meets the requirements for minimum respirator performance in this appendix.

(3) A record of the test shall be kept on file, assuming the fit test was successful. The record must contain the test subject's name; overall fit factor; make, model, style, and size of respirator used; and date tested.

4. Controlled Negative Pressure (CNP) Quantitative Fit-Testing Protocol

The CNP protocol provides an alternative to aerosol fit-test methods. The CNP fit-test method technology is based on exhausting air from a temporarily sealed respirator facepiece to generate and then maintain a constant negative-pressure inside the facepiece. The rate of air exhaust is controlled so that a constant negative pressure is maintained in the respirator during the fit test. The level of pressure is selected to replicate the mean inspiratory pressure that causes leakage into the respirator under normal use conditions. With pressure held constant, air flow out of the respirator is equal to air flow into the respirator. Therefore, measurement of the exhaust stream that is required to hold the pressure in the temporarily sealed respirator constant yields a direct measure of leakage air flow into the respirator. The CNP fit-test method measures leak rates through the facepiece as a method for determining the facepiece fit for negative-pressure respirators. The CNP instrument manufacturer Dynatech Nevada also provides attachments (sampling manifolds) that replace the filter cartridges to permit fit-testing in an employee's own respirator. To perform the test, the test subject closes his or her mouth and holds his or her breath, after which an air pump removes air from the respirator facepiece at a preselected constant pressure. The facepiece fit is expressed as the leak rate through the facepiece, expressed as milliliters per minute. The quality and validity of the CNP fit tests are determined by the degree to which the in-mask pressure tracks the test pressure during the system measurement time of approximately 5 s. Instantaneous feedback in the form of a real-time pressure trace of the in-mask pressure is provided and used to determine test validity and quality. A minimum fit factor pass level of 100 is necessary for a half-mask respirator and a minimum fit factor of at least 500 is required for a full-facepiece respirator. The entire screening and testing procedure shall be explained to the test subject prior to the conduct of the screening test.

(a) CNP Fit-Test Requirements.

(1) The instrument shall have a nonadjustable test pressure of 15.0 mm water pressure.

(2) The CNP system defaults selected for test pressure shall be set at −15 mm of water (−0.58 in. of water) and the modeled inspiratory flow rate shall be 53.8 l/min for performing fit tests.

[**Note:** CNP systems have built-in capability to conduct fit testing that is specific to unique work rate, mask, and gender situations that might apply in a specific work-place. Use of system default values, which were selected to represent respirator wear with medium cartridge resistance at a low to moderate work rate, will allow intertest comparison of the respirator fit.]

(3) The individual who conducts the CNP fit testing shall be thoroughly trained to perform the test.

(4) The respirator filter or cartridge needs to be replaced with the CNP test manifold. The inhalation valve downstream from the manifold either needs to be temporarily removed or propped open.

(5) The test subject shall be trained to hold his or her breath for at least 20 s.

(6) The test subject shall don the test respirator without any assistance from the individual who conducts the CNP fit test.

(7) The QNFT protocol shall be followed according to Section I.C.1 of this appendix with an exception for the CNP test exercises.

(b) CNP Test Exercises.

(1) Normal breathing. In a normal standing position, without talking, the subject shall breathe normally for 1 min. After the normal breathing exercise, the subject needs to hold his or her head straight ahead and hold his or her breath for 10 s during the test measurement.

(2) Deep breathing. In a normal standing position, the subject shall breathe slowly and deeply for 1 min, being careful not to hyperventilate. After the deep breathing exercise, the subject shall hold his or her head straight ahead and hold his or her breath for 10 s during test measurement.

(3) Turning head side to side. Standing in place, the subject shall slowly turn his or her head from side to side between the extreme positions on each side for 1 min. The head shall be held at each extreme momentarily so the subject can inhale at each side. After the turning head side to side exercise, the subject needs to hold his or her head full left and hold his or her breath for 10 s during test measurement. Next, the subject needs to hold his or her head full right and hold his or her breath for 10 s during test measurement.

(4) Moving head up and down. Standing in place, the subject shall slowly move his or her head up and down for 1 min. The subject shall be instructed to inhale in the up position (i.e., when looking toward the ceiling). After the moving head up and down exercise, the subject shall hold his or her head full up and hold his or her breath for 10 s during test measurement. Next, the subject shall hold his or her head full down and hold his or her breath for 10 s during test measurement.

(5) Talking. The subject shall talk out loud slowly and loudly enough to be heard clearly by the test conductor. The subject can read from a prepared text such as the Rainbow Passage, count backward from 100, or recite a memorized poem or song for 1 min. After the talking exercise, the subject shall hold his or her head straight ahead and hold his or her breath for 10 s during the test measurement.

(6) Grimace. The test subject shall grimace by smiling or frowning for 15 s.

(7) Bending over. The test subject shall bend at the waist as if he or she were to touch his or her toes for 1 min. Jogging in place shall be substituted for this exercise in those test environments such as shroud-type QNFT units that prohibit bending at the waist. After the bending-over exercise, the subject shall hold his or her head straight ahead and hold his or her breath for 10 s during the test measurement.

(8) Normal breathing. The test subject shall remove and re-don the respirator within a 1-min period. Then, in a normal standing position, without talking, the subject shall breathe normally for 1 min. After the normal breathing exercise, the subject shall hold his or her head straight ahead and hold his or her breath for 10 s during the test measurement. After the test exercises, the test subject shall be questioned by the test conductor regarding the comfort of the respirator upon completion of the protocol. If it has become unacceptable, another model of respirator shall be tried.

(c) CNP Test Instrument.

(1) The test instrument shall have an effective audio warning device when the test subject fails to hold his or her breath during the test. The test shall be terminated whenever the test subject failed to hold his or her breath. The test subject may be refitted and retested.

(2) A record of the test shall be kept on file, assuming the fit test was successful. The record must contain the test subject's name; overall fit factor; make, model, style, and size of respirator used; and date tested.

Part II. New Fit-Test Protocols

A. Any person may submit to OSHA an application for approval of a new fit-test protocol. If the application meets the following criteria, OSHA will initiate a rule-making proceeding under Section 6(b)(7) of the OSHA Act to determine whether to list the new protocol as an approved protocol in this Appendix A.

B. The application must include a detailed description of the proposed new fit-test protocol. This application must be supported by either:

1. A test report prepared by an independent government research laboratory (e.g., Lawrence Livermore National Laboratory, Los Alamos National Laboratory, the National Institute for Standards and Technology) stating that the laboratory has tested the protocol and had found it to be accurate and reliable; or

2. An article that has been published in a peer-reviewed industrial hygiene journal describing the protocol and explaining how test data support the accuracy and reliability of the protocol.

C. If OSHA determines that additional information is required before the Agency commences a rulemaking proceeding under this section, OSHA will so notify the applicant and afford the applicant the opportunity to submit the supplemental information. Initiation of a rulemaking proceeding will be deferred until OSHA has received and evaluated the supplemental information.

Appendix B-1 to §1910.134: User Seal Check Procedures (Mandatory)

The individual who uses a tight-fitting respirator is to perform a user seal check to ensure that an adequate seal is achieved each time the respirator is put on. Either the positive- and negative-pressure checks listed in this Appendix or the respirator manufacturer's recommended user seal check method shall be used. User seal checks are not substitutes for qualitative or quantitative fit tests.

I. Facepiece Positive and/or Negative Pressure Checks

A. *Positive pressure check.* Close off the exhalation valve and exhale gently into the facepiece. The face fit is considered satisfactory if a slight positive pressure can be built up inside the facepiece without any evidence of outward leakage of air at the seal. For most respirators this method of leak testing requires the wearer first to remove the exhalation valve cover before closing off the exhalation valve and then carefully replacing it after the test.

B. *Negative pressure check.* Close off the inlet opening of the canister or cartridge(s) by covering with the palm of the hand(s) or by replacing the filter seal(s), inhale gently so that the facepiece collapses slightly, and hold the breath for 10 s. The design of the inlet opening of some cartridges cannot be effectively covered with the palm of the hand. The test can be performed by covering the inlet opening of the cartridge with a thin latex or nitrile glove. If the facepiece remains in its slightly collapsed condition and no inward leakage of air is detected, the tightness of the respirator is considered satisfactory.

II. Manufacturer's Recommended User Seal Check Procedures

The respirator manufacturer's recommended procedures for performing a user seal check may be used instead of the positive- and/or negative-pressure check procedures provided that the employer demonstrates that the manufacturer's procedures are equally effective.

APPENDIX B-2 TO §1910.134: RESPIRATOR CLEANING PROCEDURES (MANDATORY)

These procedures are provided for employer use when cleaning respirators. They are general in nature, and the employer as an alternative may use the cleaning recommendations provided by the manufacturer of the respirators used by its employees, provided such procedures are as effective as those listed here in Appendix B-2. Equivalent effectiveness simply means that the procedures used must accomplish the objectives set forth in Appendix B-2, i.e., must ensure that the respirator is properly cleaned and disinfected in a manner that prevents damage to the respirator and does not cause harm to the user.

I. Procedures for Cleaning Respirators

A. Remove filters, cartridges, or canisters. Disassemble facepieces by removing speaking diaphragms, demand and pressure-demand valve assemblies, hoses, or any components recommended by the manufacturer. Discard or repair any defective parts.

B. Wash components in warm (43°C; 110°F maximum) water with a mild detergent or with a cleaner recommended by the manufacturer. A stiff bristle (not wire) brush may be used to facilitate the removal of dirt.

C. Rinse components thoroughly in clean, warm (43°C; 110°F maximum), preferably running water. Drain.

D. When the cleaner used does not contain a disinfecting agent, respirator components should be immersed for 2 min in one of the following:

1. Hypochlorite solution (50 ppm of chlorine) made by adding approximately 1 ml of laundry bleach to 1 liter of water at 43°C (110°F); or

2. Aqueous solution of iodine (50 ppm iodine) made by adding approximately 0.8 ml of tincture of iodine (6 to 8 g ammonium and/or potassium iodide/100 cc of 45% alcohol) to 1 liter of water at 43°C (110°F); or

3. Other commercially available cleansers of equivalent disinfectant quality when used as directed, if their use is recommended or approved by the respirator manufacturer.

E. Rinse components thoroughly in clean, warm (43°C; 110°F maximum), preferably running water. Drain. The importance of thorough rinsing cannot be overemphasized. Detergents or disinfectants that dry on facepieces may result in dermatitis. In addition, some disinfectants may cause deterioration of rubber or corrosion of metal parts if not completely removed.

F. Components should be hand-dried with a clean, lint-free cloth or air-dried.

G. Reassemble facepiece, replacing filters, cartridges, and canisters where necessary.

H. Test the respirator to ensure that all components work properly.

APPENDIX C TO §1910.134: OSHA RESPIRATOR MEDICAL EVALUATION QUESTIONNAIRE (MANDATORY)

To the employer: Answers to questions in Section 1, and to question 9 in Section 2 of Part A, do not require a medical examination.

To the employee:

Can you read (circle one): Yes/No

Your employer must allow you to answer this questionnaire during normal working hours, or at a time and place that is convenient to you. To maintain your confidentiality, your employer or supervisor must not look at or review your answers, and your employer must tell you how to deliver or send this questionnaire to the healthcare professional who will review it.

Part A. Section 1. (Mandatory) The following information must be provided by every employee who has been selected to use any type of respirator (please print).

1. Today's date: _____
2. Your name: _____
3. Your age (to nearest year): _____
4. Sex (circle one): Male/Female
5. Your height: _____ ft _____ in.
6. Your weight: _____ lb
7. Your job title: _____
8. A phone number where you can be reached by the health-care professional who reviews this questionnaire (include the area code): _____
9. The best time to call you at this number: _____

10. Has your employer told you how to contact the health-care professional who will review this questionnaire? (circle one) Yes/No

11. Check the type of respirator you will use (you can check more than one category):

   a. _____ N, R, or P disposable respirator (filter-mask, noncartridge type only).

   b. _____ Other type (for example, half- or full-facepiece type, powered-air purifying, supplied-air, self-contained breathing apparatus).

12. Have you worn a respirator? (circle one) Yes/No

   If "yes," what type(s): _____

_____

Part A. Section 2. (Mandatory) Questions 1 through 9 below must be answered by every employee who has been selected to use any type of respirator (please circle "yes" or "no").

   1. Do you currently smoke tobacco, or have you smoked tobacco in the last month? Yes/No

   2. Have you ever had any of the following conditions?

      a. Seizures (fits): Yes/No

      b. Diabetes (sugar disease): Yes/No

      c. Allergic reactions that interfere with your breathing: Yes/No

      d. Claustrophobia (fear of closed-in places): Yes/No

      e. Trouble smelling odors? Yes/No

   3. Have you ever had any of the following pulmonary or lung problems?

      a. Asbestosis: Yes/No

      b. Asthma: Yes/No

      c. Chronic bronchitis: Yes/No

      d. Emphysema: Yes/No

      e. Pneumonia: Yes/No

      f. Tuberculosis: Yes/No

      g. Silicosis: Yes/No

      h. Pneumothorax (collapsed lung): Yes/No

      i. Lung cancer: Yes/No

      j. Broken ribs: Yes/No

      k. Any chest injuries or surgeries: Yes/No

      l. Any other lung problem that you've been told about: Yes/No

   4. Do you currently have any of the following symptoms of pulmonary or lung illness?

      a. Shortness of breath: Yes/No

      b. Shortness of breath when walking fast on level ground or walking up a slight hill or incline: Yes/No

      c. Shortness of breath when walking with other people at an ordinary pace on level ground: Yes/No

      d. Have to stop for breath when walking at your own pace on level ground: Yes/No

      e. Shortness of breath when washing or dressing yourself: Yes/No

    f. Shortness of breath that interferes with your job: Yes/No

    g. Coughing that produces phlegm (thick sputum): Yes/No

    h. Coughing that wakes you early in the morning: Yes/No

    i. Coughing that occurs mostly when you are lying down: Yes/No

    j. Coughing up blood in the last month: Yes/No

    k. Wheezing: Yes/No

    l. Wheezing that interferes with your job: Yes/No

    m. Chest pain when you breathe deeply: Yes/No

    n. Any other symptoms that you think may be related to lung problems: Yes/No

5. Have you ever had any of the following cardiovascular or heart problems?

    a. Heart attack: Yes/No

    b. Stroke: Yes/No

    c. Angina: Yes/No

    d. Heart failure: Yes/No

    e. Swelling in your legs or feet (not caused by walking): Yes/No

    f. Heart arrhythmia (heart beating irregularly): Yes/No

    g. High blood pressure: Yes/No

    h. Any other heart problem that you have been told about: Yes/No

6. Have you ever had any of the following cardiovascular or heart symptoms?

    a. Frequent pain or tightness in your chest: Yes/No

    b. Pain or tightness in your chest during physical activity: Yes/No

    c. Pain or tightness in your chest that interferes with your job: Yes/No

    d. In the past 2 years, have you noticed your heart skipping or missing a beat: Yes/No

    e. Heartburn or indigestion that is not related to eating: Yes/No

    f. Any other symptoms that you think may be related to heart or circulation problems: Yes/No

7. Do you currently take medication for any of the following problems?

    a. Breathing or lung problems: Yes/No

    b. Heart trouble: Yes/No

    c. Blood pressure: Yes/No

    d. Seizures (fits): Yes/No

8. If you have used a respirator, have you ever had any of the following problems? (If you have never used a respirator, check the following space and go to question 9:) _____

    a. Eye irritation: Yes/No

    b. Skin allergies or rashes: Yes/No

    c. Anxiety: Yes/No

    d. General weakness or fatigue: Yes/No

    e. Any other problem that interferes with your use of a respirator: Yes/No

9. Would you like to talk to the health-care professional who will review this questionnaire about your answers to this questionnaire? Yes/No

Questions 10 to 15 below must be answered by every employee who has been selected to use either a full-facepiece respirator or a self-contained breathing apparatus (SCBA). For employees who have been selected to use other types of respirators, answering these questions is voluntary.

10. Have you ever lost vision in either eye (temporarily or permanently)? Yes/No
11. Do you currently have any of the following vision problems?
    a. Wear contact lenses: Yes/No
    b. Wear glasses: Yes/No
    c. Color blind: Yes/No
    d. Any other eye or vision problem: Yes/No
12. Have you ever had an injury to your ears, including a broken ear drum? Yes/No
13. Do you currently have any of the following hearing problems?
    a. Difficulty hearing: Yes/No
    b. Wear a hearing aid: Yes/No
    c. Any other hearing or ear problem: Yes/No
14. Have you ever had a back injury? Yes/No
15. Do you currently have any of the following musculoskeletal problems?
    a. Weakness in any of your arms, hands, legs, or feet: Yes/No
    b. Back pain: Yes/No
    c. Difficulty fully moving your arms and legs: Yes/No
    d. Pain or stiffness when you lean forward or backward at the waist: Yes/No
    e. Difficulty fully moving your head up or down: Yes/No
    f. Difficulty fully moving your head side to side: Yes/No
    g. Difficulty bending at your knees: Yes/No
    h. Difficulty squatting to the ground: Yes/No
    i. Difficulty climbing a flight of stairs or a ladder carrying more than 25 lb: Yes/No
    j. Any other muscle or skeletal problem that interferes with using a respirator: Yes/No

Part B. Any of the following questions, and other questions not listed, may be added to the questionnaire at the discretion of the health-care professional who will review the questionnaire.

1. In your present job, are you working at high altitudes (over 5000 ft) or in a place that has lower-than-normal amounts of oxygen? Yes/No
   If "yes," do you have feelings of dizziness, shortness of breath, pounding in your chest, or other symptoms when you are working under these conditions: Yes/No
2. At work or at home, have you ever been exposed to hazardous solvents, hazardous airborne chemicals (e.g., gases, fumes, or dust), or have you come into skin contact with hazardous chemicals? Yes/No
   If "yes," name the chemicals if you know them: _____
   _____
   _____

3. Have you ever worked with any of the materials, or under any of the conditions, listed below?
   a. Asbestos: Yes/No
   b. Silica (e.g., in sandblasting): Yes/No
   c. Tungsten/cobalt (e.g., grinding or welding this material): Yes/No
   d. Beryllium: Yes/No
   e. Aluminum: Yes/No
   f. Coal (for example, mining): Yes/No
   g. Iron: Yes/No
   h. Tin: Yes/No
   i. Dusty environments: Yes/No
   j. Any other hazardous exposures: Yes/No
   If "yes," describe these exposures: _____
   _____
   _____

4. List any second jobs or side businesses you have:_____
   _____
   _____

5. List your previous occupations: _____

6. List your current and previous hobbies: _____
   _____

7. Have you been in the military services? Yes/No
   If "yes," were you exposed to biological or chemical agents (either in training or combat): Yes/No
8. Have you ever worked on a HAZMAT team? Yes/No
9. Other than medications for breathing and lung problems, heart trouble, blood pressure, and seizures mentioned earlier in this questionnaire, are you taking any other medications for any reason (including over-the-counter medications)? Yes/No
   If "yes," name the medications if you know them: _____
10. Will you be using any of the following items with your respirator(s)?
    a. HEPA filters: Yes/No
    b. Canisters (for example, gas masks): Yes/No
    c. Cartridges: Yes/No
11. How often are you expected to use the respirator(s) (circle "yes" or "no" for all answers that apply to you)?
    a. Escape only (no rescue): Yes/No
    b. Emergency rescue only: Yes/No
    c. Less than 5 h/week: Yes/No
    d. Less than 2 h/day: Yes/No
    e. 2 to 4 h/day: Yes/No
    f. Over 4 h/day: Yes/No

12. During the period you are using the respirator(s), is your work effort:
   a. Light (less than 200 kcal per hour)? Yes/No
      If "yes," how long does this period last during the average shift:
      _____ hr _____ min
      Examples of a light work effort are sitting while writing, typing, draft-
      ing, or performing light assembly work; or standing while operating a
      drill press (1 to 3 lb) or controlling machines.
   b. Moderate (200 to 350 kcal/h)? Yes/No
      If "yes," how long does this period last during the average shift:
      _____ hr _____ min
      Examples of moderate work effort are sitting while nailing or filing;
      driving a truck or bus in urban traffic; standing while drilling, nailing,
      performing assembly work, or transferring a moderate load (about
      35 lb) at trunk level; walking on a level surface about 2 mph or down
      a 5° grade about 3 mph; or pushing a wheelbarrow with a heavy load
      (about 100 lb) on a level surface.
   c. Heavy (above 350 kcal/h)? Yes/No
      If "yes," how long does this period last during the average shift:
      _____ hr _____ min
      Examples of heavy work are lifting a heavy load (about 50 lb) from
      the floor to your waist or shoulder; working on a loading dock; shov-
      eling; standing while bricklaying or chipping castings; walking up an
      8° grade about 2 mph; climbing stairs with a heavy load (about 50 lb).
13. Will you be wearing protective clothing and/or equipment (other than the
    respirator) when you are using your respirator? Yes/No
    If "yes," describe this protective clothing and/or equipment: _____
    _____

14. Will you be working under hot conditions (temperature exceeding 77°F)?
    Yes/No
15. Will you be working under humid conditions? Yes/No
16. Describe the work you will be doing while you are using your respirator(s):
17. Describe any special or hazardous conditions you might encounter when
    you are using your respirator(s) (for example, confined spaces, life-threat-
    ening gases):
18. Provide the following information, if you know it, for each toxic substance
    that you will be exposed to when you are using your respirator(s):
    Name of the first toxic substance: _____
    Estimated maximum exposure level per shift: _____
    Duration of exposure per shift: _____
    Name of the second toxic substance: _____
    Estimated maximum exposure level per shift: _____
    Duration of exposure per shift: _____
    Name of the third toxic substance: _____
    Estimated maximum exposure level per shift: _____
    Duration of exposure per shift: _____

The name of any other toxic substances that you will be exposed to while using your respirator:

_____

_____

19. Describe any special responsibilities you will have while using your respirator(s) that may affect the safety and well-being of others (for example, rescue, security):

_____

APPENDIX D TO §1910.134 (MANDATORY) INFORMATION FOR EMPLOYEES USING RESPIRATORS WHEN NOT REQUIRED UNDER THE STANDARD

Respirators are an effective method of protection against designated hazards when properly selected and worn. Respirator use is encouraged, even when exposures are below the exposure limit, to provide an additional level of comfort and protection for workers. However, if a respirator is used improperly or not kept clean, the respirator itself can become a hazard to the worker. Sometimes, workers may wear respirators to avoid exposures to hazards, even if the amount of hazardous substance does not exceed the limits set by OSHA standards. If your employer provides respirators for your voluntary use, of if you provide your own respirator, you need to take certain precautions to be sure that the respirator itself does not present a hazard.

You should do the following:

1. Read and heed all instructions provided by the manufacturer on use, maintenance, cleaning, and care, and warnings regarding the limitations of the respirators.
2. Choose respirators certified for use to protect against the contaminant of concern. NIOSH, the National Institute for Occupational Safety and Health of the U.S. Department of Health and Human Services, certifies respirators. A label or statement of certification should appear on the respirator or respirator packaging. It will tell you what the respirator is designed for and how much it will protect you.
3. Do not wear your respirator into atmospheres containing contaminants for which your respirator is not designed to protect against. For example, a respirator designed to filter dust particles will not protect you against gases, vapors, or very small solid particles of fumes or smoke.
4. Keep track of your respirator so that you do not mistakenly use someone else's respirator.

3.4 HEAD PROTECTION

29 CFR 1910.135

(a)
General requirements.

(a)(1)
The employer shall ensure that each affected employee wears a protective helmet when working in areas where there is a potential for injury to the head from falling objects.

(a)(2)
The employer shall ensure that a protective helmet designed to reduce electrical shock hazard is worn by each such affected employee when near exposed electrical conductors that could contact the head.

(b)
Criteria for protective helmets.

(b)(1)
Protective helmets purchased after July 5, 1994 shall comply with ANSI Z89.1-1986, "American National Standard for Personnel Protection—Protective Headwear for Industrial Workers—Requirements," which is incorporated by reference as specified in Section 1910.6, or shall be demonstrated to be equally effective.

(b)(2)
Protective helmets purchased before July 5, 1994 shall comply with the ANSI standard "American National Standard Safety Requirements for Industrial Head Protection," ANSI Z89.1-1969, which is incorporated by reference as specified in Section 1910.6, or shall be demonstrated by the employer to be equally effective.

3.5 FOOT PROTECTION

29 CFR 1910.136

(a)
General requirements. The employer shall ensure that each affected employee uses protective footwear when working in areas where there is a danger of foot injuries due to falling or rolling objects, or objects piercing the sole, and where such employee's feet are exposed to electrical hazards.

(b)
Criteria for protective footwear.

(b)(1)
Protective footwear purchased after July 5, 1994 shall comply with ANSI Z41-1991, "American National Standard for Personal Protection—Protective Footwear," which is incorporated by reference as specified in Section 1910.6, or shall be demonstrated by the employer to be equally effective.

(b)(2)
Protective footwear purchased before July 5, 1994 shall comply with the ANSI standard "USA Standard for Men's Safety — Toe Footwear," Z41.1-1967, which is incorporated by reference as specified in Section 1910.6, or shall be demonstrated by the employer to be equally effective.

3.6 FIRE PROTECTION

29 CFR 1910.155

(a)
Scope. This subpart contains requirements for fire brigades, and all portable and fixed fire suppression equipment, fire detection systems, and fire or employee alarm systems installed to meet the fire protection requirements of 29 CFR Part 1910.

(b)
Application. This subpart applies to all employments except for maritime, construction, and agriculture.

(c)
Definitions applicable to this subpart.

(c)(1)
"After-flame" means the time a test specimen continues to flame after the flame source has been removed.

(c)(2)
"Aqueous film forming foam" (AFFF) means a fluorinated surfactant with a foam stabilizer that is diluted with water to act as a temporary barrier to exclude air from mixing with the fuel vapor by developing an aqueous film on the fuel surface of some hydrocarbons that is capable of suppressing the generation of fuel vapors.

(c)(3)
"Approved" means acceptable to the Assistant Secretary under the following criteria:

(c)(3)(i)
If it is accepted, or certified, or listed, or labeled or otherwise determined to be safe by a nationally recognized testing laboratory; or

(c)(3)(ii)
With respect to an installation or equipment of a kind that no nationally recognized testing laboratory accepts, certifies, lists, labels, or determines to be safe, if it is inspected or tested by another federal agency and found in compliance with the provisions of the applicable National Fire Protection Association Fire Code; or

(c)(3)(iii)
With respect to custom-made equipment or related installations that are designed, fabricated for, and intended for use by its manufacturer on the basis of test data that the employer keeps and makes available for inspection to the Assistant Secretary.

(c)(3)(iv)
For the purposes of paragraph (c)(3) of this section:

(c)(3)(iv)(A)
Equipment is listed if it is of a kind mentioned in a list that is published by a nationally recognized testing laboratory that makes periodic inspections of the production of such equipment and that states that such equipment meets nationally recognized standards or has been tested and found safe for use in a specified manner;

(c)(3)(iv)(B)
Equipment is labeled if there is attached to it a label, symbol, or other identifying mark of a nationally recognized testing laboratory that makes periodic inspections of the production of such equipment, and whose labeling indicates compliance with nationally recognized standards or tests to determine safe use in a specified manner;

(c)(3)(iv)(C)
Equipment is accepted if it has been inspected and found by a nationally recognized testing laboratory to conform to specified plans or to procedures of applicable codes; and

(c)(3)(iv)(D)
Equipment is certified if it has been tested and found by a nationally recognized testing laboratory to meet nationally recognized standards or to be safe for use in a specified manner or is of a kind whose production is periodically inspected by a nationally recognized testing laboratory, and if it bears a label, tag, or other record of certification.

(c)(4)
"Assistant Secretary" means the Assistant Secretary of Labor for Occupational Safety and Health or designee.

(c)(5)
"Automatic fire detection device" means a device designed to automatically detect the presence of fire by heat, flame, light, smoke, or other products of combustion.

(c)(6)
"Buddy-breathing device" means an accessory to self-contained breathing apparatus that permits a second person to share the same air supply as that of the wearer of the apparatus.

(c)(7)
"Carbon dioxide" means a colorless, odorless, electrically nonconductive inert gas (chemical formula CO_2) that is a medium for extinguishing fires by reducing the concentration of oxygen or fuel vapor in the air to the point where combustion is impossible.

(c)(8)
"Class A fire" means a fire involving ordinary combustible materials such as paper, wood, cloth, and some rubber and plastic materials.

(c)(9)
"**Class B fire**" means a fire involving flammable or combustible liquids, flammable gases, greases and similar materials, and some rubber and plastic materials.

(c)(10)
"**Class C fire**" means a fire involving energized electrical equipment where safety to the employee requires the use of electrically nonconductive extinguishing media.

(c)(11)
"**Class D fire**" means a fire involving combustible metals such as magnesium, titanium, zirconium, sodium, lithium, and potassium.

(c)(12)
"**Dry chemical**" means an extinguishing agent composed of very small particles of chemicals such as, but not limited to, sodium bicarbonate, potassium bicarbonate, urea-based potassium bicarbonate, potassium chloride, or monoammonium phosphate supplemented by special treatment to provide resistance to packing and moisture absorption (caking) as well as to provide proper flow capabilities. Dry chemical does not include dry powders.

(c)(13)
"**Dry powder**" means a compound used to extinguish or control Class D fires.

(c)(14)
"**Education**" means the process of imparting knowledge or skill through systematic instruction. It does not require formal classroom instruction.

(c)(15)
"**Enclosed structure**" means a structure with a roof or ceiling and at least two walls that may present fire hazards to employees, such as accumulations of smoke, toxic gases, and heat, similar to those found in buildings.

(c)(16)
"**Extinguisher classification**" means the letter classification given an extinguisher to designate the class or classes of fire on which an extinguisher will be effective.

(c)(17)
"**Extinguisher rating**" means the numerical rating given to an extinguisher that indicates the extinguishing potential of the unit based on standardized tests developed by Underwriters' Laboratories, Inc.

(c)(18)
"**Fire brigade**" (private fire department, industrial fire department) means an organized group of employees who are knowledgeable, trained, and skilled in at least basic fire-fighting operations.

(c)(19)
"**Fixed extinguishing system**" means a permanently installed system that either extinguishes or controls a fire at the location of the system.

(c)(20)
"Flame resistance" is the property of materials or combinations of component materials to retard ignition and restrict the spread of flame.

(c)(21)
"Foam" means a stable aggregation of small bubbles that flow freely over a burning liquid surface and form a coherent blanket that seals combustible vapors and thereby extinguishes the fire.

(c)(22)
"Gaseous agent" is a fire-extinguishing agent that is in the gaseous state at normal room temperature and pressure. It has low viscosity, can expand or contract with changes in pressure and temperature, and has the ability to diffuse readily and to distribute itself uniformly throughout an enclosure.

(c)(23)
"Halon 1211" means a colorless, faintly sweet-smelling, electrically nonconductive liquefied gas (chemical formula $CBrClF_2$) that is a medium for extinguishing fires by inhibiting the chemical chain reaction of fuel and oxygen. It is also known as bromochlorodifluoromethane.

(c)(24)
"Halon 1301" means a colorless, odorless, electrically nonconductive gas (chemical formula $CBrF_3$), which is a medium for extinguishing fires by inhibiting the chemical chain reaction of fuel and oxygen. It is also known as bromotrifluoromethane.

(c)(25)
"Helmet" is a head protective device consisting of a rigid shell, energy absorption system, and chin strap intended to be worn to provide protection for the head or portions thereof, against impact, flying or falling objects, electric shock, penetration, heat, and flame.

(c)(26)
"Incipient stage fire" means a fire that is in the initial or beginning stage and that can be controlled or extinguished by portable fire extinguishers, Class II standpipe, or small hose systems without the need for protective clothing or breathing apparatus.

(c)(27)
"Inspection" means a visual check of fire protection systems and equipment to ensure that they are in place, charged, and ready for use in the event of a fire.

(c)(28)
"Interior structural fire fighting" means the physical activity of fire suppression, rescue, or both, inside buildings or enclosed structures that are involved in a fire situation beyond the incipient stage.

(c)(29)
"Lining" means a material permanently attached to the inside of the outer shell of a garment for the purpose of thermal protection and padding.

(c)(30)

"Local application system" means a fixed fire suppression system that has a supply of extinguishing agent, with nozzles arranged to discharge automatically an extinguishing agent directly on the burning material to extinguish or control a fire.

(c)(31)

"Maintenance" means the performance of services on fire protection equipment and systems to assure that they will perform as expected in the event of a fire. Maintenance differs from inspection in that maintenance requires the checking of internal fittings, devices, and agent supplies.

(c)(32)

"Multipurpose dry chemical" means a dry chemical that is approved for use on Class A, Class B, and Class C fires.

(c)(33)

"Outer shell" is the exterior layer of material on the fire coat and protective trousers that forms the outermost barrier between the firefighter and the environment. It is attached to the vapor barrier and liner and is usually constructed with a storm flap, suitable closures, and pockets.

(c)(34)

"Positive-pressure breathing apparatus" means a self-contained breathing apparatus in which the pressure in the breathing zone is positive in relation to the immediate environment during inhalation and exhalation.

(c)(35)

"Pre-discharge employee alarm" means an alarm that will sound at a set time prior to actual discharge of an extinguishing system so that employees may evacuate the discharge area prior to system discharge.

(c)(36)

"Quick disconnect valve" means a device that starts the flow of air by inserting the hose (which leads from the facepiece) into the regulator of self-contained breathing apparatus, and stops the flow of air by disconnecting the hose from the regulator.

(c)(37)

"Sprinkler alarm" means an approved device installed so that any water flow from a sprinkler system equal to or greater than that from a single automatic sprinkler will result in an audible alarm signal on the premises.

(c)(38)

"Sprinkler system" means a system of piping designed in accordance with fire protection engineering standards and installed to control or extinguish fires. The system includes an adequate and reliable water supply, and a network of specially sized piping and sprinklers, which are interconnected. The system also includes a control valve and a device for actuating an alarm when the system is in operation.

(c)(39)(i)
"Class I standpipe system" means a 2 1/2 in. (6.3 cm) hose connection for use by fire departments and those trained in handling heavy fire streams.

(c)(39)(ii)
"Class II standpipe system" means a 1 1/2 in. (3.8 cm) hose system that provides a means for the control or extinguishment of incipient stage fires.

(c)(39)(iii)
"Class III standpipe system" means a combined system of hose that is intended for the use of employees trained in the use of hose operations and that is capable of furnishing effective water discharge during the more-advanced stages of fire (beyond the incipient stage) in the interior of workplaces. Hose outlets are available for both 1 1/2 in. (3.8 cm) and 2 1/2 in. (6.3 cm) hose.

(c)(39)(iv)
"Small hose system" means a system of hose ranging in diameter from 5/8 in. (1.6 cm) up to 1 1/2 in. (3.8 cm) that is for the use of employees and that provides a means for the control and extinguishment of incipient stage fires.

(c)(40)
"Total flooding system" means a fixed suppression system that is arranged to discharge automatically a predetermined concentration of agent into an enclosed space for the purpose of fire extinguishment or control.

(c)(41)
"Training" means the process of making proficient through instruction and hands-on practice in the operation of equipment, including respiratory protection equipment, that is expected to be used and in the performance of assigned duties.

(c)(42)
"Vapor barrier" means that material used to prevent or inhibit substantially the transfer of water, corrosive liquids, and steam or other hot vapors from the outside of a garment to the wearer's body.

3.7 FIRE EXTINGUISHERS
29 CFR 1910.157

(a)
Scope and application. The requirements of this section apply to the placement, use, maintenance, and testing of portable fire extinguishers provided for the use of employees. Paragraph (d) of this section does not apply to extinguishers provided for employee use on the outside of workplace buildings or structures. Where extinguishers are provided but are not intended for employee use and the employer has an emergency action plan and a fire prevention plan that meet the requirements of 1910.38, then only the requirements of paragraphs (e) and (f) of this section apply.

(b)
Exemptions.

(b)(1)
Where the employer has established and implemented a written fire safety policy that requires the immediate and total evacuation of employees from the workplace upon the sounding of a fire alarm signal and that includes an emergency action plan and a fire prevention plan that meet the requirements of 1910.38, and when extinguishers are not available in the workplace, the employer is exempt from all requirements of this section unless a specific standard in Part 1910 requires that a portable fire extinguisher be provided.

(b)(2)
Where the employer has an emergency action plan meeting the requirements of 1910.38 that designates certain employees to be the only employees authorized to use the available portable fire extinguishers and that requires all other employees in the fire area to evacuate immediately the affected work area upon the sounding of the fire alarm, the employer is exempt from the distribution requirements in paragraph (d) of this section.

(c)
General requirements.

(c)(1)
The employer shall provide portable fire extinguishers and shall mount, locate, and identify them so that they are readily accessible to employees without subjecting the employees to possible injury.

(c)(2)
Only approved portable fire extinguishers shall be used to meet the requirements of this section.

(c)(3)
The employer shall not provide or make available in the workplace portable fire extinguishers using carbon tetrachloride or chlorobromomethane extinguishing agents.

(c)(4)
The employer shall assure that portable fire extinguishers are maintained in a fully charged and operable condition and kept in their designated places at all times except during use.

(c)(5)
The employer shall remove from service all soldered or riveted shell self-generating soda acid or self-generating foam or gas cartridge water type portable fire extinguishers that are operated by inverting the extinguisher to rupture the cartridge or to initiate an uncontrollable pressure-generating chemical reaction to expel the agent.

(d)
Selection and distribution.

(d)(1)
Portable fire extinguishers shall be provided for employee use and selected and distributed based on the classes of anticipated workplace fires and on the size and degree of hazard that would affect their use.

(d)(2)
The employer shall distribute portable fire extinguishers for use by employees on Class A fires so that the travel distance for employees to any extinguisher is 75 ft (22.9 m) or less.

(d)(3)
The employer may use uniformly spaced standpipe systems or hose stations connected to a sprinkler system installed for emergency use by employees instead of Class A portable fire extinguishers, provided that such systems meet the respective requirements of 1910.158 or 1910.159, that they provide total coverage of the area to be protected, and that employees are trained at least annually in their use.

(d)(4)
The employer shall distribute portable fire extinguishers for use by employees on Class B fires so that the travel distance from the Class B hazard area to any extinguisher is 50 ft (15.2 m) or less.

(d)(5)
The employer shall distribute portable fire extinguishers used for Class C hazards on the basis of the appropriate pattern for the existing Class A or Class B hazards.

(d)(6)
The employer shall distribute portable fire extinguishers or other containers of Class D extinguishing agent for use by employees so that the travel distance from the combustible metal working area to any extinguishing agent is 75 ft (22.9 m) or less. Portable fire extinguishers for Class D hazards are required in those combustible metal working areas where combustible metal powders, flakes, shavings, or similarly sized products are generated at least once every 2 weeks.

(e)
Inspection, maintenance, and testing.

(e)(1)
The employer shall be responsible for the inspection, maintenance, and testing of all portable fire extinguishers in the workplace.

(e)(2)
Portable extinguishers or hose used in lieu thereof under paragraph (d)(3) of this section shall be visually inspected monthly.

(e)(3)
The employer shall assure that portable fire extinguishers are subjected to an annual maintenance check. Stored pressure extinguishers do not require an internal

examination. The employer shall record the annual maintenance date and retain this record for 1 year after the last entry or the life of the shell, whichever is less. The record shall be available to the Assistant Secretary upon request.

(e)(4)

The employer shall assure that stored pressure dry chemical extinguishers that require a 12-year hydrostatic test are emptied and subjected to applicable maintenance procedures every 6 years. Dry chemical extinguishers having nonrefillable disposable containers are exempt from this requirement. When recharging or hydrostatic testing is performed, the 6-year requirement begins from that date.

(e)(5)

The employer shall assure that alternative equivalent protection is provided when portable fire extinguishers are removed from service for maintenance and recharging.

(f)

Hydrostatic testing.

(f)(1)

The employer shall assure that hydrostatic testing is performed by trained persons with suitable testing equipment and facilities.

(f)(2)

The employer shall assure that portable extinguishers are hydrostatically tested at the intervals listed in Table L-1 of this section, except under any of the following conditions:

(f)(2)(i)

When the unit has been repaired by soldering, welding, brazing, or use of patching compounds;

(f)(2)(ii)

When the cylinder or shell threads are damaged;

(f)(2)(iii)

When there is corrosion that has caused pitting, including corrosion under removable name plate assemblies;

(f)(2)(iv)

When the extinguisher has been burned in a fire; or

(f)(2)(v)

When a calcium chloride extinguishing agent has been used in a stainless steel shell.

(f)(3)

In addition to an external visual examination, the employer shall assure that an internal examination of cylinders and shells to be tested is made prior to the hydrostatic tests.

TABLE L-1

Type of Extinguisher	Test Interval (years)
Soda acid (stainless steel shell)	5
Cartridge operated water and/or antifreeze	5
Stored pressure water and/or antifreeze	5
Wetting agent	5
Foam (stainless steel shell)	5
Aqueous film forming foam (AFFF)	5
Loaded stream	5
Dry chemical with stainless steel	5
Carbon dioxide	5
Dry chemical, stored pressure, with mild steel, brazed brass, or aluminum shells	12
Dry chemical, cartridge or cylinder operated, with mild steel shells	12
Halon 1211	12
Halon 1301	12
Dry powder, cartridge or cylinder operated, with mild steel shells	12

(f)(4)

The employer shall assure that portable fire extinguishers are hydrostatically tested whenever they show new evidence of corrosion or mechanical injury, except under the conditions listed in paragraphs (f)(2)(i) to (v) of this section.

(f)(5)

The employer shall assure that hydrostatic tests are performed on extinguisher hose assemblies that are equipped with a shutoff nozzle at the discharge end of the hose. The test interval shall be the same as specified for the extinguisher on which the hose is installed.

(f)(6)

The employer shall assure that carbon dioxide hose assemblies with a shutoff nozzle are hydrostatically tested at 1250 psi (8620 kPa).

(f)(7)

The employer shall assure that dry chemical and dry powder hose assemblies with a shutoff nozzle are hydrostatically tested at 300 psi (2070 kPa).

(f)(8)

Hose assemblies passing a hydrostatic test do not require any type of recording or stamping.

(f)(9)

The employer shall assure that hose assemblies for carbon dioxide extinguishers that require a hydrostatic test are tested within a protective cage device.

(f)(10)

The employer shall assure that carbon dioxide extinguishers and nitrogen or carbon dioxide cylinders used with wheeled extinguishers are tested every 5 years at 5/3 of the service pressure as stamped into the cylinder. Nitrogen cylinders that comply with 49 CFR 173.34(e)(15) may be hydrostatically tested every 10 years.

(f)(11)

The employer shall assure that all stored pressure and Halon 1211 types of extinguishers are hydrostatically tested at the factory test pressure not to exceed two times the service pressure.

(f)(12)

The employer shall assure that acceptable self-generating type soda acid and foam extinguishers are tested at 350 psi (2410 kPa).

(f)(13)

Air or gas pressure may not be used for hydrostatic testing.

(f)(14)

Extinguisher shells, cylinders, or cartridges that fail a hydrostatic pressure test, or that are not fit for testing shall be removed from service and from the workplace.

(f)(15)
(f)(15)(i)

The equipment for testing compressed gas type cylinders shall be of the water jacket type. The equipment shall be provided with an expansion indicator that operates with an accuracy within 1% of the total expansion or 0.1 cc (0.1 ml) of liquid.

(f)(15)(ii)

The equipment for testing noncompressed gas type cylinders shall consist of the following:

(f)(15)(ii)(A)

A hydrostatic test pump, hand or power operated, capable of producing not less than 150% of the test pressure, which shall include appropriate check valves and fittings;

(f)(15)(ii)(B)

A flexible connection for attachment to fittings to test through the extinguisher nozzle, test bonnet, or hose outlet, as is applicable; and

(f)(15)(ii)(C)

A protective cage or barrier for personal protection of the tester, designed to provide visual observation of the extinguisher under test.

(f)(16)

The employer shall maintain and provide upon request to the Assistant Secretary evidence that the required hydrostatic testing of fire extinguishers has been performed at the time intervals shown in Table L-1. Such evidence shall be in the form of a certification record that includes the date of the test, the signature of the person who performed the test, and the serial number, or other identifier, of the fire extinguisher that was tested. Such records shall be kept until the extinguisher is hydrostatically retested at the time interval specified in Table L-1 or until the extinguisher is taken out of service, whichever comes first.

(g)
Training and education.

(g)(1)
Where the employer has provided portable fire extinguishers for employee use in the workplace, the employer shall also provide an educational program to familiarize employees with the general principles of fire extinguisher use and the hazards involved with incipient-stage firefighting.

(g)(2)
The employer shall provide the education required in paragraph (g)(1) of this section upon initial employment and at least annually thereafter.

(g)(3)
The employer shall provide employees who have been designated to use fire-fighting equipment as part of an emergency action plan with training in the use of the appropriate equipment.

(g)(4)
The employer shall provide the training required in paragraph (g)(3) of this section upon initial assignment to the designated group of employees and at least annually thereafter.

3.8 CONFINED SPACE ENTRY

29 CFR 1910.146

SubPart Number: J

SubPart Title: General Environmental Controls

(a)
Scope and application. This section contains requirements for practices and procedures to protect employees in general industry from the hazards of entry into permit-required confined spaces. This section does not apply to agriculture, to construction, or to shipyard employment (Parts 1928, 1926, and 1915 of this chapter, respectively).

(b)
Definitions.

"Acceptable entry conditions" means the conditions that must exist in a permit space to allow entry and to ensure that employees involved with a permit-required confined space entry can safely enter into and work within the space.

"Attendant" means an individual stationed outside one or more permit spaces who monitors the authorized entrants and who performs all attendant's duties assigned in the employer's permit space program.

"Authorized entrant" means an employee who is authorized by the employer to enter a permit space.

"Blanking or blinding" means the absolute closure of a pipe, line, or duct by the fastening of a solid plate (such as a spectacle blind or a skillet blind) that completely covers the bore and that is capable of withstanding the maximum pressure of the pipe, line, or duct with no leakage beyond the plate.

"Confined space" means a space that:

(1) Is large enough and so configured that an employee can bodily enter and perform assigned work; and

(2) Has limited or restricted means for entry or exit (for example, tanks, vessels, silos, storage bins, hoppers, vaults, and pits are spaces that may have limited means of entry); and

(3) Is not designed for continuous employee occupancy.

"Double block and bleed" means the closure of a line, duct, or pipe by closing and locking or tagging two in-line valves and by opening and locking or tagging a drain or vent valve in the line between the two closed valves.

"Emergency" means any occurrence (including any failure of hazard control or monitoring equipment) or event internal or external to the permit space that could endanger entrants.

"Engulfment" means the surrounding and effective capture of a person by a liquid or finely divided (flowable) solid substance that can be aspirated to cause death by filling or plugging the respiratory system or that can exert enough force on the body to cause death by strangulation, constriction, or crushing.

"Entry" means the action by which a person passes through an opening into a permit-required confined space. Entry includes ensuing work activities in that space and is considered to have occurred as soon as any part of the entrant's body breaks the plane of an opening into the space.

"Entry permit (permit)" means the written or printed document that is provided by the employer to allow and control entry into a permit space and that contains the information specified in paragraph (f) of this section.

"Entry supervisor" means the person (such as the employer, foreman, or crew chief) responsible for determining if acceptable entry conditions are present at a permit space where entry is planned, for authorizing entry and overseeing entry operations, and for terminating entry as required by this section.

Note: An entry supervisor also may serve as an attendant or as an authorized entrant, as long as that person is trained and equipped as required by this section for each role he or she fills. Also, the duties of entry supervisor may be passed from one individual to another during the course of an entry operation.

"Hazardous atmosphere" means an atmosphere that may expose employees to the risk of death, incapacitation, impairment of ability to self-rescue (that is, escape unaided from a permit space), injury, or acute illness from one or more of the following causes:

(1) Flammable gas, vapor, or mist in excess of 10% of its lower flammable limit (LFL);

(2) Airborne combustible dust at a concentration that meets or exceeds its LFL;

Note: This concentration may be approximated as a condition in which the dust obscures vision at a distance of 5 ft (1.52 m) or less.

(3) Atmospheric oxygen concentration below 19.5% or above 23.5%;

(4) Atmospheric concentration of any substance for which a dose or a permissible exposure limit is published in Subpart G, Occupational Health and Environmental Control, or in Subpart Z, Toxic and Hazardous Substances, of this Part and which could result in employee exposure in excess of its dose or permissible exposure limit;

Note: An atmospheric concentration of any substance that is not capable of causing death, incapacitation, impairment of ability to self-rescue, injury, or acute illness due to its health effects is not covered by this provision.

(5) Any other atmospheric condition that is immediately dangerous to life or health.

Note: For air contaminants for which OSHA has not determined a dose or permissible exposure limit, other sources of information, such as Material Safety Data Sheets that comply with the Hazard Communication Standard, Sec. 1910.1200 of this Part, published information, and internal documents can provide guidance in establishing acceptable atmospheric conditions.

"Hot work permit" means the employer's written authorization to perform operations (for example, riveting, welding, cutting, burning, and heating) capable of providing a source of ignition.

"Immediately dangerous to life or health (IDLH)" means any condition that poses an immediate or delayed threat to life or that would cause irreversible adverse health effects or that would interfere with an individual's ability to escape unaided from a permit space.

Note: Some materials — hydrogen fluoride gas and cadmium vapor, for example — may produce immediate transient effects that, even if severe, may pass without medical attention, but are followed by sudden, possibly fatal collapse 12 to 72 hours after exposure. The victim "feels normal" from recovery from transient effects until collapse. Such materials in hazardous quantities are considered to be "immediately" dangerous to life or health.

"Inerting" means the displacement of the atmosphere in a permit space by a noncombustible gas (such as nitrogen) to such an extent that the resulting atmosphere is noncombustible.

Note: This procedure produces an IDLH oxygen-deficient atmosphere.

"Isolation" means the process by which a permit space is removed from service and completely protected against the release of energy and material into the space by such means as blanking or blinding; misaligning or removing sections of lines, pipes, or ducts; a double block and bleed system; lockout or tagout of all sources of energy; or blocking or disconnecting all mechanical linkages.

"Line breaking" means the intentional opening of a pipe, line, or duct that is or has been carrying flammable, corrosive, or toxic material, an inert gas, or any fluid at a volume, pressure, or temperature capable of causing injury.

"Non-permit confined space" means a confined space that does not contain or, with respect to atmospheric hazards, have the potential to contain any hazard capable of causing death or serious physical harm.

"Oxygen-deficient atmosphere" means an atmosphere containing less than 19.5% oxygen by volume.

"Oxygen-enriched atmosphere" means an atmosphere containing more than 23.5% oxygen by volume.

"Permit-required confined space (permit space)" means a confined space that has one or more of the following characteristics:

(1) Contains or has a potential to contain a hazardous atmosphere;

(2) Contains a material that has the potential for engulfing an entrant;

(3) Has an internal configuration such that an entrant could be trapped or asphyxiated by inwardly converging walls or by a floor that slopes downward and tapers to a smaller cross section; or

(4) Contains any other recognized serious safety or health hazard.

"Permit-required confined space program (permit space program)" means the employer's overall program for controlling and, where appropriate, for protecting employees from permit space hazards and for regulating employee entry into permit spaces.

"Permit system" means the employer's written procedure for preparing and issuing permits for entry and for returning the permit space to service following termination of entry.

"Prohibited condition" means any condition in a permit space that is not allowed by the permit during the period when entry is authorized.

"Rescue service" means the personnel designated to rescue employees from permit spaces.

"Retrieval system" means the equipment (including a retrieval line, chest or full-body harness, wristlets, if appropriate, and a lifting device or anchor) used for nonentry rescue of persons from permit spaces.

"Testing" means the process by which the hazards that may confront entrants of a permit space are identified and evaluated. Testing includes specifying the tests that are to be performed in the permit space.

Note: Testing enables employers both to devise and implement adequate control measures for the protection of authorized entrants and to determine if acceptable entry conditions are present immediately prior to, and during, entry.

(c)
General requirements.

(c)(1)
The employer shall evaluate the workplace to determine if any spaces are permit-required confined spaces.

(c)(2)
If the workplace contains permit spaces, the employer shall inform exposed employees, by posting danger signs or by any other equally effective means, of the existence and location of and the danger posed by the permit spaces.

Note: A sign reading DANGER — PERMIT-REQUIRED CONFINED SPACE, DO NOT ENTER or using other similar language would satisfy the requirement for a sign.

(c)(3)
If the employer decides that its employees will not enter permit spaces, the employer shall take effective measures to prevent its employees from entering the permit spaces and shall comply with paragraphs (c)(1), (c)(2), (c)(6), and (c)(8) of this section.

(c)(4)
If the employer decides that its employees will enter permit spaces, the employer shall develop and implement a written permit space program that complies with this section. The written program shall be available for inspection by employees and their authorized representatives.

(c)(5)
An employer may use the alternate procedures specified in paragraph (c)(5)(ii) of this section for entering a permit space under the conditions set forth in paragraph (c)(5)(i) of this section.

(c)(5)(i)
An employer whose employees enter a permit space need not comply with paragraphs (d) through (f) and (h) through (k) of this section, provided that:

(c)(5)(i)(A)
The employer can demonstrate that the only hazard posed by the permit space is an actual or potential hazardous atmosphere;

(c)(5)(i)(B)
The employer can demonstrate that continuous forced-air ventilation alone is sufficient to maintain that permit space safe for entry;

(c)(5)(i)(C)
The employer develops monitoring and inspection data that support the demonstrations required by paragraphs (c)(5)(i)(A) and (c)(5)(i)(B) of this section;

(c)(5)(i)(D)
If an initial entry of the permit space is necessary to obtain the data required by paragraph (c)(5)(i)(C) of this section, the entry is performed in compliance with paragraphs (d) through (k) of this section;

(c)(5)(i)(E)
The determinations and supporting data required by paragraphs (c)(5)(i)(A), (c)(5)(i)(B), and (c)(5)(i)(C) of this section are documented by the employer and are made available to each employee who enters the permit space under the terms of paragraph (c)(5) of this section or to that employee's authorized representative; and

(c)(5)(i)(F)
Entry into the permit space under the terms of paragraph (c)(5)(i) of this section is performed in accordance with the requirements of paragraph (c)(5)(ii) of this section.

(c)(5)(ii)
The following requirements apply to entry into permit spaces that meet the conditions set forth in paragraph (c)(5)(i) of this section.

(c)(5)(ii)(A)
Any conditions making it unsafe to remove an entrance cover shall be eliminated before the cover is removed.

(c)(5)(ii)(B)
When entrance covers are removed, the opening shall be promptly guarded by a railing, temporary cover, or other temporary barrier that will prevent an accidental fall through the opening and that will protect each employee working in the space from foreign objects entering the space.

(c)(5)(ii)(C)
Before an employee enters the space, the internal atmosphere shall be tested, with a calibrated direct-reading instrument, for oxygen content, for flammable gases and vapors, and for potential toxic air contaminants, in that order. Any employee who enters the space, or that employee's authorized representative, shall be provided an opportunity to observe the preentry testing required by this paragraph.

(c)(5)(ii)(C)(1)
Oxygen content,

(c)(5)(ii)(C)(2)
Flammable gases and vapors, and

(c)(5)(ii)(C)(3)
Potential toxic air contaminants.

(c)(5)(ii)(D)
There may be no hazardous atmosphere within the space whenever any employee is inside the space.

(c)(5)(ii)(E)
Continuous forced-air ventilation shall be used, as follows:

(c)(5)(ii)(E)(1)
An employee may not enter the space until the forced-air ventilation has eliminated any hazardous atmosphere;

(c)(5)(ii)(E)(2)
The forced-air ventilation shall be so directed to ventilate the immediate areas where an employee is or will be present within the space and shall continue until all employees have left the space;

(c)(5)(ii)(E)(3)
The air supply for the forced-air ventilation shall be from a clean source and may not increase the hazards in the space.

(c)(5)(ii)(F)
The atmosphere within the space shall be periodically tested as necessary to ensure that the continuous forced-air ventilation is preventing the accumulation of a hazardous atmosphere. Any employee who enters the space, or that employee's authorized representative, shall be provided with an opportunity to observe the periodic testing required by this paragraph.

(c)(5)(ii)(G)
If a hazardous atmosphere is detected during entry:

(c)(5)(ii)(G)(1)
Each employee shall leave the space immediately;

(c)(5)(ii)(G)(2)
The space shall be evaluated to determine how the hazardous atmosphere developed; and

(c)(5)(ii)(G)(3)
Measures shall be implemented to protect employees from the hazardous atmosphere before any subsequent entry takes place.

(c)(5)(ii)(H)
The employer shall verify that the space is safe for entry and that the preentry measures required by paragraph (c)(5)(ii) of this section have been taken, through a written certification that contains the date, the location of the space, and the signature of the person providing the certification. The certification shall be made before entry and shall be made available to each employee entering the space or to that employee's authorized representative.

(c)(6)

When there are changes in the use or configuration of a nonpermit confined space that might increase the hazards to entrants, the employer shall reevaluate that space and, if necessary, reclassify it as a permit-required confined space.

(c)(7)

A space classified by the employer as a permit-required confined space may be reclassified as a nonpermit confined space under the following procedures:

(c)(7)(i)

If the permit space poses no actual or potential atmospheric hazards and if all hazards within the space are eliminated without entry into the space, the permit space may be reclassified as a nonpermit confined space for as long as the nonatmospheric hazards remain eliminated.

(c)(7)(ii)

If it is necessary to enter the permit space to eliminate hazards, such entry shall be performed under paragraphs (d) through (k) of this section. If testing and inspection during that entry demonstrate that the hazards within the permit space have been eliminated, the permit space may be reclassified as a nonpermit confined space for as long as the hazards remain eliminated.

Note: Control of atmospheric hazards through forced-air ventilation does not constitute elimination of the hazards. Paragraph (c)(5) covers permit space entry where the employer can demonstrate that forced-air ventilation alone will control all hazards in the space.

(c)(7)(iii)

The employer shall document the basis for determining that all hazards in a permit space have been eliminated, through a certification that contains the date, the location of the space, and the signature of the person making the determination. The certification shall be made available to each employee entering the space or to that employee's authorized representative.

(c)(7)(iv)

If hazards arise within a permit space that has been declassified to a nonpermit space under paragraph (c)(7) of this section, each employee in the space shall exit the space. The employer shall then reevaluate the space and determine whether it must be reclassified as a permit space, in accordance with other applicable provisions of this section.

(c)(8)

When an employer (host employer) arranges to have employees of another employer (contractor) perform work that involves permit space entry, the host employer shall:

(c)(8)(i)

Inform the contractor that the workplace contains permit spaces and that permit space entry is allowed only through compliance with a permit space program meeting the requirements of this section;

(c)(8)(ii)
Apprise the contractor of the elements, including the hazards identified and the host employer's experience with the space, that make the space in question a permit space;

(c)(8)(iii)
Apprise the contractor of any precautions or procedures that the host employer has implemented for the protection of employees in or near permit spaces where contractor personnel will be working;

(c)(8)(iv)
Coordinate entry operations with the contractor, when both host employer personnel and contractor personnel will be working in or near permit spaces, as required by paragraph (d)(11) of this section; and

(c)(8)(v)
Debrief the contractor at the conclusion of the entry operations regarding the permit space program followed and regarding any hazards confronted or created in permit spaces during entry operations.

(c)(9)
In addition to complying with the permit space requirements that apply to all employers, each contractor who is retained to perform permit space entry operations shall:

(c)(9)(i)
Obtain any available information regarding permit space hazards and entry operations from the host employer;

(c)(9)(ii)
Coordinate entry operations with the host employer, when both host employer personnel and contractor personnel will be working in or near permit spaces, as required by paragraph (d)(11) of this section; and

(c)(9)(iii)
Inform the host employer of the permit space program that the contractor will follow and of any hazards confronted or created in permit spaces, either through a debriefing or during the entry operation.

(d)
Permit-required confined space program (permit space program). Under the permit space program required by paragraph (c)(4) of this section, the employer shall:

(d)(1)
Implement the measures necessary to prevent unauthorized entry;

(d)(2)
Identify and evaluate the hazards of permit spaces before employees enter them;

(d)(3)
Develop and implement the means, procedures, and practices necessary for safe permit space entry operations, including, but not limited to, the following:

(d)(3)(i)
Specifying acceptable entry conditions;

(d)(3)(ii)
Providing each authorized entrant or that employee's authorized representative with the opportunity to observe any monitoring or testing of permit spaces;

(d)(3)(iii)
Isolating the permit space;

(d)(3)(iv)
Purging, inerting, flushing, or ventilating the permit space as necessary to eliminate or control atmospheric hazards;

(d)(3)(v)
Providing pedestrian, vehicle, or other barriers as necessary to protect entrants from external hazards; and

(d)(3)(vi)
Verifying that conditions in the permit space are acceptable for entry throughout the duration of an authorized entry.

(d)(4)
Provide the following equipment (specified in paragraphs (d)(4)(i) through (d)(4)(ix) of this section) at no cost to employees, maintain that equipment properly, and ensure that employees use that equipment properly:

(d)(4)(i)
Testing and monitoring equipment needed to comply with paragraph (d)(5) of this section;

(d)(4)(ii)
Ventilating equipment needed to obtain acceptable entry conditions;

(d)(4)(iii)
Communications equipment necessary for compliance with paragraphs (h)(3) and (i)(5) of this section;

(d)(4)(iv)
Personal protective equipment insofar as feasible engineering and work practice controls do not adequately protect employees;

(d)(4)(v)
Lighting equipment needed to enable employees to see well enough to work safely and to exit the space quickly in an emergency;

(d)(4)(vi)
Barriers and shields as required by paragraph (d)(3)(iv) of this section;

(d)(4)(vii)
Equipment, such as ladders, needed for safe ingress and egress by authorized entrants;

(d)(4)(viii)
Rescue and emergency equipment needed to comply with paragraph (d)(9) of this section, except to the extent that the equipment is provided by rescue services; and

(d)(4)(ix)
Any other equipment necessary for safe entry into and rescue from permit spaces.

(d)(5)
Evaluate permit space conditions as follows when entry operations are conducted:

(d)(5)(i)
Test conditions in the permit space to determine if acceptable entry conditions exist before entry is authorized to begin, except that, if isolation of the space is infeasible because the space is large or is part of a continuous system (such as a sewer), preentry testing shall be performed to the extent feasible before entry is authorized and, if entry is authorized, entry conditions shall be continuously monitored in the areas where authorized entrants are working;

(d)(5)(ii)
Test or monitor the permit space as necessary to determine if acceptable entry conditions are being maintained during the course of entry operations; and

(d)(5)(iii)
When testing for atmospheric hazards, test first for oxygen, then for combustible gases and vapors, and then for toxic gases and vapors.

(d)(5)(iv)
Provide each authorized entrant or that employee's authorized representative an opportunity to observe the preentry and any subsequent testing or monitoring of permit spaces;

(d)(5)(v)
Reevaluate the permit space in the presence of any authorized entrant or that employee's authorized representative who requests that the employer conduct such reevaluation because the entrant or representative has reason to believe that the evaluation of that space may not have been adequate;

(d)(5)(vi)
Immediately provide each authorized entrant or that employee's authorized representative with the results of any testing conducted in accord with paragraph (d) of this section.

(d)(6)
Provide at least one attendant outside the permit space into which entry is authorized for the duration of entry operations;

Note: Attendants may be assigned to monitor more than one permit space provided the duties described in paragraph (i) of this section can be effectively performed for each permit space that is monitored. Likewise, attendants may be stationed at any location outside the permit space to be monitored as long as the duties described in paragraph (i) of this section can be effectively performed for each permit space that is monitored.

(d)(7)

If multiple spaces are to be monitored by a single attendant, include in the permit program the means and procedures to enable the attendant to respond to an emergency affecting one or more of the permit spaces being monitored without distraction from the attendant's responsibilities under paragraph (i) of this section;

(d)(8)

Designate the persons who are to have active roles (as, for example, authorized entrants, attendants, entry supervisors, or persons who test or monitor the atmosphere in a permit space) in entry operations, identify the duties of each such employee, and provide each such employee with the training required by paragraph (g) of this section;

(d)(9)

Develop and implement procedures for summoning rescue and emergency services, for rescuing entrants from permit spaces, for providing necessary emergency services to rescued employees, and for preventing unauthorized personnel from attempting a rescue;

(d)(10)

Develop and implement a system for the preparation, issuance, use, and cancellation of entry permits as required by this section;

(d)(11)

Develop and implement procedures to coordinate entry operations when employees of more than one employer are working simultaneously as authorized entrants in a permit space, so that employees of one employer do not endanger the employees of any other employer;

(d)(12)

Develop and implement procedures (such as closing off a permit space and canceling the permit) necessary for concluding the entry after entry operations have been completed;

(d)(13)

Review entry operations when the employer has reason to believe that the measures taken under the permit space program may not protect employees and revise the program to correct deficiencies found to exist before subsequent entries are authorized; and

Note: Examples of circumstances requiring the review of the permit space program include any unauthorized entry of a permit space, the detection of a permit space hazard not covered by the permit, the detection of a condition prohibited by the permit, the occurrence of an injury or near-miss during entry, a change in the use or configuration of a permit space, and employee complaints about the effectiveness of the program.

(d)(14)
Review the permit space program, using the canceled permits retained under paragraph (e)(6) of this section within 1 year after each entry and revise the program as necessary, to ensure that employees participating in entry operations are protected from permit space hazards.

Note: Employers may perform a single annual review covering all entries performed during a 12-month period. If no entry is performed during a 12-month period, no review is necessary.

(e)
Permit system.

(e)(1)
Before entry is authorized, the employer shall document the completion of measures required by paragraph (d)(3) of this section by preparing an entry permit.

(e)(2)
Before entry begins, the entry supervisor identified on the permit shall sign the entry permit to authorize entry.

(e)(3)
The completed permit shall be made available at the time of entry to all authorized entrants or their authorized representatives, by posting it at the entry portal or by any other equally effective means, so that the entrants can confirm that preentry preparations have been completed.

(e)(4)
The duration of the permit may not exceed the time required to complete the assigned task or job identified on the permit in accordance with paragraph (f)(2) of this section.

(e)(5)
The entry supervisor shall terminate entry and cancel the entry permit when:

(e)(5)(i)
The entry operations covered by the entry permit have been completed; or

(e)(5)(ii)
A condition that is not allowed under the entry permit arises in or near the permit space.

(e)(6)
The employer shall retain each canceled entry permit for at least 1 year to facilitate the review of the permit-required confined space program required by paragraph (d)(14) of this section. Any problems encountered during an entry operation shall be noted on the pertinent permit so that appropriate revisions to the permit space program can be made.

(f)
Entry permit. The entry permit that documents compliance with this section and authorizes entry to a permit space shall identify:

(f)(1)
The permit space to be entered;

(f)(2)
The purpose of the entry;

(f)(3)
The date and the authorized duration of the entry permit;

(f)(4)
The authorized entrants within the permit space, by name or by such other means (for example, through the use of rosters or tracking systems) as will enable the attendant to determine quickly and accurately, for the duration of the permit, which authorized entrants are inside the permit space;

Note: This requirement may be met by inserting a reference on the entry permit regarding the means used, such as a roster or tracking system, to keep track of the authorized entrants within the permit space.

(f)(5)
The personnel, by name, currently serving as attendants;

(f)(6)
The individual, by name, currently serving as entry supervisor, with a space for the signature or initials of the entry supervisor who originally authorized entry;

(f)(7)
The hazards of the permit space to be entered;

(f)(8)

The measures used to isolate the permit space and to eliminate or control permit space hazards before entry;

Note: Those measures can include the lockout or tagging of equipment and procedures for purging, inerting, ventilating, and flushing permit spaces.

(f)(9)
The acceptable entry conditions;

(f)(10)
The results of initial and periodic tests performed under paragraph (d)(5) of this section, accompanied by the names or initials of the testers and by an indication of when the tests were performed;

(f)(11)
The rescue and emergency services that can be summoned and the means (such as the equipment to use and the numbers to call) for summoning those services;

(f)(12)
The communication procedures used by authorized entrants and attendants to maintain contact during the entry;

(f)(13)
Equipment, such as personal protective equipment, testing equipment, communications equipment, alarm systems, and rescue equipment, to be provided for compliance with this section;

(f)(14)
Any other information whose inclusion is necessary, given the circumstances of the particular confined space, to ensure employee safety; and

(f)(15)
Any additional permits, such as for hot work, that have been issued to authorize work in the permit space.

(g)
Training.

(g)(1)
The employer shall provide training so that all employees whose work is regulated by this section acquire the understanding, knowledge, and skills necessary for the safe performance of the duties assigned under this section.

(g)(2)
Training shall be provided to each affected employee:

(g)(2)(i)
Before the employee is first assigned duties under this section;

(g)(2)(ii)
Before there is a change in assigned duties;

(g)(2)(iii)
Whenever there is a change in permit space operations that presents a hazard about which an employee has not previously been trained;

(g)(2)(iv)
Whenever the employer has reason to believe either that there are deviations from the permit space entry procedures required by paragraph (d)(3) of this section or that there are inadequacies in the employee's knowledge or use of these procedures.

(g)(3)
The training shall establish employee proficiency in the duties required by this section and shall introduce new or revised procedures, as necessary, for compliance with this section.

(g)(4)
The employer shall certify that the training required by paragraphs (g)(1) through (g)(3) of this section has been accomplished. The certification shall contain each employee's name, the signatures or initials of the trainers, and the dates of training. The certification shall be available for inspection by employees and their authorized representatives.

(h)

Duties of authorized entrants. The employer shall ensure that all authorized entrants:

(h)(1)

Know the hazards that may be faced during entry, including information on the mode, signs, or symptoms, and consequences of the exposure;

(h)(2)

Properly use equipment as required by paragraph (d)(4) of this section;

(h)(3)

Communicate with the attendant as necessary to enable the attendant to monitor entrant status and to enable the attendant to alert entrants of the need to evacuate the space as required by paragraph (i)(6) of this section;

(h)(4)

Alert the attendant whenever:

(h)(4)(i)

The entrant recognizes any warning sign or symptom of exposure to a dangerous situation, or

(h)(4)(ii)

The entrant detects a prohibited condition; and

(h)(5)

Exit from the permit space as quickly as possible whenever:

(h)(5)(i)

An order to evacuate is given by the attendant or the entry supervisor,

(h)(5)(ii)

The entrant recognizes any warning sign or symptom of exposure to a dangerous situation,

(h)(5)(iii)

The entrant detects a prohibited condition, or

(h)(5)(iv)

An evacuation alarm is activated.

(i)

Duties of attendants. The employer shall ensure that each attendant:

(i)(1)

Knows the hazards that may be faced during entry, including information on the mode, signs, or symptoms, and consequences of the exposure;

(i)(2)

Is aware of possible behavioral effects of hazard exposure in authorized entrants;

(i)(3)
Continuously maintains an accurate count of authorized entrants in the permit space and ensures that the means used to identify authorized entrants under paragraph (f)(4) of this section accurately identifies who is in the permit space;

(i)(4)
Remains outside the permit space during entry operations until relieved by another attendant;

Note: When the employer's permit entry program allows attendant entry for rescue, attendants may enter a permit space to attempt a rescue if they have been trained and equipped for rescue operations as required by paragraph (k)(1) of this section and if they have been relieved as required by paragraph (i)(4) of this section.

(i)(5)
Communicates with authorized entrants as necessary to monitor entrant status and to alert entrants of the need to evacuate the space under paragraph (i)(6) of this section;

(i)(6)
Monitors activities inside and outside the space to determine if it is safe for entrants to remain in the space and orders the authorized entrants to evacuate the permit space immediately under any of the following conditions;

(i)(6)(i)
If the attendant detects a prohibited condition;

(i)(6)(ii)
If the attendant detects the behavioral effects of hazard exposure in an authorized entrant;

(i)(6)(iii)
If the attendant detects a situation outside the space that could endanger the authorized entrants; or

(i)(6)(iv)
If the attendant cannot effectively and safely perform all the duties required under paragraph (i) of this section;

(i)(7)
Summon rescue and other emergency services as soon as the attendant determines that authorized entrants may need assistance to escape from permit space hazards;

(i)(8)
Takes the following actions when unauthorized persons approach or enter a permit space while entry is under way:

(i)(8)(i)
Warn the unauthorized persons that they must stay away from the permit space;

(i)(8)(ii)
Advise the unauthorized persons that they must exit immediately if they have entered the permit space; and

(i)(8)(iii)
Inform the authorized entrants and the entry supervisor if unauthorized persons have entered the permit space;

(i)(9)
Performs nonentry rescues as specified by the employer's rescue procedure; and

(i)(10)
Performs no duties that might interfere with the attendant's primary duty to monitor and protect the authorized entrants.

(j)
Duties of entry supervisors. The employer shall ensure that each entry supervisor:

(j)(1)
Knows the hazards that may be faced during entry, including information on the mode, signs or symptoms, and consequences of the exposure;

(j)(2)
Verifies, by checking that the appropriate entries have been made on the permit, that all tests specified by the permit have been conducted and that all procedures and equipment specified by the permit are in place before endorsing the permit and allowing entry to begin;

(j)(3)
Terminates the entry and cancels the permit as required by paragraph **(e)(5)** of this section;

(j)(4)
Verifies that rescue services are available and that the means for summoning them are operable;

(j)(5)
Removes unauthorized individuals who enter or who attempt to enter the permit space during entry operations; and

(j)(6)
Determines, whenever responsibility for a permit space entry operation is transferred and at intervals dictated by the hazards and operations performed within the space, that entry operations remain consistent with terms of the entry permit and that acceptable entry conditions are maintained.

(k)
Rescue and emergency services.

(k)(1)
An employer who designates rescue and emergency services, pursuant to paragraph (d)(9) of this section, shall:

(k)(1)(i)
Evaluate a prospective rescuer's ability to respond to a rescue summons in a timely manner, considering the hazard(s) identified;

Note: What will be considered timely will vary according to the specific hazards involved in each entry. For example, Sec. 1910.134, Respiratory Protection, requires that employers provide a standby person or persons capable of immediate action to rescue employee(s) wearing respiratory protection while in work areas defined as IDLH atmospheres.

(k)(1)(ii)
Evaluate a prospective rescue service's ability, in terms of proficiency with rescue-related tasks and equipment, to function appropriately while rescuing entrants from the particular permit space or types of permit spaces identified;

(k)(1)(iii)
Select a rescue team or service from those evaluated that:

(k)(1)(iii)(A)
Has the capability to reach the victim(s) within a time frame that is appropriate for the permit space hazard(s) identified;

(k)(1)(iii)(B)
Is equipped for and proficient in performing the needed rescue services;

(k)(1)(iv)
Inform each rescue team or service of the hazards they may confront when called on to perform rescue at the site; and

(k)(1)(v)
Provide the rescue team or service selected with access to all permit spaces from which rescue may be necessary so that the rescue service can develop appropriate rescue plans and practice rescue operations.

(k)(2)
An employer whose employees have been designated to provide permit space rescue and emergency services shall take the following measures:

(k)(2)(i)
Provide affected employees with the personal protective equipment (PPE) needed to conduct permit space rescues safely and train affected employees so they are proficient in the use of that PPE, at no cost to those employees;

(k)(2)(ii)
Train affected employees to perform assigned rescue duties. The employer must ensure that such employees successfully complete the training required to establish proficiency as an authorized entrant, as provided by paragraphs (g) and (h) of this section;

(k)(2)(iii)
Train affected employees in basic first aid and cardiopulmonary resuscitation (CPR). The employer shall ensure that at least one member of the rescue team or service holding a current certification in first aid and CPR is available; and

(k)(2)(iv)

Ensure that affected employees practice making permit space rescues at least once every 12 months, by means of simulated rescue operations in which they remove dummies, manikins, or actual persons from the actual permit spaces or from representative permit spaces. Representative permit spaces shall, with respect to opening size, configuration, and accessibility, simulate the types of permit spaces from which rescue is to be performed.

(k)(3)

To facilitate nonentry rescue, retrieval systems or methods shall be used whenever an authorized entrant enters a permit space, unless the retrieval equipment would increase the overall risk of entry or would not contribute to the rescue of the entrant. Retrieval systems shall meet the following requirements.

(k)(3)(i)

Each authorized entrant shall use a chest or full-body harness, with a retrieval line attached at the center of the entrant's back near the shoulder level, above the entrant's head, or at another point that the employer can establish presents a profile small enough for the successful removal of the entrant. Wristlets may be used in lieu of the chest or full-body harness if the employer can demonstrate that the use of a chest or full-body harness is infeasible or creates a greater hazard and that the use of wristlets is the safest and most effective alternative.

(k)(3)(ii)

The other end of the retrieval line shall be attached to a mechanical device or fixed point outside the permit space in such a manner that rescue can begin as soon as the rescuer becomes aware that rescue is necessary. A mechanical device shall be available to retrieve personnel from vertical-type permit spaces more than 5 ft (1.52 m) deep.

(k)(4)

If an injured entrant is exposed to a substance for which a Material Safety Data Sheet (MSDS) or other similar written information is required to be kept at the worksite, that MSDS or written information shall be made available to the medical facility treating the exposed entrant.

(l)

Employee participation.

(l)(1)

Employers shall consult with affected employees and their authorized representatives on the development and implementation of all aspects of the permit space program required by paragraph (c) of this section.

(l)(2)

Employers shall make available to affected employees and their authorized representatives all information required to be developed by this section.

3.9 CONTROL OF HAZARDOUS ENERGY (LOCKOUT/TAGOUT)

CFR 1910.147

SubPart Number: J

SubPart Title: General Environmental Controls

(a)
Scope, application, and purpose.

(a)(1)
Scope.

(a)(1)(i)
This standard covers the servicing and maintenance of machines and equipment in which the unexpected energization or startup of the machines or equipment or release of stored energy could cause injury to employees. This standard establishes minimum performance requirements for the control of such hazardous energy.

(a)(1)(ii)
This standard does not cover the following:

(a)(1)(ii)(A)
Construction, agriculture, and maritime employment;

(a)(1)(ii)(B)
Installations under the exclusive control of electric utilities for the purpose of power generation, transmission, and distribution, including related equipment for communication or metering; and

(a)(1)(ii)(C)
Exposure to electrical hazards from work on, near, or with conductors or equipment in electric utilization installations, which is covered by Subpart S of this part; and

(a)(1)(ii)(D)
Oil and gas well drilling and servicing.

(a)(2)
Application.

(a)(2)(i)
This standard applies to the control of energy during servicing and/or maintenance of machines and equipment.

(a)(2)(ii)
Normal production operations are not covered by this standard. Servicing and/or maintenance that takes place during normal production operations is covered by this standard only if:

(a)(2)(ii)(A)
An employee is required to remove or bypass a guard or other safety device; or

(a)(2)(ii)(B)
An employee is required to place any part of his or her body into an area on a machine or piece of equipment where work is actually performed upon the material being processed (point of operation) or where an associated danger zone exists during a machine operating cycle.

Note: Exception to paragraph (a)(2)(ii): Minor tool changes and adjustments, and other minor servicing activities, which take place during normal production operations, are not covered by this standard if they are routine, repetitive, and integral to the use of the equipment for production, provided that the work is performed using alternative measures that provide effective protection.

(a)(2)(iii)
This standard does not apply to the following:

(a)(2)(iii)(A)
Work on cord- and plug-connected electric equipment for which exposure to the hazards of unexpected energization or start-up of the equipment is controlled by the unplugging of the equipment from the energy source and by the plug being under the exclusive control of the employee performing the servicing or maintenance.

(a)(2)(iii)(B)
Hot tap operations involving transmission and distribution systems for substances such as gas, steam, water, or petroleum products when they are performed on pressurized pipelines, provided that the employer demonstrates that

(a)(2)(iii)(B)(1)
continuity of service is essential;

(a)(2)(iii)(B)(2)
shutdown of the system is impractical; and

(a)(2)(iii)(B)(3)
documented procedures are followed, and special equipment is used that will provide proven effective protection for employees.

(a)(3)
Purpose.

(a)(3)(i)
This section requires employers to establish a program and utilize procedures for affixing appropriate lockout devices or tagout devices to energy-isolating devices, and to otherwise disable machines or equipment to prevent unexpected energization, start-up, or release of stored energy to prevent injury to employees.

(a)(3)(ii)
When other standards in this part require the use of lockout or tagout, they shall be used and supplemented by the procedural and training requirements of this section.

(b)
Definitions applicable to this section.

"Affected employee." An employee whose job requires him or her to operate or use a machine or equipment on which servicing or maintenance is being performed under lockout or tagout, or whose job requires him or her to work in an area in which such servicing or maintenance is being performed.

"Authorized employee." A person who locks out or tags out machines or equipment to perform servicing or maintenance on that machine or equipment. An affected employee becomes an authorized employee when that employee's duties include performing servicing or maintenance covered under this section.

"Capable of being locked out." An energy-isolating device is capable of being locked out if it has a hasp or other means of attachment to which, or through which, a lock can be affixed, or it has a locking mechanism built into it. Other energy-isolating devices are capable of being locked out, if lockout can be achieved without the need to dismantle, rebuild, or replace the energy-isolating device or permanently alter its energy control capability.

"Energized." Connected to an energy source or containing residual or stored energy.

"Energy-isolating device." A mechanical device that physically prevents the transmission or release of energy, including but not limited to the following: a manually operated electrical circuit breaker; a disconnect switch; a manually operated switch by which the conductors of a circuit can be disconnected from all ungrounded supply conductors and, in addition, no pole can be operated independently; a line valve; a block; and any similar device used to block or isolate energy. Push-buttons, selector switches, and other control circuit-type devices are not energy-isolating devices.

"Energy source." Any source of electrical, mechanical, hydraulic, pneumatic, chemical, thermal, or other energy.

"Hot tap." A procedure used in the repair, maintenance, and service activities that involves welding on a piece of equipment (pipelines, vessels, or tanks) under pressure to install connections or appurtenances. It is commonly used to replace or add sections of pipeline without interruption of service for air, gas, water, steam, and petrochemical distribution systems.

"Lockout." The placement of a lockout device on an energy-isolating device, in accordance with an established procedure, ensuring that the energy-isolating device and the equipment being controlled cannot be operated until the lockout device is removed.

"Lockout device." A device that utilizes a positive means such as a lock, either key or combination type, to hold an energy-isolating device in the safe position and prevent the energizing of a machine or equipment. Included are blank flanges and bolted slip blinds.

"Normal production operations." The utilization of a machine or equipment to perform its intended production function.

"Servicing and/or maintenance." Workplace activities such as constructing, installing, setting up, adjusting, inspecting, modifying, and maintaining and/or servicing machines or equipment. These activities include lubrication, cleaning, or unjamming of machines or equipment and making adjustments or tool changes, where the employee may be exposed to the unexpected energization or startup of the equipment or release of hazardous energy.

"Setting up." Any work performed to prepare a machine or equipment to perform its normal production operation.

"Tagout." The placement of a tagout device on an energy-isolating device, in accordance with an established procedure, to indicate that the energy-isolating device and the equipment being controlled may not be operated until the tagout device is removed.

"Tagout device." A prominent warning device, such as a tag and a means of attachment, which can be securely fastened to an energy-isolating device in accordance with an established procedure, to indicate that the energy-isolating device and the equipment being controlled may not be operated until the tagout device is removed.

(c)
General.

(c)(1)
Energy control program. The employer shall establish a program consisting of energy control procedures, employee training, and periodic inspections to ensure that before any employee performs any servicing or maintenance on a machine or equipment where the unexpected energizing, startup, or release of stored energy could occur and cause injury, the machine or equipment shall be isolated from the energy source and rendered inoperative.

(c)(2)
Lockout/tagout.

(c)(2)(i)
If an energy-isolating device is not capable of being locked out, the employer's energy control program under paragraph (c)(1) of this section shall utilize a tagout system.

(c)(2)(ii)
If an energy-isolating device is capable of being locked out, the employer's energy control program under paragraph (c)(1) of this section shall utilize lockout, unless the employer can demonstrate that the utilization of a tagout system will provide full employee protection as set forth in paragraph (c)(3) of this section.

(c)(2)(iii)
After January 2, 1990, whenever replacement or major repair, renovation, or modification of a machine or equipment is performed, and whenever new machines or equipment are installed, energy-isolating devices for such machine or equipment shall be designed to accept a lockout device.

(c)(3)
Full employee protection.

(c)(3)(i)
When a tagout device is used on an energy-isolating device that is capable of being locked out, the tagout device shall be attached at the same location that the lockout device would have been attached, and the employer shall demonstrate that the tagout program will provide a level of safety equivalent to that obtained by using a lockout program.

(c)(3)(ii)
In demonstrating that a level of safety is achieved in the tagout program that is equivalent to the level of safety obtained by using a lockout program, the employer shall demonstrate full compliance with all tagout-related provisions of this standard together with such additional elements as are necessary to provide the equivalent safety available from the use of a lockout device. Additional means to be considered as part of the demonstration of full employee protection shall include the implementation of additional safety measures such as the removal of an isolating circuit element, blocking of a controlling switch, opening of an extra disconnecting device, or the removal of a valve handle to reduce the likelihood of inadvertent energization.

(c)(4)
Energy control procedure.

(c)(4)(i)
Procedures shall be developed, documented, and utilized for the control of potentially hazardous energy when employees are engaged in the activities covered by this section.

Note: Exception: The employer need not document the required procedure for a particular machine or equipment, when all of the following elements exist: (1) The machine or equipment has no potential for stored or residual energy or reaccumulation of stored energy after shutdown that could endanger employees; (2) the machine or equipment has a single energy source that can be readily identified and isolated; (3) the isolation and locking out of that energy source will completely deenergize and deactivate the machine or equipment; (4) the machine or equipment is isolated from that energy source and locked out during servicing or maintenance; (5) a single lockout device will achieve a locked-out condition; (6) the lockout device is under the exclusive control of the authorized employee performing the servicing or maintenance; (7) the servicing or maintenance does not create hazards for other employees; and (8) the employer, in utilizing this exception, has had no accidents involving the unexpected activation or reenergization of the machine or equipment during servicing or maintenance.

(c)(4)(ii)
The procedures shall clearly and specifically outline the scope, purpose, authorization, rules, and techniques to be utilized for the control of hazardous energy, and the means to enforce compliance including, but not limited to, the following:

(c)(4)(ii)(A)
A specific statement of the intended use of the procedure;

(c)(4)(ii)(B)
Specific procedural steps for shutting down, isolating, blocking, and securing machines or equipment to control hazardous energy;

(c)(4)(ii)(C)
Specific procedural steps for the placement, removal, and transfer of lockout devices or tagout devices and the responsibility for them; and

(c)(4)(ii)(D)
Specific requirements for testing a machine or equipment to determine and verify the effectiveness of lockout devices, tagout devices, and other energy control measures.

(c)(5)
Protective materials and hardware.

(c)(5)(i)
Locks, tags, chains, wedges, key blocks, adapter pins, self-locking fasteners, or other hardware shall be provided by the employer for isolating, securing, or blocking of machines or equipment from energy sources.

(c)(5)(ii)
Lockout devices and tagout devices shall be singularly identified; shall be the only devices(s) used for controlling energy; shall not be used for other purposes; and shall meet the following requirements:

(c)(5)(ii)(A)
Durable.

(c)(5)(ii)(A)(1)
Lockout and tagout devices shall be capable of withstanding the environment to which they are exposed for the maximum period of time that exposure is expected.

(c)(5)(ii)(A)(2)
Tagout devices shall be constructed and printed so that exposure to weather conditions or wet and damp locations will not cause the tag to deteriorate or the message on the tag to become illegible.

(c)(5)(ii)(A)(3)
Tags shall not deteriorate when used in corrosive environments such as areas where acid and alkali chemicals are handled and stored.

(c)(5)(ii)(B)
Standardized. Lockout and tagout devices shall be standardized within the facility in at least one of the following criteria: color, shape, or size. Additionally, in the case of tagout devices, print and format shall be standardized.

(c)(5)(ii)(C)
Substantial.

(c)(5)(ii)(C)(1)
Lockout devices. Lockout devices shall be substantial enough to prevent removal without the use of excessive force or unusual techniques, such as with the use of bolt cutters or other metal-cutting tools.

(c)(5)(ii)(C)(2)
Tagout devices. Tagout devices, including their means of attachment, shall be substantial enough to prevent inadvertent or accidental removal. Tagout device attachment means shall be of a nonreusable type, attachable by hand, self-locking, and nonreleasable with a minimum unlocking strength of no less than 50 lb and having the general design and basic characteristics of being at least equivalent to a one-piece, all-environment-tolerant nylon cable tie.

(c)(5)(ii)(D)
Identifiable. Lockout devices and tagout devices shall indicate the identity of the employee applying the device(s).

(c)(5)(iii)
Tagout devices shall warn against hazardous conditions if the machine or equipment is energized and shall include a legend such as the following: Do Not Start. Do Not Open. Do Not Close. Do Not Energize. Do Not Operate.

(c)(6)
Periodic inspection.

(c)(6)(i)
The employer shall conduct a periodic inspection of the energy control procedure at least annually to ensure that the procedure and the requirements of this standard are being followed.

(c)(6)(i)(A)
The periodic inspection shall be performed by an authorized employee other than the ones(s) utilizing the energy control procedure being inspected.

(c)(6)(i)(B)
The periodic inspection shall be conducted to correct any deviations or inadequacies identified.

(c)(6)(i)(C)
Where lockout is used for energy control, the periodic inspection shall include a review, between the inspector and each authorized employee, of that employee's responsibilities under the energy control procedure being inspected.

(c)(6)(i)(D)
Where tagout is used for energy control, the periodic inspection shall include a review, between the inspector and each authorized and affected employee, of that employee's responsibilities under the energy control procedure being inspected, and the elements set forth in paragraph (c)(7)(ii) of this section.

(c)(6)(ii)

The employer shall certify that the periodic inspections have been performed. The certification shall identify the machine or equipment on which the energy control procedure was being utilized, the date of the inspection, the employees included in the inspection, and the person performing the inspection.

(c)(7)

Training and communication.

(c)(7)(i)

The employer shall provide training to ensure that the purpose and function of the energy control program are understood by employees and that the knowledge and skills required for the safe application, usage, and removal of the energy controls are acquired by employees. The training shall include the following:

(c)(7)(i)(A)

Each authorized employee shall receive training in the recognition of applicable hazardous energy sources, the type and magnitude of the energy available in the workplace, and the methods and means necessary for energy isolation and control.

(c)(7)(i)(B)

Each affected employee shall be instructed in the purpose and use of the energy control procedure.

(c)(7)(i)(C)

All other employees whose work operations are or may be in an area where energy control procedures may be utilized, shall be instructed about the procedure, and about the prohibition relating to attempts to restart or reenergize machines or equipment that are locked out or tagged out.

(c)(7)(ii)

When tagout systems are used, employees shall also be trained in the following limitations of tags:

(c)(7)(ii)(A)

Tags are essentially warning devices affixed to energy-isolating devices, and do not provide the physical restraint on those devices that is provided by a lock.

(c)(7)(ii)(B)

When a tag is attached to an energy-isolating means, it is not to be removed without authorization of the authorized person responsible for it, and it is never to be bypassed, ignored, or otherwise defeated.

(c)(7)(ii)(C)

Tags must be legible and understandable by all authorized employees, affected employees, and all other employees whose work operations are or may be in the area, in order to be effective.

(c)(7)(ii)(D)

Tags and their means of attachment must be made of materials that will withstand the environmental conditions encountered in the workplace.

(c)(7)(ii)(E)
Tags may evoke a false sense of security, and their meaning needs to be understood as part of the overall energy control program.

(c)(7)(ii)(F)
Tags must be securely attached to energy-isolating devices so that they cannot be inadvertently or accidentally detached during use.

(c)(7)(iii)
Employee retraining.

(c)(7)(iii)(A)
Retraining shall be provided for all authorized and affected employees whenever there is a change in their job assignments, a change in machines, equipment, or processes that present a new hazard, or when there is a change in the energy control procedures.

(c)(7)(iii)(B)
Additional retraining shall also be conducted whenever a periodic inspection under paragraph (c)(6) of this section reveals, or whenever the employer has reason to believe that there are deviations from or inadequacies in the employee's knowledge or use of the energy control procedures.

(c)(7)(iii)(C)
The retraining shall reestablish employee proficiency and introduce new or revised control methods and procedures, as necessary.

(c)(7)(iv)
The employer shall certify that employee training has been accomplished and is being kept up-to-date. The certification shall contain each employee's name and dates of training.

(c)(8)
Energy isolation. Lockout or tagout shall be performed only by the authorized employees who are performing the servicing or maintenance.

(c)(9)
Notification of employees. Affected employees shall be notified by the employer or authorized employee of the application and removal of lockout devices or tagout devices. Notification shall be given before the controls are applied, and after they are removed from the machine or equipment.

(d)
Application of control. The established procedures for the application of energy control (the lockout or tagout procedures) shall cover the following elements and actions and shall be done in the following sequence:

(d)(1)
Preparation for shutdown. Before an authorized or affected employee turns off a machine or equipment, the authorized employee shall have knowledge of the type

and magnitude of the energy, the hazards of the energy to be controlled, and the method or means to control the energy.

(d)(2)

Machine or equipment shutdown. The machine or equipment shall be turned off or shut down using the procedures established for the machine or equipment. An orderly shutdown must be utilized to avoid any additional or increased hazard(s) to employees as a result of the equipment stoppage.

(d)(3)

Machine or equipment isolation. All energy-isolating devices that are needed to control the energy to the machine or equipment shall be physically located and operated in such a manner as to isolate the machine or equipment from the energy source(s).

(d)(4)

Lockout or tagout device application.

(d)(4)(i)

Lockout or tagout devices shall be affixed to each energy-isolating device by authorized employees.

(d)(4)(ii)

Lockout devices, where used, shall be affixed in a manner that will hold the energy-isolating devices in a "safe" or "off" position.

(d)(4)(iii)

Tagout devices, where used, shall be affixed in such a manner as will clearly indicate that the operation or movement of energy-isolating devices from the "safe" or "off" position is prohibited.

(d)(4)(iii)(A)

Where tagout devices are used with energy-isolating devices designed with the capability of being locked, the tag attachment shall be fastened at the same point at which the lock would have been attached.

(d)(4)(iii)(B)

Where a tag cannot be affixed directly to the energy-isolating device, the tag shall be located as close as safely possible to the device, in a position that will be immediately obvious to anyone attempting to operate the device.

(d)(5)

Stored energy.

(d)(5)(i)

Following the application of lockout or tagout devices to energy-isolating devices, all potentially hazardous stored or residual energy shall be relieved, disconnected, restrained, and otherwise rendered safe.

(d)(5)(ii)
If there is a possibility of reaccumulation of stored energy to a hazardous level, verification of isolation shall be continued until the servicing or maintenance is completed, or until the possibility of such accumulation no longer exists.

(d)(6)
Verification of isolation. Prior to starting work on machines or equipment that have been locked out or tagged out, the authorized employee shall verify that isolation and deenergization of the machine or equipment have been accomplished.

(e)
Release from lockout or tagout. Before lockout or tagout devices are removed and energy is restored to the machine or equipment, procedures shall be followed and actions taken by the authorized employee(s) to ensure the following:

(e)(1)
The machine or equipment. The work area shall be inspected to ensure that nonessential items have been removed and to ensure that machine or equipment components are operationally intact.

(e)(2)
Employees.

(e)(2)(i)
The work area shall be checked to ensure that all employees have been safely positioned or removed.

(e)(2)(ii)
After lockout or tagout devices have been removed and before a machine or equipment is started, affected employees shall be notified that the lockout or tagout device(s) have been removed.

(e)(3)
Lockout or tagout devices removal. Each lockout or tagout device shall be removed from each energy-isolating device by the employee who applied the device. Exception to paragraph (e)(3): When the authorized employee who applied the lockout or tagout device is not available to remove it, that device may be removed under the direction of the employer, provided that specific procedures and training for such removal have been developed, documented, and incorporated into the employer's energy control program. The employer shall demonstrate that the specific procedure provides equivalent safety to the removal of the device by the authorized employee who applied it. The specific procedure shall include at least the following elements:

(e)(3)(i)
Verification by the employer that the authorized employee who applied the device is not at the facility:

(e)(3)(ii)
Making all reasonable efforts to contact the authorized employee to inform him or her that his or her lockout or tagout device has been removed; and

(e)(3)(iii)

Ensuring that the authorized employee has this knowledge before he or she resumes work at that facility.

(f)

Additional requirements.

(f)(1)

Testing or positioning of machines, equipment, or components thereof. In situations in which lockout or tagout devices must be temporarily removed from the energy-isolating device and the machine or equipment energized to test or position the machine, equipment, or component thereof, the following sequence of actions shall be followed:

(f)(1)(i)

Clear the machine or equipment of tools and materials in accordance with paragraph (e)(1) of this section;

(f)(1)(ii)

Remove employees from the machine or equipment area in accordance with paragraph (e)(2) of this section;

(f)(1)(iii)

Remove the lockout or tagout devices as specified in paragraph (e)(3) of this section;

(f)(1)(iv)

Energize and proceed with testing or positioning;

(f)(1)(v)

Deenergize all systems and reapply energy control measures in accordance with paragraph (d) of this section to continue the servicing and/or maintenance.

(f)(2)

Outside personnel (contractors, etc.).

(f)(2)(i)

Whenever outside servicing personnel are to be engaged in activities covered by the scope and application of this standard, the on-site employer and the outside employer shall inform each other of their respective lockout or tagout procedures.

(f)(2)(ii)

The on-site employer shall ensure that his or her employees understand and comply with the restrictions and prohibitions of the outside employer's energy control program.

(f)(3)

Group lockout or tagout.

(f)(3)(i)

When servicing and/or maintenance is performed by a crew, craft, department, or other group, it shall utilize a procedure that affords the employees a level of protection equivalent to that provided by the implementation of a personal lockout or tagout device.

(f)(3)(ii)
Group lockout or tagout devices shall be used in accordance with the procedures required by paragraph (c)(4) of this section including, but not necessarily limited to, the following specific requirements:

(f)(3)(ii)(A)
Primary responsibility is vested in an authorized employee or a set number of employees working under the protection of a group lockout or tagout device (such as an operations lock);

(f)(3)(ii)(B)
Provision for the authorized employee to ascertain the exposure status of individual group members with regard to the lockout or tagout of the machine or equipment; and

(f)(3)(ii)(C)
When more than one crew, craft, department, etc. is involved, assignment of overall job-associated lockout or tagout control responsibility to an authorized employee designated to coordinate affected work forces and ensure continuity of protection; and

(f)(3)(ii)(D)
Each authorized employee shall affix a personal lockout or tagout device to the group lockout device, group lockbox, or comparable mechanism when he or she begins work, and shall remove those devices when he or she stops working on the machine or equipment being serviced or maintained.

(f)(4)
Shift or personnel changes. Specific procedures shall be utilized during shift or personnel changes to ensure the continuity of lockout or tagout protection, including provision for the orderly transfer of lockout or tagout device protection between off-going and oncoming employees, to minimize exposure to hazards from the unexpected energization or startup of the machine or equipment, or the release of stored energy.

3.10 HAZARDOUS COMMUNICATIONS

CRF 1910.1200

SubPart Number: Z

(a)
Purpose.

(a)(1)
The purpose of this section is to ensure that the hazards of all chemicals produced or imported are evaluated, and that information concerning their hazards is transmitted to employers and employees. This transmittal of information is to be accomplished by means of comprehensive hazard communication programs, which are to include container labeling and other forms of warning, material safety data sheets, and employee training.

(a)(2)

This occupational safety and health standard is intended to address comprehensively the issue of evaluating the potential hazards of chemicals, and communicating information concerning hazards and appropriate protective measures to employees, and to preempt any legal requirements of a state, or political subdivision of a state, pertaining to this subject. Evaluating the potential hazards of chemicals, and communicating information concerning hazards and appropriate protective measures to employees, may include, for example, but is not limited to, provisions for developing and maintaining a written hazard communication program for the workplace, including lists of hazardous chemicals present; labeling of containers of chemicals in the workplace, as well as of containers of chemicals being shipped to other workplaces; preparation and distribution of material safety data sheets to employees and downstream employers; and development and implementation of employee training programs regarding hazards of chemicals and protective measures. Under section 18 of the Act, no state or political subdivision of a state may adopt or enforce, through any court or agency, any requirement relating to the issue addressed by this federal standard, except pursuant to a federally approved state plan.

(b)

Scope and application.

(b)(1)

This section requires chemical manufacturers or importers to assess the hazards of chemicals that they produce or import, and all employers to provide information to their employees about the hazardous chemicals to which they are exposed, by means of a hazard communication program, labels and other forms of warning, material safety data sheets, and information and training. In addition, this section requires distributors to transmit the required information to employers. (Employers who do not produce or import chemicals need only focus on those parts of this rule that deal with establishing a workplace program and communicating information to their workers.)

(b)(2)

This section applies to any chemical that is known to be present in the workplace in such a manner that employees may be exposed under normal conditions of use or in a foreseeable emergency.

(b)(3)

This section applies to laboratories only as follows:

(b)(3)(i)

Employers shall ensure that labels on incoming containers of hazardous chemicals are not removed or defaced;

(b)(3)(ii)

Employers shall maintain any material safety data sheets that are received with incoming shipments of hazardous chemicals, and ensure that they are readily accessible during each workshift to laboratory employees when they are in their work areas;

(b)(3)(iii)
Employers shall ensure that laboratory employees are provided information and training in accordance with paragraph (h) of this section, except for the location and availability of the written hazard communication program under paragraph (h)(2)(iii) of this section; and,

(b)(3)(iv)
Laboratory employers that ship hazardous chemicals are considered to be either a chemical manufacturer or a distributor under this rule, and thus must ensure that any containers of hazardous chemicals leaving the laboratory are labeled in accordance with paragraph (f)(1) of this section, and that a material safety data sheet is provided to distributors and other employers in accordance with paragraphs (g)(6) and (g)(7) of this section.

(b)(4)
In work operations where employees only handle chemicals in sealed containers that are not opened under normal conditions of use (such as are found in marine cargo handling, warehousing, or retail sales), this section applies to these operations only as follows:

(b)(4)(i)
Employers shall ensure that labels on incoming containers of hazardous chemicals are not removed or defaced;

(b)(4)(ii)
Employers shall maintain copies of any material safety data sheets that are received with incoming shipments of the sealed containers of hazardous chemicals, shall obtain a material safety data sheet as soon as possible for sealed containers of hazardous chemicals received without a material safety data sheet if an employee requests the material safety data sheet, and shall ensure that the material safety data sheets are readily accessible during each work shift to employees when they are in their work area(s); and

(b)(4)(iii)
Employers shall ensure that employees are provided with information and training in accordance with paragraph (h) of this section (except for the location and availability of the written hazard communication program under paragraph (h)(2)(iii) of this section), to the extent necessary to protect them in the event of a spill or leak of a hazardous chemical from a sealed container.

(b)(5)
This section does not require labeling of the following chemicals:

(b)(5)(i)
Any pesticide as such term is defined in the Federal Insecticide, Fungicide, and Rodenticide Act (7 U.S.C. 136 et seq.), when subject to the labeling requirements of that Act and labeling regulations issued under that Act by the Environmental Protection Agency;

(b)(5)(ii)
Any chemical substance or mixture as such terms are defined in the Toxic Substances Control Act (15 U.S.C. 2601 et seq.), when subject to the labeling requirements of that Act and labeling regulations issued under that Act by the Environmental Protection Agency;

(b)(5)(iii)
Any food, food additive, color additive, drug, cosmetic, or medical or veterinary device or product, including materials intended for use as ingredients in such products (e.g., flavors and fragrances), as such terms are defined in the Federal Food, Drug, and Cosmetic Act (21 U.S.C. 301 et seq.) or the Virus-Serum-Toxin Act of 1913 (21 U.S.C. 151 et seq.), and regulations issued under those Acts, when they are subject to the labeling requirements under those Acts by either the Food and Drug Administration or the Department of Agriculture;

(b)(5)(iv)
Any distilled spirits (beverage alcohols), wine, or malt beverage intended for non-industrial use, as such terms are defined in the Federal Alcohol Administration Act (27 U.S.C. 201 et seq.) and regulations issued under that Act, when subject to the labeling requirements of that Act and labeling regulations issued under that Act by the Bureau of Alcohol, Tobacco, and Firearms;

(b)(5)(v)
Any consumer product or hazardous substance as those terms are defined in the Consumer Product Safety Act (15 U.S.C. 2051 et seq.), and Federal Hazardous Substances Act (15 U.S.C. 1261 et seq.) respectively, when subject to a consumer product safety standard or labeling requirement of those Acts, or regulations issued under those Acts by the Consumer Product Safety Commission; and

(b)(5)(vi)
Agricultural or vegetable seed treated with pesticides and labeled in accordance with the Federal Seed Act (7 U.S.C. 1551 et seq.) and the labeling regulations issued under that Act by the Department of Agriculture.

(b)(6)
This section does not apply to:

(b)(6)(i)
Any hazardous waste as such term is defined by the Solid Waste Disposal Act, as amended by the Resource Conservation and Recovery Act of 1976, as amended (42 U.S.C. 6901 et seq.), when subject to regulations issued under that Act by the Environmental Protection Agency;

(b)(6)(ii)
Any hazardous substance as such term is defined by the Comprehensive Environmental Response, Compensation and Liability ACT (CERCLA) (42 U.S.C. 9601 et seq.) when the hazardous substance is the focus of remedial or removal action being conducted under CERCLA in accordance with the Environmental Protection Agency regulations.

(b)(6)(iii)
Tobacco or tobacco products;

(b)(6)(iv)
Wood or wood products, including lumber that will not be processed, where the chemical manufacturer or importer can establish that the only hazard they pose to employees is the potential for flammability or combustibility (wood or wood products that have been treated with a hazardous chemical covered by this standard, and wood that may be subsequently sawed or cut, generating dust, are not exempted);

(b)(6)(v)
Articles (as that term is defined in paragraph (c) of this section);

(b)(6)(vi)
Food or alcoholic beverages that are sold, used, or prepared in a retail establishment (such as a grocery store, restaurant, or drinking place), and foods intended for personal consumption by employees while in the workplace;

(b)(6)(vii)
Any drug, as that term is defined in the Federal Food, Drug, and Cosmetic Act (21 U.S.C. 301 et seq.), when it is in solid, final form for direct administration to the patient (e.g., tablets or pills); drugs that are packaged by the chemical manufacturer for sale to consumers in a retail establishment (e.g., over-the-counter drugs); and drugs intended for personal consumption by employees while in the workplace (e.g., first-aid supplies);

(b)(6)(viii)

Cosmetics that are packaged for sale to consumers in a retail establishment, and cosmetics intended for personal consumption by employees while in the workplace;

(b)(6)(ix)
Any consumer product or hazardous substance, as those terms are defined in the Consumer Product Safety Act (15 U.S.C. 2051 et seq.) and Federal Hazardous Substances Act (15 U.S.C. 1261 et seq.), respectively, where the employer can show that it is used in the workplace for the purpose intended by the chemical manufacturer or importer of the product, and the use results in a duration and frequency of exposure that is not greater than the range of exposures that could reasonably be experienced by consumers when used for the purpose intended;

(b)(6)(x)
Nuisance particulates where the chemical manufacturer or importer can establish that they do not pose any physical or health hazard covered under this section;

(b)(6)(xi)
Ionizing and nonionizing radiation; and

(b)(6)(xii)
Biological hazards.

(c)
Definitions.

"Article" means a manufactured item other than a fluid or particle: (i) which is formed to a specific shape or design during manufacture; (ii) which has end use function(s) dependent in whole or in part upon its shape or design during end use; and (iii) which under normal conditions of use does not release more than very small quantities, e.g., minute or trace amounts of a hazardous chemical (as determined under paragraph (d) of this section), and does not pose a physical hazard or health risk to employees.

"Assistant Secretary" means the Assistant Secretary of Labor for Occupational Safety and Health, U.S. Department of Labor, or designee.

"Chemical" means any element, chemical compound, or mixture of elements and/or compounds.

"Chemical manufacturer" means an employer with a workplace where chemical(s) are produced for use or distribution.

"Chemical name" means the scientific designation of a chemical in accordance with the nomenclature system developed by the International Union of Pure and Applied Chemistry (IUPAC) or the Chemical Abstracts Service (CAS) rules of nomenclature, or a name that will clearly identify the chemical for the purpose of conducting a hazard evaluation.

"Combustible liquid" means any liquid having a flash point at or above 100°F (37.8°C), but below 200°F (93.3°C), except any mixture having components with flash points of 200°F (93.3°C), or higher, the total volume of which make up 99% or more of the total volume of the mixture.

"Commercial account" means an arrangement whereby a retail distributor sells hazardous chemicals to an employer, generally in large quantities over time and/or at costs that are below the regular retail price.

"Common name" means any designation or identification such as code name, code number, trade name, brand name, or generic name used to identify a chemical other than by its chemical name.

"Compressed gas" means (i) a gas or mixture of gases having, in a container, an absolute pressure exceeding 40 psi at 70°F (21.1°C); or (ii) a gas or mixture of gases having, in a container, an absolute pressure exceeding 104 psi at 130°F (54.4°C) regardless of the pressure at 70°F (21.1°C); or (iii) a liquid having a vapor pressure exceeding 40 psi at 100°F (37.8°C) as determined by ASTM D-323-72.

"Container" means any bag, barrel, bottle, box, can, cylinder, drum, reaction vessel, storage tank, or the like that contains a hazardous chemical. For purposes of this section, pipes or piping systems, and engines, fuel tanks, or other operating systems in a vehicle are not considered containers.

"Designated representative" means any individual or organization to whom an employee gives written authorization to exercise such employee's rights under this section. A recognized or certified collective bargaining agent shall be treated automatically as a designated representative without regard to written employee authorization.

"Director" means the Director, National Institute for Occupational Safety and Health, U.S. Department of Health and Human Services, or designee.

"Distributor" means a business, other than a chemical manufacturer or importer, that supplies hazardous chemicals to other distributors or to employers.

"Employee" means a worker who may be exposed to hazardous chemicals under normal operating conditions or in foreseeable emergencies. Workers such as office workers or bank tellers who encounter hazardous chemicals only in nonroutine, isolated instances are not covered.

"Employer" means a person engaged in a business where chemicals are used or distributed, or are produced for use or distribution, including a contractor or subcontractor.

"Explosive" means a chemical that causes a sudden, almost instantaneous release of pressure, gas, and heat when subjected to sudden shock, pressure, or high temperature.

"Exposure or exposed" means that an employee is subjected in the course of employment to a chemical that is a physical or health hazard, and includes potential (e.g., accidental or possible) exposure. "Subjected" in terms of health hazards includes any route of entry (e.g., inhalation, ingestion, skin contact, or absorption.)

"Flammable" means a chemical that falls into one of the following categories:

(i) **"Aerosol, flammable"** means an aerosol that, when tested by the method described in 16 CFR 1500.45, yields a flame projection exceeding 18 in. at full valve opening, or a flashback (a flame extending back to the valve) at any degree of valve opening; (ii) **"Gas, flammable"** means (A) a gas that, at ambient temperature and pressure, forms a flammable mixture with air at a concentration of 13% by volume or less; or (B) a gas that, at ambient temperature and pressure, forms a range of flammable mixtures with air wider than 12% by volume, regardless of the lower limit; (iii) **"Liquid, flammable"** means any liquid having a flash point below 100°F (37.8°C), except any mixture having components with flash points of 100°F (37.8°C) or higher, the total of which make up 99% or more of the total volume of the mixture; (iv) **"Solid, flammable"** means a solid, other than a blasting agent or explosive as defined in 1910.109(a), that is liable to cause fire through friction, absorption of moisture, spontaneous chemical change, or retained heat from manufacturing or processing, or which can be ignited readily and when ignited burns so vigorously and persistently that it creates a serious hazard. A chemical shall be considered to be a flammable solid if, when tested by the method described in 16 CFR 1500.44, it ignites and burns with a self-sustained flame at a rate greater than one tenth of an inch per second along its major axis.

"Flash point" means the minimum temperature at which a liquid gives off a vapor in sufficient concentration to ignite when tested as follows: (i) Tagliabue Closed Tester [See American National Standard Method of Test for Flash Point by Tag Closed Tester, Z11.24-1979 (ASTM D 56-79)] for liquids with a viscosity of less than 45 Saybolt Universal Seconds (SUS) at 100°F (37.8°C), that do not contain suspended solids and do not have a tendency to form a surface film under test; or (ii) Pensky-Martens Closed Tester [see American National Standard Method of Test for Flash Point by Pensky-Martens Closed Tester, Z11.7-1979 (ASTM D 93-79)] for liquids with a viscosity equal to or greater than 45 SUS at 100°F (37.8°C), or that contain suspended solids, or that have a tendency to form a surface film under test; or (iii) Setaflash Closed Tester [see American National Standard Method of Test for Flash Point by Setaflash Closed Tester (ASTM D 3278-78)]. Organic peroxides, which undergo autoaccelerating thermal decomposition, are excluded from any of the flash point determination methods specified above.

"Foreseeable emergency" means any potential occurrence such as, but not limited to, equipment failure, rupture of containers, or failure of control equipment that could result in an uncontrolled release of a hazardous chemical into the workplace.

"Hazardous chemical" means any chemical that is a physical hazard or a health hazard.

"Hazard warning" means any words, pictures, symbols, or combination thereof appearing on a label or other appropriate form of warning that convey the specific physical and health hazard(s), including target organ effects, of the chemical(s) in the container(s). (*See* the definitions for "physical hazard" and "health hazard" to determine the hazards that must be covered.)

"Health hazard" means a chemical for which there is statistically significant evidence based on at least one study conducted in accordance with established scientific principles that acute or chronic health effects may occur in exposed employees. The term *health hazard* includes chemicals that are carcinogens, toxic or highly toxic agents, reproductive toxins, irritants, corrosives, sensitizers, hepatotoxins, nephrotoxins, neurotoxins, agents that act on the hematopoietic system, and agents that damage the lungs, skin, eyes, or mucous membranes.

"Identity" means any chemical or common name that is indicated on the material safety data sheet (MSDS) for the chemical. The identity used shall permit crossreferences to be made among the required list of hazardous chemicals, the label, and the MSDS.

"Immediate use" means that the hazardous chemical will be under the control of and used only by the person who transfers it from a labeled container and only within the work shift in which it is transferred.

"Importer" means the first business with employees within the Customs Territory of the United States that receives hazardous chemicals produced in other countries for the purpose of supplying them to distributors or employers within the United States.

"Label" means any written, printed, or graphic material displayed on or affixed to containers of hazardous chemicals.

"Material safety data sheet" (MSDS) means written or printed material concerning a hazardous chemical that is prepared in accordance with paragraph (g) of this section.

"Mixture" means any combination of two or more chemicals if the combination is not, in whole or in part, the result of a chemical reaction.

"Organic peroxide" means an organic compound that contains the bivalent –O–O–structure and that may be considered to be a structural derivative of hydrogen peroxide where one or both of the hydrogen atoms have been replaced by an organic radical.

"Oxidizer" means a chemical other than a blasting agent or explosive as defined in 1910.109(a), that initiates or promotes combustion in other materials, thereby causing fire either of itself or through the release of oxygen or other gases.

"Physical hazard" means a chemical for which there is scientifically valid evidence that it is a combustible liquid, a compressed gas, explosive, flammable, an organic peroxide, an oxidizer, pyrophoric, unstable (reactive), or water reactive.

"Produce" means to manufacture, process, formulate, blend, extract, generate, emit, or repackage.

"Pyrophoric" means a chemical that will ignite spontaneously in air at a temperature of 130°F (54.4°C) or below.

"Responsible party" means someone who can provide additional information on the hazardous chemical and appropriate emergency procedures, if necessary.

"Specific chemical identity" means the chemical name, Chemical Abstracts Service (CAS) Registry Number, or any other information that reveals the precise chemical designation of the substance.

"Trade secret" means any confidential formula, pattern, process, device, information, or compilation of information that is used in an employer's business, and that gives the employer an opportunity to obtain an advantage over competitors who do not know or use it.

"Unstable (reactive)" means a chemical that in the pure state, or as produced or transported, will vigorously polymerize, decompose, condense, or will become self-reactive under conditions of shock, pressure, or temperature.

"Use" means to package, handle, react, emit, extract, generate as a by-product, or transfer.

"Water-reactive" means a chemical that reacts with water to release a gas that is either flammable or presents a health hazard.

"Work area" means a room or defined space in a workplace where hazardous chemicals are produced or used, and where employees are present.

"Workplace" means an establishment, job site, or project, at one geographical location containing one or more work areas.

(d)
Hazard determination.

(d)(1)
Chemical manufacturers and importers shall evaluate chemicals produced in their workplaces or imported by them to determine if they are hazardous. Employers are not required to evaluate chemicals unless they choose not to rely on the evaluation performed by the chemical manufacturer or importer for the chemical to satisfy this requirement.

(d)(2)
Chemical manufacturers, importers, or employers evaluating chemicals shall identify and consider the available scientific evidence concerning such hazards. For health hazards, evidence that is statistically significant and that is based on at least one positive study conducted in accordance with established scientific principles is considered to be sufficient to establish a hazardous effect if the results of the study meet the definitions of health hazards in this section.

(d)(3)
The chemical manufacturer, importer, or employer evaluating chemicals shall treat the following sources as establishing that the chemicals listed in them are hazardous:

(d)(3)(i)
29 CFR part 1910, subpart Z, Toxic and Hazardous Substances, Occupational Safety and Health Administration (OSHA); or

(d)(3)(ii)
"Threshold Limit Values for Chemical Substances and Physical Agents in the Work Environment," American Conference of Governmental Industrial Hygienists (ACGIH) (latest edition). The chemical manufacturer, importer, or employer is still responsible for evaluating the hazards associated with the chemicals in these source lists in accordance with the requirements of this standard.

(d)(4)
Chemical manufacturers, importers, and employers evaluating chemicals shall treat the following sources as establishing that a chemical is a carcinogen or potential carcinogen for hazard communication purposes:

(d)(4)(i)
National Toxicology Program (NTP), "Annual Report on Carcinogens" (latest edition);

(d)(4)(ii)
International Agency for Research on Cancer (IARC) "Monographs" (latest editions); or

(d)(4)(iii)
29 CFR part 1910, subpart Z, Toxic and Hazardous Substances, Occupational Safety and Health Administration.

Note: The "Registry of Toxic Effects of Chemical Substances" published by the National Institute for Occupational Safety and Health indicates whether a chemical has been found by NTP or IARC to be a potential carcinogen.

(d)(5)
The chemical manufacturer, importer, or employer shall determine the hazards of mixtures of chemicals as follows:

(d)(5)(i)
If a mixture has been tested as a whole to determine its hazards, the results of such testing shall be used to determine whether the mixture is hazardous;

(d)(5)(ii)
If a mixture has not been tested as a whole to determine whether the mixture is a health hazard, the mixture shall be assumed to present the same health hazards as do the components that comprise 1% (by weight or volume) or greater of the mixture, except that the mixture shall be assumed to present a carcinogenic hazard if it contains a component in concentrations of 0.1% or greater that is considered to be a carcinogen under paragraph (d)(4) of this section;

(d)(5)(iii)
If a mixture has not been tested as a whole to determine whether the mixture is a physical hazard, the chemical manufacturer, importer, or employer may use whatever scientifically valid data are available to evaluate the physical hazard potential of the mixture; and,

(d)(5)(iv)
If the chemical manufacturer, importer, or employer has evidence to indicate that a component present in the mixture in concentrations of less than 1% (or in the case of carcinogens, less than 0.1%) could be released in concentrations that would exceed an established OSHA permissible exposure limit or ACGIH Threshold Limit Value, or could present a health risk to employees in those concentrations, the mixture shall be assumed to present the same hazard.

(d)(6)
Chemical manufacturers, importers, or employers evaluating chemicals shall describe in writing the procedures they use to determine the hazards of the chemical they evaluate. The written procedures are to be made available, upon request, to employees, their designated representatives, the Assistant Secretary, and the Director. The written description may be incorporated into the written hazard communication program required under paragraph **(e)** of this section.

(e)
Written hazard communication program.

(e)(1)

Employers shall develop, implement, and maintain at each workplace, a written hazard communication program that at least describes how the criteria specified in paragraphs (f), (g), and (h) of this section for labels and other forms of warning, material safety data sheets, and employee information and training will be met, and that also includes the following:

(e)(1)(i)

A list of the hazardous chemicals known to be present using an identity that is referenced on the appropriate material safety data sheet (the list may be compiled for the workplace as a whole or for individual work areas); and

(e)(1)(ii)

The methods the employer will use to inform employees of the hazards of nonroutine tasks (for example, the cleaning of reactor vessels), and the hazards associated with chemicals contained in unlabeled pipes in their work areas.

(e)(2)

Multiemployer workplaces. Employers who produce, use, or store hazardous chemicals at a workplace in such a way that the employees of other employer(s) may be exposed (for example, employees of a construction contractor working on-site) shall additionally ensure that the hazard communication programs developed and implemented under this paragraph (e) include the following:

(e)(2)(i)

The methods the employer will use to provide the other employer(s) on-site access to material safety data sheets for each hazardous chemical the other employer(s)' employees may be exposed to while working;

(e)(2)(ii)

The methods the employer will use to inform the other employer(s) of any precautionary measures that need to be taken to protect employees in the normal operating conditions of the workplace and in foreseeable emergencies; and

(e)(2)(iii)

The methods the employer will use to inform the other employer(s) of the labeling system used in the workplace.

(e)(3)

The employer may rely on an existing hazard communication program to comply with these requirements, provided that it meets the criteria established in this paragraph **(e)**.

(e)(4)

The employer shall make the written hazard communication program available, upon request, to employees, their designated representatives, the Assistant Secretary and the Director, in accordance with the requirements of 29 CFR 1910.1020(e).

(e)(5)
Where employees must travel between workplaces during a work shift, i.e., their work is carried out at more than one geographical location, the written hazard communication program may be kept at the primary workplace facility.

(f)
Labels and other forms of warning.

(f)(1)
The chemical manufacturer, importer, or distributor shall ensure that each container of hazardous chemicals leaving the workplace is labeled, tagged, or marked with the following information:

(f)(1)(i)
Identity of the hazardous chemical(s);

(f)(1)(ii)
Appropriate hazard warnings; and

(f)(1)(iii)
Name and address of the chemical manufacturer, importer, or other responsible party.

(f)(2)
(f)(2)(i)
For solid metal (such as a steel beam or a metal casting), solid wood, or plastic items that are not exempted as articles due to their downstream use, or shipments of whole grain, the required label may be transmitted to the customer at the time of the initial shipment, and need not be included with subsequent shipments to the same employer unless the information on the label changes;

(f)(2)(ii)
The label may be transmitted with the initial shipment itself, or with the material safety data sheet that is to be provided prior to or at the time of the first shipment; and

(f)(2)(iii)
This exception to requiring labels on every container of hazardous chemicals is only for the solid material itself, and does not apply to hazardous chemicals used in conjunction with, or known to be present with, the material and to which employees handling the items in transit may be exposed (for example, cutting fluids or pesticides in grains).

(f)(3)
Chemical manufacturers, importers, or distributors shall ensure that each container of hazardous chemicals leaving the workplace is labeled, tagged, or marked in accordance with this section in a manner that does not conflict with the requirements of the Hazardous Materials Transportation Act (49 U.S.C. 1801 et seq.) and regulations issued under that Act by the Department of Transportation.

(f)(4)
If the hazardous chemical is regulated by OSHA in a substance-specific health standard, the chemical manufacturer, importer, distributor, or employer shall ensure

that the labels or other forms of warning used are in accordance with the requirements of that standard.

(f)(5)
Except as provided in paragraphs (f)(6) and (f)(7) of this section, the employer shall ensure that each container of hazardous chemicals in the workplace is labeled, tagged, or marked with the following information:

(f)(5)(i)
Identity of the hazardous chemical(s) contained therein; and

(f)(5)(ii)
Appropriate hazard warnings or, alternatively, words, pictures, symbols, or combination thereof, that provide at least general information regarding the hazards of the chemicals, and that, in conjunction with the other information immediately available to employees under the hazard communication program, will provide employees with the specific information regarding the physical and health hazards of the hazardous chemical.

(f)(6)
The employer may use signs, placards, process sheets, batch tickets, operating procedures, or other such written materials in lieu of affixing labels to individual stationary process containers, as long as the alternative method identifies the containers to which it is applicable and conveys the information required by paragraph (f)(5) of this section to be on a label. The written materials shall be readily accessible to the employees in their work area throughout each work shift.

(f)(7)
The employer is not required to label portable containers into which hazardous chemicals are transferred from labeled containers, and which are intended only for the immediate use of the employee who performs the transfer. For purposes of this section, drugs that are dispensed by a pharmacy to a health-care provider for direct administration to a patient are exempted from labeling.

(f)(8)
The employer shall not remove or deface existing labels on incoming containers of hazardous chemicals, unless the container is immediately marked with the required information.

(f)(9)
The employer shall ensure that labels or other forms of warning are legible, in English, and prominently displayed on the container, or readily available in the work area throughout each work shift. Employers having employees who speak other languages may add the information in their language to the material presented, as long as the information is presented in English as well.

(f)(10)
The chemical manufacturer, importer, distributor, or employer need not affix new labels to comply with this section if existing labels already convey the required information.

(f)(11)

Chemical manufacturers, importers, distributors, or employers who become newly aware of any significant information regarding the hazards of a chemical shall revise the labels for the chemical within 3 months of becoming aware of the new information. Labels on containers of hazardous chemicals shipped after that time shall contain the new information. If the chemical is not currently produced or imported, the chemical manufacturer, importers, distributor, or employer shall add the information to the label before the chemical is shipped or introduced into the workplace again.

(g)

Material safety data sheets.

(g)(1)

Chemical manufacturers and importers shall obtain or develop a material safety data sheet for each hazardous chemical they produce or import. Employers shall have a material safety data sheet in the workplace for each hazardous chemical they use.

(g)(2)

Each material safety data sheet shall be in English (although the employer may maintain copies in other languages as well), and shall contain at least the following information:

(g)(2)(i)

The identity used on the label and, except as provided for in paragraph (i) of this section, on trade secrets:

(g)(2)(i)(A)

If the hazardous chemical is a single substance, its chemical and common name(s);

(g)(2)(i)(B)

If the hazardous chemical is a mixture that has been tested as a whole to determine its hazards, the chemical and common name(s) of the ingredients that contribute to these known hazards, and the common name(s) of the mixture itself; or

(g)(2)(i)(C)

If the hazardous chemical is a mixture that has not been tested as a whole:

(g)(2)(i)(C)(1)

The chemical and common name(s) of all ingredients that have been determined to be health hazards, and that comprise 1% or greater of the composition, except that chemicals identified as carcinogens under paragraph (d) of this section shall be listed if the concentrations are 0.1% or greater; and

(g)(2)(i)(C)(2)

The chemical and common name(s) of all ingredients that have been determined to be health hazards, and that comprise less than 1% (0.1% for carcinogens) of the mixture, if there is evidence that the ingredient(s) could be released from the mixture in concentrations that would exceed an established OSHA permissible exposure limit or ACGIH Threshold Limit Value, or could present a health risk to employees; and

(g)(2)(i)(C)(3)
The chemical and common name(s) of all ingredients that have been determined to present a physical hazard when present in the mixture;

(g)(2)(ii)
Physical and chemical characteristics of the hazardous chemical (such as vapor pressure, flash point);

(g)(2)(iii)
The physical hazards of the hazardous chemical, including the potential for fire, explosion, and reactivity;

(g)(2)(iv)
The health hazards of the hazardous chemical, including signs and symptoms of exposure, and any medical conditions that are generally recognized as being aggravated by exposure to the chemical;

(g)(2)(v)
The primary route(s) of entry;

(g)(2)(vi)
The OSHA permissible exposure limit, ACGIH Threshold Limit Value, and any other exposure limit used or recommended by the chemical manufacturer, importer, or employer preparing the material safety data sheet, where available;

(g)(2)(vii)
Whether the hazardous chemical is listed in the National Toxicology Program (NTP) Annual Report on Carcinogens (latest edition) or has been found to be a potential carcinogen in the International Agency for Research on Cancer (IARC) Monographs (latest editions), or by OSHA;

(g)(2)(viii)
Any generally applicable precautions for safe handling and use that are known to the chemical manufacturer, importer, or employer preparing the material safety data sheet, including appropriate hygienic practices, protective measures during repair and maintenance of contaminated equipment, and procedures for cleanup of spills and leaks;

(g)(2)(ix)
Any generally applicable control measures that are known to the chemical manufacturer, importer, or employer preparing the material safety data sheet, such as appropriate engineering controls, work practices, or personal protective equipment;

(g)(2)(x)
Emergency and first-aid procedures;

(g)(2)(xi)
The date of preparation of the material safety data sheet or the last change to it; and

(g)(2)(xii)
The name, address, and telephone number of the chemical manufacturer, importer, employer, or other responsible party preparing or distributing the material safety data sheet, who can provide additional information on the hazardous chemical and appropriate emergency procedures, if necessary.

(g)(3)
If no relevant information is found for any given category on the material safety data sheet, the chemical manufacturer, importer, or employer preparing the material safety data sheet shall mark it to indicate that no applicable information was found.

(g)(4)
Where complex mixtures have similar hazards and contents (i.e., the chemical ingredients are essentially the same, but the specific composition varies from mixture to mixture), the chemical manufacturer, importer, or employer may prepare one material safety data sheet to apply to all of these similar mixtures.

(g)(5)
The chemical manufacturer, importer, or employer preparing the material safety data sheet shall ensure that the information recorded accurately reflects the scientific evidence used in making the hazard determination. If the chemical manufacturer, importer, or employer preparing the material safety data sheet becomes newly aware of any significant information regarding the hazards of a chemical, or ways to protect against the hazards, this new information shall be added to the material safety data sheet within 3 months. If the chemical is not currently being produced or imported, the chemical manufacturer or importer shall add the information to the material safety data sheet before the chemical is introduced into the workplace again.

(g)(6)
(g)(6)(i)
Chemical manufacturers or importers shall ensure that distributors and employers are provided an appropriate material safety data sheet with their initial shipment, and with the first shipment after a material safety data sheet is updated;

(g)(6)(ii)
The chemical manufacturer or importer shall either provide material safety data sheets with the shipped containers or send them to the distributor or employer prior to or at the time of the shipment;

(g)(6)(iii)
If the material safety data sheet is not provided with a shipment that has been labeled as a hazardous chemical, the distributor or employer shall obtain one from the chemical manufacturer or importer as soon as possible; and

(g)(6)(iv)
The chemical manufacturer or importer shall also provide distributors or employers with a material safety data sheet upon request.

(g)(7)
(g)(7)(i)
Distributors shall ensure that material safety data sheets, and updated information, are provided to other distributors and employers with their initial shipment and with the first shipment after a material safety data sheet is updated;

(g)(7)(ii)
The distributor shall either provide material safety data sheets with the shipped containers, or send them to the other distributor or employer prior to or at the time of the shipment;

(g)(7)(iii)
Retail distributors selling hazardous chemicals to employers having a commercial account shall provide a material safety data sheet to such employers upon request, and shall post a sign or otherwise inform them that a material safety data sheet is available;

(g)(7)(iv)
Wholesale distributors selling hazardous chemicals to employers over-the-counter may also provide material safety data sheets upon the request of the employer at the time of the over-the-counter purchase, and shall post a sign or otherwise inform such employers that a material safety data sheet is available;

(g)(7)(v)
If an employer without a commercial account purchases a hazardous chemical from a retail distributor not required to have material safety data sheets on file (i.e., the retail distributor does not have commercial accounts and does not use the materials), the retail distributor shall provide the employer, upon request, with the name, address, and telephone number of the chemical manufacturer, importer, or distributor from which a material safety data sheet can be obtained;

(g)(7)(vi)
Wholesale distributors shall also provide material safety data sheets to employers or other distributors upon request; and

(g)(7)(vii)
Chemical manufacturers, importers, and distributors need not provide material safety data sheets to retail distributors that have informed them that the retail distributor does not sell the product to commercial accounts or open the sealed container to use it in their own workplaces.

(g)(8)
The employer shall maintain in the workplace copies of the required material safety data sheets for each hazardous chemical, and shall ensure that they are readily accessible during each work shift to employees when they are in their work area(s). (Electronic access, microfiche, and other alternatives to maintaining paper copies of the material safety data sheets are permitted as long as no barriers to immediate employee access in each workplace are created by such options.)

(g)(9)
Where employees must travel between workplaces during a work shift, i.e., their work is carried out at more than one geographical location, the material safety data sheets may be kept at the primary workplace facility. In this situation, the employer shall ensure that employees can immediately obtain the required information in an emergency.

(g)(10)
Material safety data sheets may be kept in any form, including operating procedures, and may be designed to cover groups of hazardous chemicals in a work area where it may be more appropriate to address the hazards of a process rather than individual hazardous chemicals. However, the employer shall ensure that in all cases the required information is provided for each hazardous chemical, and is readily accessible during each work shift to employees when they are in their work area(s).

(g)(11)
Material safety data sheets shall also be made readily available, upon request, to designated representatives and to the Assistant Secretary, in accordance with the requirements of 29 CFR 1910.1020(**e**). The Director shall also be given access to material safety data sheets in the same manner.

(h)
Employee information and training.

(h)(1)
Employers shall provide employees with effective information and training on hazardous chemicals in their work area at the time of their initial assignment, and whenever a new physical or health hazard the employees have not previously been trained about is introduced into their work area. Information and training may be designed to cover categories of hazards (e.g., flammability, carcinogenicity) or specific chemicals. Chemical-specific information must always be available through labels and material safety data sheets.

(h)(2)
Information. Employees shall be informed of:

(h)(2)(i)
The requirements of this section;

(h)(2)(ii)
Any operations in their work area where hazardous chemicals are present; and

(h)(2)(iii)
The location and availability of the written hazard communication program, including the required list(s) of hazardous chemicals, and material safety data sheets required by this section.

(h)(3)
Training. Employee training shall include at least:

(h)(3)(i)
Methods and observations that may be used to detect the presence or release of a hazardous chemical in the work area (such as monitoring conducted by the employer, continuous monitoring devices, visual appearance or odor of hazardous chemicals when being released, etc.);

(h)(3)(ii)
The physical and health hazards of the chemicals in the work area;

(h)(3)(iii)
The measures employees can take to protect themselves from these hazards, including specific procedures the employer has implemented to protect employees from exposure to hazardous chemicals, such as appropriate work practices, emergency procedures, and personal protective equipment to be used; and

(h)(3)(iv)
The details of the hazard communication program developed by the employer, including an explanation of the labeling system and the material safety data sheet, and how employees can obtain and use the appropriate hazard information.

(i)
Trade secrets.

(i)(1)
The chemical manufacturer, importer, or employer may withhold the specific chemical identity, including the chemical name and other specific identification of a hazardous chemical, from the material safety data sheet, provided that:

(i)(1)(i)
The claim that the information withheld is a trade secret can be supported;

(i)(1)(ii)
Information contained in the material safety data sheet concerning the properties and effects of the hazardous chemical is disclosed;

(i)(1)(iii)
The material safety data sheet indicates that the specific chemical identity is being withheld as a trade secret; and

(i)(1)(iv)
The specific chemical identity is made available to health professionals, employees, and designated representatives in accordance with the applicable provisions of this paragraph.

(i)(2)
Where a treating physician or nurse determines that a medical emergency exists and the specific chemical identity of a hazardous chemical is necessary for emergency or first-aid treatment, the chemical manufacturer, importer, or employer shall immediately disclose the specific chemical identity of a trade secret chemical to that treating physician or nurse, regardless of the existence of a written statement of need or a confidentiality agreement. The chemical manufacturer, importer, or employer

may require a written statement of need and confidentiality agreement, in accordance with the provisions of paragraphs (i)(3) and (4) of this section, as soon as circumstances permit.

(i)(3)

In nonemergency situations, a chemical manufacturer, importer, or employer shall, upon request, disclose a specific chemical identity, otherwise permitted to be withheld under paragraph (i)(1) of this section, to a health professional (i.e., physician, industrial hygienist, toxicologist, epidemiologist, or occupational health nurse) providing medical or other occupational health services to exposed employee(s), and to employees or designated representatives, if:

(i)(3)(i)

The request is in writing;

(i)(3)(ii)

The request describes with reasonable detail one or more of the following occupational health needs for the information:

(i)(3)(ii)(A)

To assess the hazards of the chemicals to which employees will be exposed;

(i)(3)(ii)(B)

To conduct or assess sampling of the workplace atmosphere to determine employee exposure levels;

(i)(3)(ii)(C)

To conduct preassignment or periodic medical surveillance of exposed employees;

(i)(3)(ii)(D)

To provide medical treatment to exposed employees;

(i)(3)(ii)(E)

To select or assess appropriate personal protective equipment for exposed employees;

(i)(3)(ii)(F)

To design or assess engineering controls or other protective measures for exposed employees; and

(i)(3)(ii)(G)

To conduct studies to determine the health effects of exposure.

(i)(3)(iii)

The request explains in detail why the disclosure of the specific chemical identity is essential and that, in lieu thereof, the disclosure of the following information to the health professional, employee, or designated representative would not satisfy the purposes described in paragraph (i)(3)(ii) of this section:

(i)(3)(iii)(A)

The properties and effects of the chemical;

(i)(3)(iii)(B)
Measures for controlling workers' exposure to the chemical;

(i)(3)(iii)(C)
Methods of monitoring and analyzing worker exposure to the chemical; and

(i)(3)(iii)(D)
Methods of diagnosing and treating harmful exposures to the chemical;

(i)(3)(iv)
The request includes a description of the procedures to be used to maintain the confidentiality of the disclosed information; and

(i)(3)(v)
The health professional and the employer or contractor of the services of the health professional (i.e., downstream employer, labor organization, or individual employee), employee, or designated representative agree in a written confidentiality agreement that the health professional, employee, or designated representative will not use the trade secret information for any purpose other than the health need(s) asserted and agree not to release the information under any circumstances other than to OSHA, as provided in paragraph (i)(6) of this section, except as authorized by the terms of the agreement or by the chemical manufacturer, importer, or employer.

(i)(4)
The confidentiality agreement authorized by paragraph (i)(3)(iv) of this section:

(i)(4)(i)
May restrict the use of the information to the health purposes indicated in the written statement of need;

(i)(4)(ii)
May provide for appropriate legal remedies in the event of a breach of the agreement, including stipulation of a reasonable preestimate of likely damages; and

(i)(4)(iii)
May not include requirements for the posting of a penalty bond.

(i)(5)
Nothing in this standard is meant to preclude the parties from pursuing noncontractual remedies to the extent permitted by law.

(i)(6)
If the health professional, employee, or designated representative receiving the trade secret information decides that there is a need to disclose it to OSHA, the chemical manufacturer, importer, or employer who provided the information shall be informed by the health professional, employee, or designated representative prior to, or at the same time as, such disclosure.

(i)(7)
If the chemical manufacturer, importer, or employer denies a written request for disclosure of a specific chemical identity, the denial must:

(i)(7)(i)
Be provided to the health professional, employee, or designated representative, within 30 days of the request;

(i)(7)(ii)
Be in writing;

(i)(7)(iii)
Include evidence to support the claim that the specific chemical identity is a trade secret;

(i)(7)(iv)
State the specific reasons the request is being denied; and

(i)(7)(v)
Explain in detail how alternative information may satisfy the specific medical or occupational health need without revealing the specific chemical identity.

(i)(8)

The health professional, employee, or designated representative whose request for information is denied under paragraph (i)(3) of this section may refer the request and the written denial of the request to OSHA for consideration.

(i)(9)
When a health professional, employee, or designated representative refers the denial to OSHA under paragraph (i)(8) of this section, OSHA shall consider the evidence to determine if:

(i)(9)(i)
The chemical manufacturer, importer, or employer has supported the claim that the specific chemical identity is a trade secret;

(i)(9)(ii)
The health professional, employee, or designated representative has supported the claim that there is a medical or occupational health need for the information; and

(i)(9)(iii)
The health professional, employee, or designated representative has demonstrated adequate means to protect the confidentiality.

(i)(10)
(i)(10)(i)
If OSHA determines that the specific chemical identity requested under paragraph (i)(3) of this section is not a "bona fide" trade secret, or that it is a trade secret, but the requesting health professional, employee, or designated representative has a legitimate medical or occupational health need for the information, has executed a written confidentiality agreement, and has shown adequate means to protect the confidentiality of the information, the chemical manufacturer, importer, or employer will be subject to citation by OSHA.

(i)(10)(ii)

If a chemical manufacturer, importer, or employer demonstrates to OSHA that the execution of a confidentiality agreement would not provide sufficient protection against the potential harm from the unauthorized disclosure of a trade secret specific chemical identity, the Assistant Secretary may issue such orders or impose such additional limitations or conditions upon the disclosure of the requested chemical information as may be appropriate to assure that the occupational health services are provided without an undue risk of harm to the chemical manufacturer, importer, or employer.

(i)(11)

If a citation for a failure to release specific chemical identity information is contested by the chemical manufacturer, importer, or employer, the matter will be adjudicated before the Occupational Safety and Health Review Commission in accordance with the enforcement scheme of the Act and the applicable commission rules of procedure. In accordance with the Commission rules, when a chemical manufacturer, importer, or employer continues to withhold the information during the contest, the Administrative Law Judge may review the citation and supporting documentation "in camera" or issue appropriate orders to protect the confidentiality of such matters.

(i)(12)

Notwithstanding the existence of a trade secret claim, a chemical manufacturer, importer, or employer shall, upon request, disclose to the Assistant Secretary any information that this section requires the chemical manufacturer, importer, or employer to make available. Where there is a trade secret claim, such claim shall be made no later than at the time the information is provided to the Assistant Secretary so that suitable determinations of trade secret status can be made and the necessary protections can be implemented.

(i)(13)

Nothing in this paragraph shall be construed as requiring the disclosure under any circumstances of process or percentage of mixture information that is a trade secret.

4 Chemical Profiles

Hazardous Communication (HazCom) is among the most frequently cited violations in the OSHA standard. It is amazing that, after all these years, HazCom would be so heavily cited. What is it that makes this standard so impossible to comply with? How does one even begin to complete the enormous amount of training on hundreds or thousands of chemicals? What is really expected of supervisors? Managers? Employees? This section hopes to provide the resources to answer these questions.

To start, it is important to distinguish between the different regulations related to chemicals. Although they may sound similar and are often confused, there are specific differences that are important to understand. There are three main chemical safety regulations that affect almost all companies regardless of how small or large. Those three regulations are 29 CFR 1910.1200 — Hazardous Communications, 29 CFR 1910.120 — Hazardous Waste Operations (including emergency response), and 49 CFR Part 172 — Hazardous Material Transportation. Do not fall victim to complacency or naiveté and automatically think, "our facility doesn't fall under that regulation." This is the first reason OSHA cites so heavily in this area. Understand the differences in these major regulations and learn how to comply.

This section should make complying with 29 CFR 1910.1200 — Hazardous Communications standard much more straightforward and achievable.

The first step in getting a handle on HazCom is to make **a comprehensive list of chemicals** that employees could be exposed to in the workplace. This list should include chemicals purchased from vendors, chemicals manufactured in the facility, by-products such as fumes, gases, or vapors, and any chemicals that are transported or stored by the facility. In small facilities this can easily be hundreds of chemicals, whereas large facilities frequently have several thousand chemicals to list. The list should include the chemical name, manufacturer, the manufacturer's phone number or contact address, the area where the chemical is stored or used, and when the latest material safety data sheets (MSDS) were obtained.

With a thorough chemical inventory to work from, proceed to prepare and implement a **Written Hazardous Communication program.** This does not have to be a cumbersome or difficult task. Sample programs are available on the Web and through many vendors. This process can be as simple as filling in the blanks applicable to the facility. One important aspect of a written program is the designation of a person(s) who will be responsible for chemical labeling of containers and who will review and update the labeling information.

Another aspect of the written program is identifying what method will be used to **provide all employees with current MSDS.** With hundreds or thousands of MSDS to keep current and available this is an almost impossible undertaking. Electronic data management tools have made this task much easier. These are easily found in trade magazines, professional publications, as well as on the Internet. Internet-based retrieval systems can make employee retrieval of MSDS easy and relatively inexpensive. Many universities offer large databases of MSDSs. These are often very helpful and quick to access. Remember the standard stipulates every employee must have ready access to applicable MSDSs in their work areas during their work shift.

Probably the most troublesome aspect of the HazCom regulation is the criteria that **all employees must be trained on every chemical they are exposed to in the workplace.** Where there are only a few chemicals to which an employee is exposed, this may not be too burdensome, whereas in facilities with hundreds or thousands of chemicals this part of the standard can be overwhelming. The National Toxicology Program (NTP) has collected chemical data on more than 2000 chemicals suitable for training. Training material can be accessed at

http://ntp-server.niehs.nih.gov/Main_Pages/Chem-HS.html

Another site providing information suitable for individual chemical training is Scorecard. Scorecard provides detailed information on more than 6800 chemicals, including most chemicals used in large amounts in the United States and all the chemicals regulated under major environmental laws.

http://www.scorecard.org/chemical-profiles/

The regulations offer a more manageable method to address this potential training monster. Training can be conducted by "category of hazard" (e.g., flammables, carcinogens, acids, etc.) This method of training can cut thousands of chemicals down, to, say, 20 "categories of hazards." This type of chemical profile categories of hazards is available at

starmark@arn.net

or Starmark Engineering at 806-273-3830. Examples of these "chemical profiles" follow.

The following pages are devoted to providing some of the basic chemical category training needed to bring a HazCom program into compliance. This does not represent all the chemical categories, but provides a good start, a good template to use in developing additional and customized training modules, and resources of vendors that might help you as well. These chemical profiles include:

Aerosols
Compressed gases
Corrosives
Flammable liquids
Fuels
Insulation

Paint
Pesticides
Reactives
Solvents
Toluene or xylene
Waste

4.1 AEROSOLS

4.1.1 AEROSOL INFORMATION SOURCES

This Hazardous Communication safety information is intended to provide general safety information only. Its intent is to serve as an easy-to-understand reference for the different aspects of handling aerosols. One should always review the product labels and Material Safety Data Sheet (MSDS) for the specific aerosols in use. In some cases, it may even be necessary to consult the manufacturer of specific aerosols before handling them.

4.1.2 AEROSOL SYNONYMS, COMMON NAMES, AND SIMILAR PRODUCTS

Spray Can, Spray, Propellant, Pressurized Spray Can, Pump Spray.

4.1.3 WHAT ARE AEROSOLS?

Thousands of different kinds of products come as aerosols. In general, any product that can be sprayed from a container is an aerosol. These might include fine mists or solids, such as foam or creams. Many different agents are used as propellants for the product being disbursed. These propellants can be user-friendly and environmentally safe, or they can be dangerous to the user and bystanders and damaging to the environment. It is important to understand what kind of aerosol is in use and the proper safeguards, both from a personal perspective and an environmental perspective.

Hazardous aerosol propellants include, but are not limited to, the following materials: chlorofluorocarbons (CFCs), volatile organic compounds (VOCs), or other hydrocarbons that are defined as hazardous by federal and state regulations. Since 1978, CFCs have been used only for a very small percentage of aerosols and only those specifically approved by the government. Some foreign-supplied aerosols still contain CFCs.

Nonhazardous and ozone-safe propellants are good alternatives. They include aerosols that use a gas propellant, e.g., nitrogen, nitrous oxide, carbon dioxide, which are nonhazardous and nonreactive with the ozone layer; products that are packaged in nonaerosols (pump-spray) containers; and products that can be applied with an application system that uses compressed air and reusable containers. To the maximum extent possible, minimize or eliminate the use of products that generate hazardous waste by selecting nonhazardous substitutes.

4.1.4 COMMON CHARACTERISTICS

The most common aerosols come in the form of a spray can. A typical example would be a can of spray paint. Some kind of propellant is always used to force the product out of the aerosol container. Typical propellants would include, but not be limited to, carbon dioxide, compressed air, methane, propane, and nitrogen.

4.1.5 PERSONAL HAZARDS

All aerosols can be dangerous if not used properly. Even if the product content and propellant are not hazardous, the container can still pose a hazard because the container is pressurized. Pressurized containers can release tremendous amounts of energy when punctured, burned, or otherwise damaged.

It is important to remember that aerosol hazards must be considered from at least two different perspectives: (1) the type of product and (2) the type of propellant. An aerosol product may be nonhazardous but may be dispensed with a hazardous propellant. Likewise, a nonhazardous propellant may be used to dispense a hazardous product. Careful review of the specific MSDS is important.

Eyes	Use of aerosols can present significant eye irritation from both the product and the propellant, as well as hazards related to the pressure that the aerosol is under.
Skin	Skin should always be protected when using an aerosol. Use of some aerosols will require personal protective equipment (PPE). If the product gets on skin, it should be washed off immediately. Prolonged contact with skin of any aerosol can be dangerous and should be avoided. See MSDS for proper PPE.
Inhalation	Inhalation can represent a major hazard while using aerosols because the product and propellant are airborne. Use adequate ventilation. Consult the specific MSDS for proper PPE and exposure limits.
Ingestion	Typically, ingestion of significant amounts of a product via aerosol is not a typical route of exposure, although it is important to remember ingestion of a product or propellant via the suspended particles in the air is possible. Ingestion of any propellent or product being sprayed should be avoided.

4.1.6 PERSONAL PROTECTIVE EQUIPMENT

The MSDS for each aerosol in use will also provide guidance for the appropriate protective equipment. PPE will vary according to the particular job and specific substances involved.

Eyes	Always wear eye protection while using aerosols. Safety glasses and, in some cases, chemical safety goggles, which have indirect ventilation ports, or face shields may be required.

Skin   Check the MSDS to see if gloves are required and, if so, use those
            specified for handling that specific aerosol. In some cases, fire-
            retardant clothing, apron, boots, and/or full-cover work clothes
            may be required.

Inhalation Respiratory protection should always be worn when using aero-
            sols. In many cases, a P-100 dust filter will be adequate protection,
            whereas in other cases, a specific cartridge or air-supplied respi-
            rator may be needed. Always use NIOSH-approved respirators.
            Consult the specific MSDS or required respiratory protection
            guide. Wear proper protective clothing with self-contained breath-
            ing apparatus when necessary. When a respirator is necessary, it
            must meet company requirements, MSDS recommendations, and
            OSHA guidelines.

Note: If employees wear any type of respirator, they must be fit-tested and trained
on proper respirator use prior to beginning use.

4.1.7 STORAGE AND DISPOSAL

Special precautions necessary to prevent a hazardous situation are also noted on the
MSDS. These precautions may identify storage procedures, such as storing in a cool,
dry place, storing away from sources of heat, and storing aerosols separately. In
addition, these precautions may include ventilation requirements, as well as special
tools and any other pertinent precautions.

Use common sense when storing or disposing of aerosols. Follow company
procedures and those listed on the MSDS. If there are questions, ask the supervisor.
Always store aerosols away from excessive heat or open flame.

In case of a spill, contain and collect the free liquid. Contain spills with absor-
bents such as sand or clay. Do not allow spills to enter waterways or drainage
systems.

Aerosol containers that no longer deliver sufficient product for normal use may
still have a small amount of product or propellant remaining in the container.
Whenever any chemicals in the product or propellant are regulated materials, the
container must be considered hazardous waste. Review the MSDS for the product
to determine if the aerosol product contains any regulated materials. Do not reuse,
cut, or puncture containers.

An aerosol container that meets both of the following criteria can be managed
as nonhazardous waste:

1. The spray mechanism was not defective, and the contents and propellant
   were discharged to the maximum extent practical under normal use.
2. The container did not contain an extremely or acutely hazardous material.

Dispose of waste in accordance with local, state, and federal EPA regulations.
Treat waste as outlined on the MSDS. To the maximum extent possible, minimize
or eliminate the use of products that generate hazardous waste by selecting

nonhazardous substitutes. Many aerosol containers can be recycled. When possible, this should be done.

4.1.8 EMERGENCY RESPONSE

Since it is impossible to predict when an emergency will occur, take the time to learn the emergency procedure before working with aerosols. Consult the emergency and first-aid procedure section of the MSDS for the specific substances in use.

Victims of aerosol emergencies can be seriously hurt. If seriously hurt, send for medical help as soon as possible, and perform first aid while waiting for emergency personnel. This may include bringing the victim to an eyewash station, safety shower, or fresh air source or removing clothing from the victim. Follow appropriate guidelines in administering first-aid procedures.

Eyes	Flush well with water.
Skin	Wash well with soap and water. Remove contaminated clothing and launder. Seek medical attention if irritation occurs.
Inhalation	Move victim to fresh air. If not breathing, use artificial respiration. If breathing is difficult, give oxygen, if available. Seek medical attention.
Ingestion	Low oral toxicity. Do **not** induce vomiting. Consult a physician.

4.1.9 HELPFUL HANDLING HINTS

- Know the location of nearest eyewash stations, safety showers, fresh air sources, and fire extinguishers. Use the right kind of extinguisher for the chemical in use — the wrong kind can spread the fire.
- Keep aerosols away from heat and flame.
- Consult the MSDS for the specifics on aerosols handled in the work area. Read the label on the container.
- Keep incompatible chemicals separate. Never mix or store them together.
- Label all containers, even temporary ones. Make sure that seals, screens, caps, and containers are working properly and do not leak.
- Never interchange spray can nozzles.
- Never try to use an aerosol can as a torch.
- If an aerosol can becomes dented, discard according to MSDS instructions.
- Make sure all aerosol cans have proper Hazardous Material Information System (HMIS) label until they are properly discarded.
- Learn first-aid skills, including cardiopulmonary resuscitation (CPR) and the company's emergency plan.
- Stay calm. Someone who is not trained to handle emergencies involving aerosols can make the situation worse. An employee who has not received training for handling emergencies involving aerosols can still help keep possible injury and damage to a minimum by remaining calm.
- Have emergency PPE on hand **before** an emergency arises.

- If PPE is necessary, be sure to wear the appropriate PPE and make sure it fits. Never wear less than the recommended PPE as outlined on the MSDS.
- For additional information on aerosols, contact the National Aerosol Association, 584 Bellerive Drive, Suite 3D, Annapolis, MD 21401.

4.2 COMPRESSED GASES

4.2.1 COMPRESSED GASES INFORMATION SOURCES

This Hazardous Communication safety information is intended to provide general safety information only. Its intent is to serve as an easy-to-understand reference for the different aspects of handling compressed gases. One should always review the product labels and Material Safety Data Sheet (MSDS) for the specific compressed gas in use. In some cases, it may even be necessary to consult the compressed gas manufacturer before handling the compressed gases.

4.2.2 COMPRESSED GASES SYNONYMS, COMMON NAMES, AND SIMILAR PRODUCTS

Oxygen, Acetylene, Nitrogen, Hydrogen, LPG, Propane.

4.2.3 WHAT ARE COMPRESSED GASES?

A material or mixture in a container with an absolute pressure of 40 psi (pounds per square inch) at 70°F, a material or mixture in a container with an absolute pressure exceeding 104 psi at 130°F, and a liquid material having a vapor pressure exceeding 40 psi absolute at 100°F are all considered compressed gases. Absolute pressure is the pressure reading on the gauge plus local atmospheric pressure (14.7 psi at sea level).

4.2.4 COMMON CHARACTERISTICS

In general, gases may be hazardous because they are under high pressure, flammable, asphyxiant (inert), oxidizing, corrosive, toxic or highly toxic, or cryogenic (extremely cold).

Compressed gases are hazardous simply because of the high pressure at which they are stored in cylinders, pressure tanks, etc. One dramatic example of what can happen is rocketing. This occurs when a high-pressure cylinder accidentally ruptures or when a valve assembly breaks off. Driven by the pressure of the contents, the cylinder becomes a missile and can blast its way right through a concrete wall. This possibility means that compressed gases should be handled with the utmost care.

Flammable gases catch fire easily and burn quickly. Hydrogen, acetylene, ethylene, propane, and natural gas are some examples. Add flammability to a compression hazard, and some extremely dangerous materials result.

Compressed gases can be corrosive, combustible, flammable, explosive, toxic, or all of these combined. So that everyone will know what type of gas is in a compressed gas cylinder, the cylinders must be legibly marked for identification purposes. Somewhere on the bottle, usually just below the cap, are the identification code and label stating the contents of the bottle. Never rely solely on the color of the bottle when trying to determine what is inside; bottles are sometimes improperly repainted.

4.2.5 PERSONAL HAZARDS

Corrosive gases attack tissue and other materials. Be aware that special personal protective equipment (PPE) and a self-contained breathing apparatus are required when handling these gases.

The degree of hazard of a compressed gas depends on its specific chemical properties. Workers must know what materials they are handling and the hazards they face. Review the MSDS to determine the hazards associated with the compressed gas in use.

4.2.6 PERSONAL PROTECTIVE EQUIPMENT

The MSDS for each compressed gas in use with will also provide guidance for the appropriate protective equipment. PPE will vary according to the particular job and specific substances involved.

PPE required while working with compressed gases will be determined by the hazards of the specific compressed gas. PPE should include safety glasses and, in some cases, chemical safety goggles, which have indirect ventilation ports, or face shields. If gloves are required, check the MSDS for the proper gloves to be worn when handling the specific compressed gases. When a respirator is necessary, it **must** meet company and MSDS standards.

4.2.7 STORAGE AND DISPOSAL

Compressed gas cylinders must be stored so that they are secure and upright. The names of the gases must be posted in the storage area. The storage area should be well ventilated without temperature extremes (especially above 125°F). Compressed gases should be kept away from fires, flames, and welding torches. Containers with removable caps must have caps on while the container is not connected to dispensing equipment. When containers are in storage, regulators, cylinder connections, hose lines, and other similar auxiliary devices must be removed. Oxygen must be stored separately from flammable gas containers and combustibles — 20-ft rule or non-combustible barrier (5 ft high with a fire-resistance rating of 1/2 hour). Mark cylinders when they are empty and avoid storing empty cylinders with full cylinders.

Special precautions necessary to prevent a hazardous situation are also noted on the MSDS. These precautions may identify storage procedures, such as storing in a cool, dry place, storing away from sources of heat, and storing compressed gases separately. In addition, these precautions may include ventilation requirements, as well as special tools and any other pertinent precautions.

Use common sense when storing or disposing of compressed gases. Follow company procedures and those listed on the MSDS. If there are questions, ask the supervisor.

4.2.8 EMERGENCY RESPONSE

Since it is impossible to predict when an emergency will occur, take the time to learn the emergency procedure before working with any compressed gas. Consult the emergency and first-aid procedure section of the MSDS for the specific substances in use.

If there are problems with compressed gases, first contact the gas supplier for assistance. In addition, emergency response advice can be found by calling the Chemical Transportation Emergency Center (CHEMTREC) at 800-424-9300.

Victims of compressed gas emergencies can be seriously hurt. Send for medical help as soon as possible, and perform first aid while waiting for emergency personnel. This may include bringing the victim to an eyewash station, safety shower, or fresh air source or removing the victim's clothing. Follow appropriate guidelines in administering first-aid procedures.

4.2.9 HELPFUL HANDLING HINTS

- Inspect containers oxidizing gases can explode violently when they react with organic and combustible materials. It is important that containers of oxidizing gases or oxygen and associated equipment be free of oils, greases, and other hydrocarbon-based materials. In addition, clothing that has been exposed to an oxygen-rich atmosphere is a fire hazard.
- Identify compressed gases by a label and not solely by the color of the cylinder. Each cylinder must have a Department of Transportation (DOT) label that identifies its contents.
- Inspect equipment used with compressed gas cylinders, such as regulators, valves, connectors, and hose lines, regularly. Damaged equipment should not be accepted when it is delivered or should be removed from service if damage is detected once in use. A damaged cylinder should never be used, and the supplier should be contacted for disposal.
- Inspect threads on valves, regulators, and similar devices for damage, dirt, grease, and oil. The threading on regulators and other auxiliary equipment must match the container valve threads. Adapters are not allowed. Connections should not be tampered with and not be forced together.
- Be aware of special threads on the cylinders for each type of gas. Never try to adapt the wrong regulator to a cylinder by use of an adapter. Never transfer pressure from one cylinder to another.
- Design regulators, pressure relief devices, valves, hoses, and other auxiliary equipment for the specific container and compressed gas to be used. Do not interchange equipment between different types of gases.

- Open container valves slowly with valve outlets pointed away from the user and other persons, using the wrench supplied by the supplier. They should not be hammered open or closed.
- When using compressed gas cylinders, always use the correct regulator for that particular bottle. Open the bottle valve slowly, and do not use tools to force open a valve. If it is difficult to open, return the bottle for a new one. Damaged or difficult-to-open cylinders should be red-tagged and returned to the storeroom or supplier.
- Never tamper with or attempt to repair defective valves or safety relief devices, or cylinders. Such cylinders should be returned to the vendor immediately.
- Do not exceed the specified pressure. Pressure relief devices and safety devices help maintain cylinder or system pressure at the desired levels. Exceeding the desired pressure could damage the cylinder or system.
- Remember that compressed gas, under control, can be extremely helpful in performing many tasks. Out of control, it can cause serious injury — even death.
- Always secure cylinders when in use, in storage, and in transport. If a cylinder is not secured, secure it or let someone know. And when a cylinder is not in use, it should have its protective cap on. If the cylinder valve is knocked off, the cylinder will take off like a missile.
- Secure the protective cap. Some cylinders are designed with a protective cap that screws over the valve at the end of the cylinder. Except when the cylinder is connected to a line or hose, the cap should be kept on the cylinder at all times. Never use the protective cap for lifting or handling the cylinder. Never use a hammer or wrench to open a cylinder valve.
- Be sure that a cylinder is always in the upright position when being used. However, it should never be left in the upright position unless properly secured by means of a substantial chain, cable, or other secure closure. Never drop gas cylinders.
- Do not drop or bang cylinders together violently. Move them only with approved hand trucks. When transporting cylinders by crane, use approved material skiffs; never use magnets or slings. Never use cylinders as rollers to move material.
- Do not drop, slide, or roll cylinders. Carts or other material-handling equipment can be used to move cylinders.
- Keep the cylinder away from all forms of fire and spark-producing operations and electric lines. A compressed gas cylinder should never be exposed to excessive heat so its outside surface exceeds 125°F.
- Never take a compressed gas cylinder into a confined space. Always set the cylinder outside the space, and run the hose or tubing into the space.
- Do not use compressed gas to blow debris off clothing. This is a serious safety violation.
- Install a check valve on the downstream side of the regulator valve whenever there is danger of material flowing back into the cylinder. Avoid placing cylinders where they might form part of an electrical circuit.

- Consult the MSDS for the specifics on the compressed gases in use in the work area. Read the label on the container.
- Keep incompatible chemicals separate. Never mix or store them together.
- Learn first-aid skills, including CPR (cardiopulmonary resuscitation) and the company's emergency plan.
- Be calm. Someone who is not trained to handle emergencies involving compressed gases can make the situation worse. An employee who has not received training for this chemical can still help keep possible injury and damage to a minimum by remaining calm.
- Have emergency PPE on hand **before** an emergency arises.
- If PPE is necessary, be sure to wear the appropriate PPE and make sure it fits. Never wear less than the recommended PPE as outlined on the MSDS.
- Know the location of nearest eyewash stations, safety showers, fresh air sources, and fire extinguishers. Use the right kind of extinguisher for the chemical in use — the wrong kind can spread the fire.

4.3 CORROSIVES

4.3.1 CORROSIVES INFORMATION SOURCES

This Hazardous Communication safety information is intended to provide general safety information only. Its intent is to serve as an easy-to-understand reference for the different aspects of handling corrosives. One should always review the product labels and Material Safety Data Sheet (MSDS) for the specific corrosives in use. In some cases, it may even be necessary to consult the manufacturer of the corrosives before handling them.

4.3.2 CORROSIVES SYNONYMS, COMMON NAMES, AND SIMILAR PRODUCTS

Sulfuric Acid, Caustic, Nitric Acid.

4.3.3 WHAT ARE CORROSIVES?

Whether they are acids such as sulfuric or bases such as lye, corrosives can be hazardous when mishandled. They are usually toxic (poisonous) to the human body, destroying skin, eyes, and other organs on contact. Corrosive vapors can harm internal organs, if they are inhaled or swallowed. If corrosives are combined or stored with the wrong chemicals, there may be an explosion, fire, or release of dangerous vapors.

Contact with a corrosive can cause different amounts of damage, depending on how the corrosive enters the body (i.e., through the skin or through breathing), the quantity of the corrosive, its strength (concentration), and other qualities of the chemical. Permissible Exposure Limits (PELs) for many corrosives have been set by the Occupational Safety and Health Administration and are listed on the appropriate MSDS.

Breathing in a small amount of corrosive mist can cause symptoms including nose, mouth, and throat irritation. Breathing in more highly concentrated or a greater quantity of corrosive mist may cause a heavy sensation in the chest, a hacking cough or chest pain, and difficulty breathing. Swallowing even a small amount of a corrosive may cause severe abdominal pain. Get medical help immediately in all these situations.

4.3.4 COMMON CHARACTERISTICS

These harsh corrosive acids and bases can explode, cause fires, or harm the human body very quickly. A rapid response to an emergency can mean the difference between a slight skin irritation and a blistering burn, or even between life and death.

Even someone who has worked with corrosives before may not know all the different kinds of hazards they represent. Harsh corrosives, which generally are acids or bases, are used in many processes such as water treatment, chemical manufacturing, and metal plating.

4.3.5 PERSONAL HAZARDS

Eyes   If a corrosive mist or other small amount gets in the eyes, symptoms can include burning, watering, irritation, or inflammation. If a corrosive liquid splashes in the eyes, the results can include cloudy scarring or even blindness. Seek emergency medical treatment immediately.

Skin   Skin exposed to small amounts or low concentrations of corrosives can become irritated, itchy, or show signs of contact dermatitis. If skin is splashed by a corrosive, the corrosive can burn, cause blisters, or may penetrate through the skin itself.

4.3.6 PERSONAL PROTECTIVE EQUIPMENT

The MSDS for each corrosive in use will also provide guidance for the appropriate personal protective equipment (PPE). PPE will vary according to the particular job and specific substances involved. Special PPE is required for handling most corrosives. Most commonly, they include safety goggles, which have indirect ventilation, face shield, chemical apron, and gloves designed for the specific corrosive. Read the MSDS and container label for the specific corrosive. Use the PPE and procedures listed on the MSDS.

Emergency-response PPE should be kept in the immediate area in a designated locker or cabinet and be clearly marked.

4.3.7 STORAGE AND DISPOSAL

Special precautions necessary to prevent a hazardous situation are also noted on the MSDS. These precautions may identify storage procedures such as storing in a cool, dry place, storing away from sources of heat, and storing corrosives separately. In addition, these precautions may include ventilation requirements, as well as special tools and any other pertinent precautions.

Use common sense when storing or disposing of corrosives. Follow company procedures and those listed on the MSDS. If there are questions, ask the supervisor.

Acids and bases are an explosive combination, so store them separately. If it is not clear whether the substance is an acid or a base, read the label or MSDS.

Store corrosives in tightly closed approved containers separate from flammable and/or combustible liquids. Dispense them only from approved nozzles and dispensers. Store large drums and containers below eye level whenever possible to avoid splashing the face or eyes.

Check storage containers regularly for leaks, and make sure that caps and spouts are in good working order. Do not smoke around corrosives. Do not carry lighters, matches, or sparking devices when handling corrosives.

Know whether contaminated clothing, PPE, rags, and other materials should be decontaminated, cleaned, or disposed of. Dispose of waste as outlined in the appropriate MSDS.

4.3.8 EMERGENCY RESPONSE

Since it is impossible to predict when an emergency will occur, take the time to learn the emergency procedure before working with corrosives. Consult the emergency and first-aid procedure section of the MSDS for the specific substances.

An employee should handle a spill, leak, fire, or other emergency **only** if it is small and if he or she is trained to do so. Otherwise, the employee might make the situation more dangerous. One should evacuate the area as quickly as possible while letting others know about the emergency.

If a corrosive contact clothing, keep gloves on while removing the clothes immediately.

For most medical emergencies involving corrosives, the first treatment will be water (for skin and eyes) or fresh air (for inhalation). Since corrosives are so harsh, immediate treatment by medical professionals is always critical.

Eyes   If a corrosive contacts the eyes, go directly to the nearest eyewash station. If none is nearby, use any low-pressure clean water source, such as a hose. Remove contact lenses, which can absorb or trap the corrosive. Flush the eyes for 15–20 min, letting the water run from the inside to the outside of the eyes. Do not apply neutralizers or ointments, which have the potential to make damage worse. Someone who has corrosives in the eyes may need to be restrained to receive first aid. Immediately after flushing, transport the person to a medical professional for further treatment.

Skin   If a corrosive contacts the skin, rinse (do not scrub) the affected area for 15–20 min. If a safety shower is not nearby, use a faucet, hose, or any clean water source. Remove any contaminated clothing and, if necessary, dispose of it. Do not put clothing back on

	until it has been decontaminated or cleaned. Do not apply burn ointments or neutralizing solutions. If possible, after rinsing, cover the burn with a sterile dressing: then seek a medical professional immediately.
Inhalation	If corrosive vapors are inhaled, symptoms can range from slight throat irritation to serious difficulty breathing. Get to fresh air immediately, and ask a coworker to obtain medical help right away.
Ingestion	If a corrosive is swallowed, ask a co-worker to obtain medical assistance immediately and to call the local Poison Control Center (dial 911). Do not eat or drink anything unless so instructed by the MSDS, the corrosive label, or a medical professional.

Victims of corrosive emergencies can be seriously hurt. Send for medical help as soon as possible, and perform first aid while waiting for emergency personnel. This may include bringing the victim to an eyewash station, safety shower, or fresh air source or removing the victim's clothing. Follow appropriate guidelines in administering first-aid procedures.

Corrosives do much of their damage quickly, but some symptoms may not appear immediately. Victims of an emergency involving corrosives should stay under medical observation until your doctor releases them.

4.3.9 HELPFUL HANDLING HINTS

- Consult the MSDS for the specifics on the corrosives handled in the work area. Read the label on the container.
- Keep incompatible chemicals separate. Never mix or store them together.
- Label all containers, even temporary ones. Make sure that seals, screens, caps, and containers are working properly and do not leak.
- Learn first-aid skills, including CPR (cardiopulmonary resuscitation), and the company's emergency plan.
- Stay calm. Someone who is not trained to handle emergencies involving corrosives can make the situation worse. An employee who has not received training for this chemical can still help keep possible injury and damage to a minimum by remaining calm.
- Have emergency PPE on hand **before** an emergency arises.
- If PPE is necessary, be sure to wear the appropriate PPE and make sure it fits. Never wear less than the recommended PPE as outlined on the MSDS.
- Know the location of nearest eyewash stations, safety showers, fresh air sources and fire extinguishers. Use the right kind of extinguisher for the chemical in use — the wrong kind can spread the fire.

4.4 FLAMMABLE LIQUIDS

4.4.1 FLAMMABLE LIQUIDS INFORMATION SOURCES

This Hazardous Communication safety information is intended to provide general safety information only. Its intent is to serve as an easy-to-understand reference for the different aspects of handling flammable liquids. One should always review the product labels and MSDS for the specific flammable liquids in use. In some cases, it may even be necessary to consult the flammable liquid manufacturer before handling the flammable liquid.

4.4.2 FLAMMABLE LIQUIDS SYNONYMS, COMMON NAMES, AND SIMILAR PRODUCTS

Gasoline, Paints, Acetone, Toluene, Solvents, Cleaners.

4.4.3 WHAT ARE FLAMMABLE LIQUIDS?

Liquids are rated as flammable or combustible based upon the temperature required for the liquid to give off enough vapor to form an ignitable mixture with the air. Flammable liquids form such a vapor at temperatures below 100°F and combustible liquids at temperatures between 100°F and 200°F. Gasoline will form a flammable mixture in temperatures as low as –50°F, while some heavy fuel oils must be heated or sprayed before they vaporize enough to form the mixture.

4.4.4 COMMON CHARACTERISTICS

Flammables are common chemicals. They are liquids (like gasoline) and gases that burn, release vapors, or even explode under what seem like safe conditions. Flammable gases and liquids burn at close to room temperature (under 100°F), when they are near a spark, flame, or even static electricity. Many of them evaporate quickly. These are called volatile. Flammables can also explode. Their MSDSs indicate when: look for the upper and lower explosive limits. Some chemicals are so flammable that they burn simply from contact with air. These are called pyrophoric.

Flammable emergencies can have a domino effect quickly. A small spark that causes gasoline to explode leads to a large fire, which then leads to a large explosion. Injury to people and damage to property can be extensive.

4.4.5 PERSONAL HAZARDS

The personal hazards involved with flammable materials are significant. Once a situation has become an incident, catastrophic injuries and property damage can take place instantly. While working with flammables, one must always protect against the worst possible scenario of fire and explosion because even the tiniest of ignition source can be extremely detrimental.

4.4.6 PERSONAL PROTECTIVE EQUIPMENT

The MSDS for each flammable liquid will also provide guidance for the appropriate personal protective equipment (PPE). PPE will vary according to the particular job and specific substances involved.

PPE should include safety glasses and, in some cases, chemical safety goggles, which have indirect ventilation ports, or face shields. If gloves are required, check the MSDS for the proper gloves to be worn when handling the specific flammable liquids. If the job deals with many flammable liquids or continual use of flammable liquid materials, fire-retardant clothing may be required because even the slightest of unstable conditions can be extremely dangerous. When a respirator is necessary, it **must** meet company and MSDS standards.

4.4.7 STORAGE AND DISPOSAL

Special precautions necessary to prevent a hazardous situation are also noted on the MSDS. These precautions may identify storage procedures, such as storing in a cool, dry place, storing away from sources of heat, and storing flammable liquids separately. In addition, these precautions may include ventilation requirements, as well as special tools and any other pertinent precautions.

Large amounts of these liquids should be stored in a special flammable storage room or cabinet. It should be well ventilated. Keep flammables far from heat or electric sources. The amount of flammable liquid kept on hand near any industrial operation should be limited to a supply for 1 day or one shift. Flammable liquid containers must be clearly identifiable. A red diamond shape with black lettering is used to designate flammables. Store flammables in a self-closing safety can with a spark arrest or in the pouring spout. Do not leave flammable liquids in open containers because the liquid can vaporize and cause an ignitable mixture to build up.

When rags or other materials are used with flammables, store the liquid-soaked rags in a metal container with a close-fitting lid. This keeps excess oxygen away from the rags and reduces the possibility of a fire. When exposed to the air, some rags can produce enough heat to cause them to ignite spontaneously.

Use common sense when disposing of flammable liquids. Never pour flammables into the drain, sewer, garbage can, or on the ground. Follow company procedures and those listed on the MSDS. If there are questions, ask the supervisor.

4.4.8 EMERGENCY RESPONSE

Since it is impossible to predict when an emergency will occur, take the time to learn the emergency procedure before working with flammable liquids. Consult the emergency and first aid procedure section of the MSDS for the specific substances in use.

Knowing what to do in an emergency can prevent an accident from becoming more serious. Because it is never known when an emergency will happen, it is a good idea to take the time now to read the MSDS, labels, and other materials. Become familiar with emergency plans, and speak to the supervisor if there are questions.

In an emergency, a quick response may prevent damage. But an employee should handle the emergency only if it is small and if trained to handle it.

Turn off any flames and equipment that can spark. Open windows and ventilate the area thoroughly. Clean up any spills using safe procedures and materials. If your clothing is contaminated, remove it immediately. Decontaminate, wash, or dispose of it properly.

In an emergency involving a flammable, obtain medical attention for the victims as soon as possible. Fast response might save their vision, lungs, or even lives. Know the location of the nearest eyewash stations, safety showers, fresh air sources, and fire extinguishers.

Special care must be taken to clean up any spilled material, and it must be properly disposed of in accordance with local, state, and federal regulations.

Victims of flammable liquid emergencies can be seriously hurt. Send for medical help as soon as possible, and perform first aid while waiting for emergency personnel. This may include bringing the victim to an eyewash station, safety shower, or fresh air source or removing the victim's clothing. Follow appropriate guidelines in administering first-aid procedures.

4.4.9 HELPFUL HANDLING HINTS

- Control all ignition sources around flammable liquid. The "no smoking" rule must be enforced and nonsparking tools may be required. Special explosion-proof electrical equipment may be required; never use standard electric power tools around flammable liquids.
- Know what causes flammable liquids to react. Never move, mix, or work with a flammable liquid until its nature is known. Some materials can be ignited by the minimal energy in a static spark; therefore, when drawing liquids from a bulk tank to a portable-use container, the containers should be bonded to the tank, which means that there should be a solid connection between the tank or barrel and the container. Self-closing valves must be used with the dispensing containers to limit spills.
- For fire protection, keep fuel sources to a minimum, limit the oxygen available to the fuel, and control heat or ignition sources.
- Keep paper and cloth away from open flames and keep matches and cigarettes away from flammable liquids such as gasoline, kerosene, or other solvents. Watch for excessive heat such as that generated by friction on machines.
- Keep flammables and reactives away from each other. Reactives are chemicals that explode, burn, or release dangerous vapors very easily.
- Work in well-ventilated areas. Use available ventilation hoods and systems.
- Check to see that all containers are labeled, even temporary ones. They should be in good working order. Check caps, screens, valves, seals, and containers for leaks. Replace or dispose of containers and parts if necessary.
- Do not mix flammables with other chemicals unless instructed to do so.

- Use grounding and bonding wires to prevent dangerous static electricity while transferring flammables from one container to another.
- Know where fire extinguishers are located. Use the right kind of extinguisher for the flammable liquid in use— the wrong kind can spread the fire.
- Turn off flames or equipment that can spark, if possible, from outside the area.
- Learn first-aid skills, including CPR (cardiopulmonary resuscitation) and the company's emergency plan.
- Have emergency PPE on hand **before** an emergency arises.
- If PPE is necessary, be sure to wear the appropriate PPE and make sure it fits. Never wear less than the recommended PPE as outlined on the MSDS.
- Know the location of nearest eyewash stations, safety showers, fresh air sources, and fire extinguishers.

4.5 FUELS

4.5.1 FUELS INFORMATION SOURCES

This Hazardous Communication safety information is intended to provide general safety information only. Its intent is to serve as an easy-to-understand reference for the different aspects of handling fuels. One should always review the product labels and Material Safety Data Sheet (MSDS) for the specific fuels in use. In some cases, it may even be necessary to consult the manufacturer of specific fuels before handling them.

4.5.2 FUELS SYNONYMS, COMMON NAMES, AND SIMILAR PRODUCTS

Gasoline, Petrol, Benzene, Diesel, Propane, Unleaded Gasoline.

4.5.3 WHAT ARE FUELS?

Fuels are highly flammable materials that are easily ignited by heat, sparks, or flames.

4.5.4 COMMON CHARACTERISTICS

Fuels are products that have very low flash points. Vapors may form explosive mixtures with air. Vapors may travel to a source of ignition and flash back. Most vapors are heavier than air. They will spread along the ground and collect in low or confined areas (sewers, basements, tanks). Vapors are explosion hazards indoors, outdoors, and in sewers.

4.5.5 PERSONAL HAZARDS

Inhalation   Inhalation or contact with material may irritate or burn skin and eyes. Vapors may cause dizziness or suffocation.

4.5.6 PERSONAL PROTECTIVE EQUIPMENT

The MSDS for each fuel will also provide guidance for the appropriate personal protective equipment (PPE). PPE will vary according to the particular job and specific substances involved.

PPE should include safety glasses and, in some cases, chemical safety goggles, which have indirect ventilation ports, or face shields. If gloves are required, check the MSDS for the proper gloves to be worn when handling specific fuels. If the job deals with many fuel materials or continual use of fuel materials, fire-retardant clothing may be required because even the slightest of unstable conditions can be extremely dangerous. When a respirator is necessary, it **must** meet company and MSDS standards. Wear positive-pressure self-contained breathing apparatus (SCBA) while fighting fires.

- All equipment used when handling the product must be grounded.

4.5.7 STORAGE AND DISPOSAL

Special precautions necessary to prevent a hazardous situation are also noted on the MSDS. These precautions may identify storage procedures, such as storing in a cool, dry place, storing away from sources of heat, and storing fuels separately. In addition, these precautions may include ventilation requirements, as well as special tools and any other pertinent precautions.

Use common sense when storing or disposing of fuels. Follow company procedures and those listed on the MSDS. If there are questions, ask the supervisor.

4.5.8 EMERGENCY RESPONSE

Since it is impossible to predict when an emergency will occur, take the time to learn the emergency procedure before working with the fuel. Consult the emergency and first-aid procedure section of the MSDS for the specific substances.

Move the victim to fresh air. Call emergency medical care. Apply artificial respiration if the victim is not breathing. Administer oxygen if breathing is difficult. Remove and isolate contaminated clothing and shoes. In case of contact with a substance, immediately flush skin or eyes with running water for at least 20 min. Wash skin with soap and water. Keep victim warm and quiet. Ensure that medical personnel are aware of the material(s) involved and take precautions to protect themselves.

- Use of water spray when fighting fire may be inefficient. Dry chemical, CO_2, water spray, or regular foam should be used to fight small fires.
- Vapors are potential explosion hazards indoors, outdoors, and in sewers.
- Runoff to sewer may create fire or explosion hazard.
- Containers may explode when heated.
- Eliminate all ignition sources (no smoking, flares, sparks or flames in immediate area).
- Do not touch or walk through spilled material to avoid spreading the fuel and becoming exposed to injury.

Spills or Leaks:

- Stop the leak if it possible to do it without risk.
- Prevent entry of material into waterways, sewers, basements, or confined areas.
- Use vapor-suppressing foam to reduce vapors.
- Absorb material or cover with dry earth, sand, or other noncombustible material and transfer to containers.
- Use clean, nonsparking tools to collect absorbed material.
- If a spill is large, dike far ahead of the liquid spill for later disposal.
- Be aware that water spray may reduce vapor, but may not prevent ignition in closed spaces.
- Isolate spill or leak area immediately for at least 25 to 50 m (80 to 160 ft) in all directions.
- Keep out of low areas.
- Ventilate closed spaces before entering.
- Stay upwind.
- Keep unauthorized personnel away.

Victims of fuel emergencies can be seriously hurt. Send for medical help as soon as possible, and perform first aid while waiting for emergency personnel. This may include bringing the victim to an eyewash station, safety shower, or fresh air source or removing clothing from the victim. Follow appropriate guidelines in administering first-aid procedures.

4.5.9 HELPFUL HANDLING HINTS

- Consult the MSDS for the specifics on fuels handled in the work area. Read the label on the container.
- Keep incompatible chemicals separate. Never mix or store them together.
- Label all containers, even temporary ones. Make sure that seals, screens, caps, and containers are working properly and do not leak.
- Learn first-aid skills, including CPR (cardiopulmonary resuscitation) and the company's emergency plan.
- Stay calm. Someone who is not trained to handle emergencies involving fuels can make the situation worse. An employee who has not received training for this chemical can still help keep possible injury and damage to a minimum by remaining calm.
- Have emergency PPE on hand **before** an emergency arises.
- If PPE is necessary, be sure to wear the appropriate PPE and make sure it fits. Never wear less than the recommended PPE as outlined on the MSDS.
- Know the location of nearest eyewash stations, safety showers, fresh air sources and fire extinguishers. Use the right kind of extinguisher for the chemical in use — the wrong kind can spread the fire.

4.6 INSULATION

4.6.1 INSULATION INFORMATION SOURCES

This Hazardous Communication safety information is intended to provide general safety information only. Its intent is to serve as an easy-to-understand reference for the different aspects of handling insulation. One should always review the product labels and Material Safety Data Sheet (MSDS) for the specific insulation in use. In some cases, it may even be necessary to consult the insulation manufacturer before handling the insulation.

4.6.2 INSULATION SYNONYMS, COMMON NAMES, AND SIMILAR PRODUCTS

Acoustical Backing Board, Duct Wrap, Attic Blanket, Duct Board, Insulation Batts, Wall Insulation, Pipe Wrap Insulation, Water Heater Blanket, Rockwool, Insulboard, Fiberglass Wool.

4.6.3 WHAT IS INSULATION?

Insulation is used for many different purposes, but generally its function can be summed up by saying it is used to keep heat contained or to keep heat away. Insulation slows heat transfer, thereby improving the efficiency of the system. For example, on process pipes, insulation is used to retain as much heat as possible in the pipes. When used in a refrigerator, insulation is used to keep the cold in and to keep heat out of the refrigerated area. Insulation also has sound absorption qualities that are often utilized. Insulation and refractory materials perform similar functions.

4.6.4 COMMON CHARACTERISTICS

Insulation can come in many different forms. Fiberglass (wool) is a common type of insulation. It is a fibrous, nondense padding-type material. Fiberglass can be in a loose form, which is installed by being blown through air hoses. Insulation can also be a more dense material such as insulboard or acoustical backing board.

4.6.5 PERSONAL HAZARDS

Primary Routes of Entry: Although insulation can prove irritating to skin and eyes, it does not typically enter the body this way. If skin or eyes become irritated by insulation, remove from source and wash affected area thoroughly.

Acute (short term): Fiberglass wool is a mechanical irritant and may cause temporary irritation of the respiratory tract, skin, and eyes.

Chronic (long term): Fiberglass wool is a possible cancer hazard. Use of these products has not been shown to cause cancer in humans, although some ingredients commonly found in insulation have been determined to be possible causes of cancer in humans (formaldehyde, for example). Consult the specific MSDS.

Medical Conditions Aggravated by Exposure: Chronic respiratory or skin conditions may temporarily worsen from exposure to these products.

Ingestion: Ingestion of this material is unlikely. If it does occur, observe the individual for several days to ensure that intestinal blockage does not occur.

The exposure limits for fibrous glass in insulation is 5 mg/m^3 (respirable dust); 15 mg/m^3 (total dust); ACGIH TLV (threshold limit valve) is 10 mg/m^3. The exposure limit for formaldehyde in insulation is 0.75 ppm TWA (time-washed average) and 2 ppm STEL (short-term exposure limit) according to OSHA PEL (permissible exposure limit), and 0.03 ppm ceiling, according to the ACGIH TLV. Consult the specific MSDS for verification of TWA, TLV, STEL, and other exposure limits. If exact exposure concentrations are not readily known, respiratory protection that would protect against maximum exposures should be worn. Industrial hygiene monitoring should be performed.

Insulation may contain binders or other additives that may be hazardous. Consult the MSDS of the particular insulation being used.

Carcinogenicity: The table below indicates whether or not the agencies listed consider typical insulation ingredients as carcinogens.

Ingredient	ACGIH	IARC	NTP	OSHA
Formaldehyde	Yes	Yes	Yes	Yes
Fiberglass Wool	No	Yes	Yes	No
Cured Resin	No	No	No	No

4.6.6 PERSONAL PROTECTIVE EQUIPMENT

The MSDS for each insulation will also provide guidance for the appropriate personal protective equipment (PPE). PPE will vary according to the particular job and specifics substances involved.

Eyes	Safety glasses, goggles, or face shield.
Skin	Loose-fitting, long-sleeved shirt, long pants, and gloves.
Inhalation	Consult the specific MSDS for the respirator to be worn. A P-100 dust mask should always be considered as the minimum respiratory protection.

PPE should include safety glasses and, in some cases, chemical safety goggles, which have indirect ventilation ports, or face shields. If gloves are required, check the MSDS for the proper gloves to be worn when handling the specific insulation. When a respirator is necessary, it **must** meet company and MSDS standards.

Engineering Controls: General dilution ventilation and/or local exhaust ventilation should be provided as necessary to maintain exposures below regulatory limits. Dust collection systems should be used in operations involving cutting or machining of insulation products and may be required in operations using power tools.

4.6.7 STORAGE AND DISPOSAL

Resource Conservation and Recovery Act (RCRA) Hazard Class:	Nonhazardous.
Toxic Substance Control Act (TSCA) Status:	Typically each ingredient is on the inventory.
Superfund Amendments and Reauthorization Act (SARA) Title III:	Classifies acute health and chronic health of inhalation as hazardous categories, whereas listing fire hazard, pressure hazard, and reactivity hazard as no hazardous categories.
Storage:	No special storage or handling procedures are required for this material.
Disposal:	
Land spill	Scoop up or vacuum material and put into a suitable container for disposal as a nonhazardous waste.
Water spill	This material will sink and disperse along the bottom of waterways and ponds. It cannot easily be removed after it is waterborne, but is generally considered nonhazardous in water.
Air release	This material will settle out of the air. It can then be scooped up or vacuumed for disposal as a nonhazardous waste.

Special precautions necessary to prevent a hazardous situation are also noted on the MSDS. These precautions may identify storage procedures, such as storing in a cool, dry place, storing away from sources of heat, and storing insulation separately. In addition, these precautions may include ventilation requirements, as well as special tools and any other pertinent precautions. Use common sense when storing or disposing of insulation. Follow company procedures and those listed on the MSDS. If there are questions, ask the supervisor.

4.6.8 EMERGENCY RESPONSE

Since it is impossible to predict when an emergency will occur, take the time to learn the emergency procedure before working with the insulation. Consult the emergency and first-aid procedure section of the MSDS for the specific substances.

Eyes	Flush eyes with running water for at least 15 min. Seek medical attention if irritation persists.
Skin	Wash with mild soap and running water. Use a washcloth to help remove fibers. To avoid further irritation, do not rub or scratch irritated areas. Rubbing or scratching may force fibers into skin. Seek medical attention if irritation persists.

Inhalation   Move person to fresh air. Administer cardiopulmonary resuscitation (CPR) if a pulse is not detectable or if the person is unable to breathe. Provide oxygen if breathing is difficult. Obtain medical assistance if irritation persists.

Consult the MSDS for the specifics on insulation handled in the work area. Read the label on the container.

Fire Fighting Instructions: Use self-contained breathing apparatus (SCBA) in a sustained fire.

Hazardous Combustion Products: Primary combustion products are carbon monoxide, carbon dioxide, ammonia, and water.

Victims of insulation emergencies can be seriously hurt. Send for medical help as soon as possible, and perform first aid while waiting for emergency personnel. This may include bringing the victim to an eyewash station, safety shower, or fresh air source or removing the victim's clothing. Follow appropriate guidelines in administering first-aid procedures.

4.6.9 Helpful Handling Hints

- Never mix, pour, stir, or otherwise work insulation without the use of a respirator (dust mask). Some insulations will require a cartridge respirator. Always consult the applicable MSDS.
- Never cut or saw insulation without the use of a respirator (dust mask). Some refractories will require a cartridge respirator. Always consult the applicable MSDS.
- Keep incompatible chemicals separate. Never mix or store them together.
- Label all containers, even temporary ones. Make sure that seals, screens, caps, and containers are working properly and do not leak.
- Learn first-aid skills, including CPR (cardiopulmonary resuscitation) and the company's emergency plan.
- Stay calm. Someone who is not trained to handle emergencies involving insulation can make the situation worse. An employee who has not received training for this insulation can still help keep possible injury and damage to a minimum by remaining calm.
- Have emergency PPE on hand **before** an emergency arises.
- If PPE is necessary, be sure to wear the appropriate PPE and make sure it fits. Never wear less than the recommended PPE as outlined on the MSDS.
- Know the location of nearest eyewash stations, safety showers, fresh air sources and fire extinguishers. Use the right kind of extinguisher for the insulation in use — the wrong kind can spread the fire.

4.7 PAINT

4.7.1 Paint Information Sources

This Hazardous Communication safety information is intended to provide general safety information only. Its intent is to serve as an easy-to-understand reference for the different aspects of handling paints. One should always review the product labels and Material Safety Data Sheet (MSDS) for the specific paint in use. In some cases, it may even be necessary to consult the paint manufacturer before handling the paint.

4.7.2   PAINT SYNONYMS, COMMON NAMES, AND SIMILAR PRODUCTS

Latex, Oil Based, Epoxy, Acrylic, Aerosol.

4.7.3   WHAT IS PAINT?

People have been using paint for thousands of years. In fact, prehistoric cave dwellers used paint to decorate the stone walls of their primitive homes. Spray painting has been known for more than 15,000 years, even though some may have thought of it as a recent invention. In those days, the paint was probably a combination of mud and colored clay. But today's paints are complex chemical mixtures of pigments to add color and binders that allow paint to adhere to various surfaces. However, the chemicals that enable modern paints to perform under a wide variety of applications can present some hazards if not handled properly. Modern paint is applied with brushes, rollers, aerosol cans, and spray guns. Rollers and sprays cover a greater area than brushes, but because they spread more paint, they also increase the amount of both solvents and pigments in the air.

4.7.4   COMMON CHARACTERISTICS

Pigments — The pigments that give paints their color can contain hazardous materials such as lead, zinc, chromium, titanium oxide, and silicates.

Binders — The binders that cause the paint to adhere to the surface being painted can contain hazardous materials, such as acyl and epoxy resins.

Solvents — These pigments and binders are dissolved in water or potentially hazardous solvents such as naphtha, ether, xylene, mineral spirits, or alcohol. When the paint is applied, these solvents evaporate leaving behind the pigment and binder.

4.7.5   PERSONAL HAZARDS

Although some of the materials in pigments, binders, and solvents in paint are considered hazardous, they can only cause harmful health effects if they enter the body.

Once a substance enters the body, two types of health effects can occur — acute and chronic. Acute effects take place soon after exposure. Chronic effects can persist for months or years. Acute or chronic effects can result from exposure to the hazardous materials in paints by any of the three routes of entry: skin, inhalation, and ingestion.

Skin   Skin contact with paints can cause skin irritation, and eye contact
            can result in serious eye injuries. For example, prolonged or
            repeated skin contact with paints containing solvents, such as xylene
            and toluene, can cause a skin condition known as dermatitis. Symp-
            toms of dermatitis include inflammation and flaking of the skin.
Inhalation Examples of chronic effects of inhaling certain paint vapors is
            permanent liver or kidney damage. Inhaling paint vapors can result
            in irritation, headaches, dizziness, mental disorientation, or uncon-
            sciousness.
Ingestion   Paints containing lead can be harmful if inhaled or ingested. Con-
            tinual exposure to even small amounts of lead can result in lead
            poisoning. Some symptoms of lead poisoning are headaches and/or
            dizziness. Ingesting paints can cause abdominal pain or nausea.

Although the materials in some paints can cause acute and chronic health effects,
it is important to understand that these health effects occur only when excessive
quantities of hazardous materials enter the body.

4.7.6 PERSONAL PROTECTIVE EQUIPMENT

The MSDS for each paint will also provide guidance for the appropriate personal
protective equipment (PPE). PPE will vary according to the particular job and the
specific substances involved.

PPE should include safety glasses and, in some cases, chemical safety goggles,
which have indirect ventilation ports, or face shields. If gloves are required, check
the MSDS for the proper gloves to be worn when handling that specific paint. Other
PPE might include aprons and barrier creams; be sure to check the MSDS. Under
certain conditions, respiratory and hearing (if noisy spray equipment is used) protection may also be necessary. Check the MSDS to determine if respiratory or hearing
protection is recommended for the paints or painting method in use. When a respirator is necessary, it **must** meet company and MSDS standards.

To help prevent overexposure from occurring, health professionals have established occupational exposure limits. The most important of these limits is the
TLV-TWA (threshold limit value – time weighted average). This is the average
concentration of material a person may be exposed to for a normal 8-hour day
without adverse health effects. The TLV-TWA can be found in the section for
hazardous ingredients in the MSDS.

4.7.7 STORAGE AND DISPOSAL

Special precautions necessary to prevent a hazardous situation are also noted on the
MSDS. Proper storage is important when using paints. It is preferable to store paints
in cabinets or rooms specially designed for flammable liquid storage. Storage areas
should be cool, dry, well ventilated, and away from heat and flames. Storage areas
and cabinets should be marked with appropriate caution signs, including, "No Smoking" signs. Carefully dispose of paints and rags in the appropriate containers when
painting is finished. Never pour paints down a sink or drain. Use common sense

when storing or disposing of paint. Follow company procedures and those listed on the MSDS. If there are questions, ask the supervisor.

4.7.8 EMERGENCY RESPONSE

Since it is impossible to predict when an emergency will occur, take the time to learn the emergency procedure before working with the paint. Consult the emergency and first-aid procedure section of the MSDS for the specific substances.

Eyes	If the eyes have been splashed, they must be flushed with running water for at least 15 min.
Inhalation	If an unconscious victim is found and there is reason to believe that the cause may be the inhalation of toxic vapors, call for help immediately. **Do not attempt a rescue.** Only qualified individuals who are wearing proper protective equipment and who have a standby observer should attempt rescue. Move the victim to fresh air and provide artificial resuscitation until medical help arrives.
Ingestion	If paint is ingested, give water to dilute the paint, but do **not** induce vomiting.

Since some paints are flammable, read the MSDS before working with them to determine what extinguishing media and fire-fighting procedures are recommended. Should a fire occur, seek help **before** taking any action.

Victims of paint emergencies can be seriously hurt. Send for medical help as soon as possible, and perform first aid while waiting on emergency personnel. This may include bringing the victim to an eyewash station, safety shower, or fresh air source or removing the victim's clothing. Follow appropriate guidelines in administering first-aid procedures.

4.7.9 HELPFUL HANDLING HINTS

- Consult the MSDS for the specifics on the paints handled in the work area. Read the label on the container.
- Be aware that, because static electricity can cause fire or explosion in areas where paint solvent vapors are present, bonding and grounding of paint containers are essential. When dispensing flammable materials from one container to another, make sure the dispensing container is grounded and that it is bonded to the receiving container.
- Provide adequate ventilation when using paints. As paint is applied and dries, it can give off vapors that are potentially hazardous. If natural ventilation is not adequate, mechanical ventilation may be necessary.
- Keep in mind that the most common physical hazards of paint are potential fire and explosion. In general, paints that are classified as flammable should be treated like any other flammable liquid.
- Because noise is also a physical hazard associated with painting operations, which create high noise level, use hearing protection.

- Wipe off splashes to the skin by a solvent-based paint with a dry cloth and then wash with soap and water.
- Wash hands thoroughly before eating or smoking to prevent ingesting hazardous ingredients.
- Label all containers, even temporary ones. Make sure that seals, screens, caps, and containers are working properly and do not leak.
- Learn first-aid skills, including CPR (cardiopulmonary resuscitation) and the company's emergency plan.
- Have emergency PPE on hand **before** an emergency arises.
- If PPE is necessary, be sure to wear the appropriate PPE and make sure it fits. Never wear less than the recommended PPE as outlined on the MSDS.
- Know the location of nearest eyewash stations, safety showers, fresh air sources, and fire extinguishers. Use the right kind of extinguisher for the paint in use — the wrong kind can spread the fire.

4.8 PESTICIDES

4.8.1 PESTICIDES INFORMATION SOURCES

This Hazardous Communication safety information is intended to provide general safety information only. Its intent is to serve as an easy-to-understand reference for the different aspects of handling pesticides. One should always review the product labels and Material Safety Data Sheet (MSDS) for the specific pesticides in use. In some cases, it may even be necessary to consult the pesticide manufacturer before handling the pesticide.

4.8.2 PESTICIDES SYNONYMS, COMMON NAMES, AND SIMILAR PRODUCTS

Rodenticide, Fungicide, Insecticide, Herbicide, Fumigant, Miticide, Repellent.

4.8.3 WHAT ARE PESTICIDES?

Pesticide is a general term used for chemicals that control or kill such pests as rats, mice, insects, bacteria, and weeds. The Federal Insecticide, Fungicide, and Rodenticide Act (FIFRA) is the law that governs the use of pesticides. FIFRA requires pesticide manufacturers to register each of their products with the U.S. Environmental Protection Agency (EPA) as either a general-use or restricted-use pesticide. Restricted-use pesticides can be applied only by certified applicators. In most cases, anyone can apply general-use pesticides according to the label without being certified.

Pesticide liquids are mixtures of powdered or liquid active ingredients combined with liquid carriers such as water or petroleum products. The common ways to apply liquid sprays are with aerosol dispensers, handheld compressed air sprayers, backpack sprayers, or larger, motorized spray units. When liquid sprays are applied, a residue of pesticide active ingredients remains on the treated surfaces and helps to manage pests over a period of time.

Gases that kill pests are known as fumigants. The process of applying fumigants, or fumigation, is much different from other forms of pesticide application and requires application by a professional pest control operator.

Dusts formulations are finely ground, dry powders that contain toxic materials. This formulation leaves visible residues on treated surfaces, which often limits its use to areas such as warehouses, attics, crawl spaces, and wall voids.

Usually granules are combined with a food substance or attractant to encourage target pests (rats, mice, etc.) to feed on them. Do not apply granules in areas where children or pets may come in contact with them.

Poisoned bait may be used to manage specific types of insects. Most baits are a combination of a pesticide and a food material. Baits are usually placed in a bait station or broadcast over the soil around the outside of a structure.

4.8.4 COMMON CHARACTERISTICS

All types of pesticides can be very dangerous if not used properly. Most pesticides have severe consequences if misused. Pesticides can be harmful regardless of their current state, i.e., liquid, solid, granule, fume, residue, etc. Although all pesticides have specific benefits, they also have specific hazards and characteristics. The specific characteristics, dangers, and precautions should be thoroughly understood before handling and before application.

Poisonous chemicals such as pesticides injure or kill people by interfering with the normal functioning of internal body organs and systems. The nature and extent of injury depend on the toxicity of the chemical as well as the dose (amount of material) that enters the body. A person's health and size may also influence the severity of injury.

Pesticides that are applied in strict accordance with their label instructions and with adherence to application rates, reentry intervals, protective equipment requirements, aeration periods, and other listed procedures generally do not leave unsafe levels of pesticide residue.

4.8.5 PERSONAL HAZARDS

Signal words are used to indicate what precautionary measures are needed for people (or animals) who may be exposed.

- **Danger** — This word signals that the pesticide is highly toxic. The product is very likely to cause acute illness from oral, dermal, or inhalation exposure, or to cause severe eye or skin irritation.
- **Posion/Skull And Crossbones** — All highly toxic pesticides that are likely to cause acute illness through oral, dermal, or inhalation exposure also will carry the word **poison** printed in red and the skull-and-crossbones symbol. Products that have the signal word **danger** because of skin and eye irritation potential will not carry the word **poison** or the skull-and-crossbones symbol.
- **Warning** — This word signals that the product is moderately likely to cause acute illness from oral, dermal, or inhalation exposure or that the product is likely to cause moderate skin or eye irritation.

- **Caution** — This word signals that the product is slightly toxic or relatively nontoxic. The product has only slight potential to cause acute illness from oral, dermal, or inhalation exposure. The skin or eye irritation it would cause, if any, is likely to be slight.

Many pesticides can cause acute effects by more than one route, so study these statements carefully. These *precautionary statements* indicate what parts of the body will need the most protection.

People who live or work in the treated area must always be protected so they are not exposed to harmful residues. Avoid using pesticides or application methods that might injure nontarget animals or plants or damage property. Pesticide use should not endanger the environment or cause contamination of groundwater, soils, air, or human and animal foods. In addition, people applying pesticides must do so in ways that avoid excessive exposure to any part of their own bodies.

Applying liquid sprays in certain areas may be extremely hazardous. For example, electric outlets, motors, or exposed wiring pose a potential threat of electrical shock to persons applying water-based pesticide sprays. Pilot lights and gas flames from heaters and appliances may ignite flammable petroleum-based pesticides.

The ingredients of some pesticides are very potent and are capable of causing poisoning at doses as small as a few drops. Regardless of the specific potential hazard, anyone working with pesticides should avoid exposure by using suitable protective clothing and application techniques.

Symptoms are abnormal conditions, feelings, or signs that indicate the presence of an injury, disease, or disorder. When a person is exposed to a large-enough dose of pesticide to cause injury or poisoning, some type of symptoms will usually appear. These symptoms may show up immediately or after several days.

The effect of an exposure can be localized (such as eye or skin irritation) or generalized when the pesticide is absorbed into the blood and distributed to other parts of the body. Examples of chronic conditions usually associated with high or prolonged levels of exposure to certain pesticides include, among others, infertility, birth defects, and cancer.

The type of symptoms may vary between chemical classes of pesticides and may also be different among pesticides within the same chemical class. Symptoms may include a skin rash, headache, or irritation of the eyes, nose, or throat. Other symptoms, which might be caused by higher levels of pesticide exposure, include any of the following: blurred vision, dizziness, heavy sweating, weakness, nausea, stomach pain, vomiting, diarrhea, extreme thirst, and blistered skin. Poisoning can also result in apprehension, restlessness, anxiety, unusual behavior, shaking, convulsions, or unconsciousness of the victim.

When using dusts, prevent their drifting into the airspace of rooms or work areas. Apply dusts only according to the instructions on the pesticide label. Wear approved respiratory protection to avoid inhaling dust particles.

Confined areas present special hazards to persons applying a pesticide. Confined areas may be attics, crawl spaces beneath buildings, storage areas, closets, small rooms, and other places that have poor ventilation. Hazards include inhaling the pesticide being applied and coming in contact with treated surfaces.

4.8.6 Personal Protective Equipment

The MSDS for each pesticide will also provide guidance for the appropriate personal protective equipment (PPE). PPE will vary according to the particular job and specific substances involved.

Avoid pesticide exposure by wearing required or recommended PPE. Carefully maintain, clean, and store PPE in order to keep it in good condition and to ensure that it provides optimal protection. Prevent skin or eye contact with spray resides or vapor. When making an application, always wear a long-sleeved shirt and full-length pants, coveralls, or lightweight spray suit. Protect hands with waterproof gloves and use a face shield or goggles to prevent spray or dust from contacting eyes. Read the pesticide label carefully for the minimum protective clothing requirements.

Whenever possible, increase ventilation in the treatment area by opening windows or using a fan to bring in fresh air. To avoid breathing fumes, wear an approved respirator for the pesticides being applied. Be sure it is in good working condition, fits well, and forms a good seal around the face.

PPE should include safety glasses and, in some cases, chemical safety goggles, which have indirect ventilation ports, or face shields. If gloves are required, check the MSDS for the proper gloves to be worn when handling the specific pesticide. When a respirator is necessary, it **must** meet company and MSDS standards.

4.8.7 Storage and Disposal

Special precautions necessary to prevent a hazardous situation during storage or disposal are noted on the MSDS. These precautions may identify storage procedures such as storing in a cool, dry place, storing away from sources of heat, and storing pesticides separately. In addition, these precautions may include ventilation requirements, as well as special tools and any other pertinent precautions.

Never drain or wash application equipment where runoff will enter sewers, sinks, sumps, or drain tile systems. Use common sense when storing or disposing of pesticides. Follow company procedures and those listed on the MSDS. If there are questions, ask the supervisor.

Residues from improper application or improper rinsing of equipment may also result in contamination of surface water or groundwater.

4.8.8 Emergency Response

Since it is impossible to predict when an emergency will occur, take the time to learn the emergency procedure before working with pesticides. Consult the emergency and first-aid procedure section of the MSDS for the specific substances.

Victims of pesticide emergencies can be seriously hurt. Send for medical help as soon as possible, and perform first aid while waiting for emergency personnel. This may include bringing the victim to an eyewash station, safety shower, or fresh air source or removing the victim's clothing. Follow appropriate guidelines in administering first-aid procedures.

4.8.9 Helpful Handling Hints

- Consult the MSDS for the specifics on the pesticides handled in the work area. Read the label on the container.
- Always apply pesticides in strict accordance with label instructions. Never use a pesticide in a building or other area unless people living or working there can be protected from exposure. This often requires that inhabitants leave the area before an application begins.
- Never apply a pesticide dissolved in oil or petroleum solvent in an enclosed area if there is any source of spark or flame, such as functioning electrical motors, wall switches, appliances, or pilot lights.
- Do not apply pesticides on or near animal food or water or dishes that are used in feeding.
- Never make an outdoor application of a liquid spray when the wind is blowing at more than 5 miles/hour.
- Do not apply a pesticide in outdoor locations where residues can be carried into a well, stream, pond, or other water body.
- Do not use a water-based spray around electric appliances, outlets, or switches.
- Use extreme caution when making pesticide applications in rooms where elderly people or children sleep or spend long periods of time and, whenever possible, avoid treating these locations.
- Keep incompatible pesticides separate. Never mix or store them together.
- Label all containers, even temporary ones. Make sure that seals, screens, caps, and containers are working properly and do not leak.
- Learn first-aid skills, including CPR (cardiopulmonary resuscitation) and the company's emergency plan.
- Have emergency PPE on hand **before** an emergency arises.
- If PPE is necessary, be sure to wear the appropriate PPE and make sure it fits. Never wear less than the recommended PPE as outlined on the MSDS.
- Know the location of nearest eyewash stations, safety showers, fresh air sources, and fire extinguishers.

4.9 REACTIVES

4.9.1 Reactives Information Sources

This Hazardous Communication safety information is intended to provide general safety information only. Its intent is to serve as an easy-to-understand reference for the different aspects of handling paints. One should always review the product labels and Material Safety Data Sheet (MSDS) for the specific reactives in use. In some cases, it may even be necessary to consult the manufacturer of the reactive before handling the reactive.

4.9.2 Reactives Synonyms, Common Names, and Similar Products

Nitric Acid, Petric Acid, Epoxy, Caustic.

4.9.3 What Are Reactives?

Reactives are "nervous" chemicals that can react violently, sometimes just by being moved. Avoid taking chances! Each time employees work with reactives, they should read the MSDS first. The information on this sheet and on container labels will also help protect them.

It takes very little to cause reactives or chemicals near them to explode, burn, or release dangerous vapors. Explosives, the most obvious reactives, can sometimes explode in the presence of a tiny spark, even from friction. Oxidizers, such as nitric acid, contain a large percentage of oxygen. They can cause other substances, like flammables, to burn.

Unstable chemicals can explode under what seem like safe conditions, such as heat or slight movements. Some chemicals, such as ether, become unstable over time. These can be especially dangerous because the usual procedures are no longer safe. Incompatible chemicals, such as acids and bases, seem stable on their own, but react strongly when they are mixed together. Polymerizing chemicals, such as epoxies, create their own chemical reaction. If this reaction happens too quickly, the result can be fire or explosion.

4.9.4 Common Characteristics

"Handle with care" should always be the guide. Before handling a reactive, read the MSDS. It will indicate what causes the chemical to react. Some examples: Unstable chemicals can explode with the slightest shock. Explosives and oxidizers can explode in the presence of the smallest spark, even friction. Incompatible chemicals are unsafe when they are in contact with each other.

4.9.5 Personal Hazards

Reactives can react so suddenly and violently that one should always play it safe. Avoid unnecessary risks. Reading the MSDS and following the list of procedures are excellent protections.

4.9.6 Personal Protective Equipment

The MSDS for each reactive will also provide guidance for the appropriate personal protective equipment (PPE). PPE will vary according to the particular job and specific substances involved.

PPE should include safety glasses and, in some cases, chemical safety goggles, which have indirect ventilation ports, or face shields. If gloves are required, check the MSDS for the proper gloves to be worn when handling the specific reactive. If the job deals with many reactive materials or continual use of reactive materials, fire-retardant clothing may be required. When a respirator is necessary, it **must** meet company and MSDS standards.

4.9.7 STORAGE AND DISPOSAL

Special precautions necessary to prevent a hazardous situation are noted on the MSDS. The safe bet is to store all reactives away from other chemicals. Keep them far from heat and electric sources. Keep oxidizers such as nitric or sulfuric acid stored separately from flammables, paper, wood, or other materials that can burn. Unstable chemicals can react violently to conditions such as movement or heat. Store them in temperature-controlled areas, which do not vibrate or receive shocks. Store incompatible chemicals such as acids and bases away from each other.

Use common sense when storing or disposing of reactives. Follow company procedures and those listed on the MSDS. If there are questions, ask the supervisor.

Always dispose of reactives in approved containers. Never pour them into a drain, sewer, or throw them into the garbage or on the ground. Never smoke around disposal sites or containers.

4.9.8 EMERGENCY RESPONSE

Since it is impossible to predict when an emergency will occur, take the time to learn the emergency procedure before working with the reactive. Consult the emergency and first-aid procedure section of the MSDS for the specific substances you use.

Emergencies involving reactives can be very serious. Explosions and fires can spread quickly, starting from a tiny spark of static electricity or from a lighted match. Take the time before beginning a task to read the appropriate MSDS.

Victims of reactive emergencies can be seriously hurt. Send for medical help as soon as possible, and perform first aid while waiting for emergency personnel. This may include bringing the victim to an eyewash station, safety shower, or fresh air source or removing the victim's clothing. Follow appropriate guidelines in administering first-aid procedures.

4.9.9 HELPFUL HANDLING HINTS

- Consult the MSDS for the specifics on the reactives handled in the work area. Read the label on the container.
- Keep incompatible chemicals separate. Never mix or store them together.
- Label all containers, even temporary ones. Make sure that seals, screens, caps, and containers are working properly and do not leak.
- Learn first-aid skills, including CPR (cardiopulmonary resuscitation) and the company's emergency plan.
- Stay calm. Someone who is not trained to handle emergencies involving reactives can make the situation worse. An employee who has not received training for this chemical can still help keep possible injury and damage to a minimum by remaining calm.

- Have emergency PPE on hand **before** an emergency arises.
- If PPE is necessary, be sure to wear the appropriate PPE and make sure it fits. Never wear less than the recommended PPE as outlined on the MSDS.
- Know the location of nearest eyewash stations, safety showers, fresh air sources, and fire extinguishers. Use the right kind of extinguisher for the reactive in use — the wrong kind can spread the fire.

4.10 REFRACTORIES

4.10.1 REFRACTORY INFORMATION SOURCES

This Hazardous Communication safety information is intended to provide general safety information only. Its intent is to serve as an easy-to-understand reference for the different aspects of handling refractories. One should always review the product labels and Material Safety Data Sheet (MSDS) for the specific refractory material in use. In some cases, it may even be necessary to consult the manufacturer of the material before handling the refractory material.

4.10.2 REFRACTORY SYNONYMS, COMMON NAMES, AND SIMILAR PRODUCTS

Fire brick, Shapes.

4.10.3 WHAT IS A REFRACTORY?

Refractories are nonmetallic materials capable of maintaining physical and chemical stability at high temperatures.

Refractories in modern practice are primarily ceramic in nature, and are widely used in a variety of industries. Wherever a process involves heat in excess of 700 or 800°F (roughly), one will find refractory materials in place, either as a lining or forming the process vessel itself. Some common process vessels using refractories are boiler combustion chambers, furnaces, forges, incinerators, many emission-control scrubbers, rotary kilns, or dryers.

4.10.4 COMMON CHARACTERISTICS

The most common refractories contain alumina and silica. Trace amounts of titanium oxides, chromium, magnesia, ferric oxide, lime, carbon, and other materials also occur in various kinds of refractory materials.

Refractories are primarily made with clay. These clays are then mixed and blended to allow specific properties to be prevalent. These mixes are then formed into a variety of products. Much of it is formed into special brick shapes, then fired again to help stabilize the size of the brick. Castable refractories are bagged and then installed much like cement: add water, pour (or gun), and fire-in-place.

4.10.5 PERSONAL HAZARDS

The personal hazards involved with refractory material vary widely depending on what type of refractory it is and what state of the refractory, i.e., brick, rubble, powder, mud, blown, etc. The same refractory material can pose different health risks at different times. For example, refractory brick poses little health risk when being installed or while in use. But the same brick could pose serious health risk when it is cut by a refractory saw and dust is generated. The refractory could also become an environmental health risk if it is disposed of in a way that will allow rain water to come in contact with it and then migrate toward a drinking water source. It is important to follow the manufacturer's recommendations for handling a refractory material from cradle to grave; that is, from the time it is created, to the time it is disposed of by an approved method.

Eyes   Whenever a refractory material is being handled, cut, sawed, ground, mixed, sprayed, or otherwise abraded or disrupted, eye protection should be used. The applicable MSDS describes the appropriate type of eye protection.

Skin   Protective clothing should be worn to limit contact of loose refractory material coming into contact with the skin. Refractory material that has come in contact with the skin should be promptly removed by the method described on the MSDS. In some cases, this may be simple washing from the skin with soap and water and in other cases it may require a specific solvent. When solvents must be used, always rewash with water.

Inhalation In most cases, the most significant personal hazard dealing with refractories is the hazard of inhalation of refractory dust generated when sawed, ground, blown, cut, or otherwise abraded. At a minimum, a dust respirator should be worn when refractory material is generating dust at any level. When suggested on the MSDS, a canister respirator should be worn. In some cases, the manufacturer will recommend using an air-supplied respirator. The manufacturer's recommendations should always be closely followed.

Ingestion   Ingestion is usually not applicable, but in heavy concentrations of refractory dust ingestion of small amounts is possible. If ingestion does occur, leave the area of exposure and follow first-aid procedures recommended in the MSDS and consult a physician or poison control.

4.10.6 PERSONAL PROTECTIVE EQUIPMENT

The MSDS for each refractory material will also provide guidance for the appropriate personal protective equipment (PPE). PPE will vary according to the particular job and specific substances involved. Handling refractory brick will be much different from working with a powder–castable refractory material.

PPE should include safety glasses and, in some cases, chemical safety goggles, which have indirect ventilation ports, or face shields. If gloves are required, check the MSDS for the proper gloves to be worn when handling the specific refractory material. If the job deals with many refractory materials or continual use of refractory materials, special respirators may be required because compounds in the refractory material can be dangerous. When a respirator is necessary, it **must** meet company and MSDS standards.

4.10.7 STORAGE AND DISPOSAL

Special precautions necessary to prevent a hazardous situation are noted on the MSDS. These precautions may identify storage procedures, such as storing in a cool, dry place, storing away from sources of heat, and storing refractories separately. In addition, these precautions may include ventilation requirements, as well as special tools and any other pertinent precautions.

Use common sense when storing or disposing of a refractory material. Follow company procedures and those listed on the MSDS. If there are questions, ask the supervisor.

Some refractory materials are considered hazardous waste because of the levels of chromium, titanium oxides, or other compounds entrapped in the refractory material. Do not be fooled by the apparent harmless look of waste refractory brick or rubble. Waste refractory material can pose a threat of groundwater contamination if disposed of improperly.

4.10.8 EMERGENCY RESPONSE

Since it is impossible to predict when an emergency will occur, take the time to learn the emergency procedure before working with the refractory material. Consult the emergency and first-aid procedure section of the MSDS for the specific substances.

Victims of refractory emergencies should receive medical help as soon as possible. Perform first aid while waiting for emergency personnel. This may include bringing the victim to an eyewash station, safety shower, or fresh air source or removing the victim's clothing. Follow appropriate guidelines in administering first-aid procedures.

4.10.9 HELPFUL HANDLING HINTS

- Never mix, pour, stir, or otherwise work castable refractory materials without the use of a respirator (dust mask). Some refractories will require a cartridge respirator. Always consult the applicable MSDS.
- Never cut or saw shapes, bricks, or other molded refractories without the use of a respirator (dust mask). Some refractories will require a cartridge respirator. Always consult the applicable MSDS.
- Consult the MSDS for the specifics on the refractory material handled in the work area. Read the label on the container.

- Keep incompatible chemicals separate. Never mix or store them together.
- Label all containers, even temporary ones. Make sure that seals, screens, caps, and containers are working properly and do not leak.
- Learn first-aid skills, including CPR (cardiopulmonary resuscitation) and the company's emergency plan.
- Have emergency PPE on hand **before** an emergency arises.
- If PPE is necessary, be sure to wear the appropriate PPE and make sure it fits. Never wear less than the recommended PPE as outlined on the MSDS.
- Know the location of nearest eyewash stations, safety showers, fresh air sources and fire extinguishers. Use the right kind of extinguisher for the material in use — the wrong kind can spread the fire.

4.11 SOLVENTS

4.11.1 SOLVENTS INFORMATION SOURCES

This Hazardous Communication safety information is intended to provide general safety information only. Its intent is to serve as an easy to understand reference for the different aspects of handling solvents. One should always review the product labels and MSDS for the specific solvents in use. In some cases, it may even be necessary to consult the solvent manufacturer before handling the solvents.

4.11.2 SOLVENTS SYNONYMS, COMMON NAMES, AND SIMILAR PRODUCTS

Naphtha, Varsol, Mineral Spirits, Paint Thinner.

4.11.3 WHAT ARE SOLVENTS?

Solvents are substances, usually liquid, that dissolve other substances. Some familiar uses include degreasing, spray painting, dry cleaning, and paint softening. Solvents are found throughout industry and include such common chemicals as paint thinners, degreasers, and industrial cleaners. Solvents can spill or leak, and the vapors can catch fire or explode.

4.11.4 COMMON CHARACTERISTICS

Every solvent is hazardous, depending on how it is used. Many organic solvents will burn. They can cause fires and explosions if misused. Many of them are toxic. Some are flammable, explosive, and toxic; however, all are useful and all can be used and worked with safely.

When a solvent is heated, vapors are produced — how much vapor will depend upon the temperature of the operation and the nature of the solvent. Some solvents evaporate very rapidly; others are slower in evaporating. The larger the area of contact between the solvent and air, the more vapor will be produced.

Know the solvent. Know if it is flammable or toxic or both. Solvents can be toxic (poisonous) to the human body and can burn, catch fire, or cause explosions. They can be especially dangerous because often they have no color or long-lasting smell. Most solvents evaporate quickly and are called volatile. With volatile solvents, there is the hazard that people can breathe in their vapors.

If a spark, flame, or static electricity is present, many solvents can explode. Their upper and lower explosive limits, which indicate when an explosion is possible, are listed on their MSDS. Some solvents have a flash point, or catch fire, at less than 100°F. They are called flammable, and are hazardous because their flash point may be below normal room temperature. Solvents with a flash point above 100°F are called combustible.

If too much of a toxic solvent is absorbed, irritated or damaged skin, eyes, lungs, and other organs can result. Permissible exposure limits (PELs) for many solvents have been set by the Occupational Safety and Health Administration and are listed on the MSDS.

4.11.5 Personal Hazards

Eyes   If a solvent splashes in the eyes, acute (short-term) symptoms can include burning, watering, irritation, and redness. Overexposure to solvent vapors or mists can eventually cause chronic (long-term) symptoms, such as blurred vision, constant irritation, or permanent vision damage. A person with solvent in the eye should seek the nearest eyewash station. If no eyewash station is available, any low-pressure, clean water source may be used. Contact lenses should be removed because they can trap or absorb the solvent. Flush the eye for 15-20 min, letting water run from the inside to the outside of the eye. Keep the affected eye turned downward to prevent the solvent from running into the other eye. Do not apply neutralizers or ointments to the eye. Someone who has solvent in the eye may need to be restrained to receive first aid.

Skin   One-time exposure, such as splashing a solvent on the skin, can cause dry, scaly skin, rashes, burning, or irritation. If a solvent enters the bloodstream through the skin, one can experience acute symptoms such as those listed below for inhalation and ingestion. Long-term overexposure to solvents can cause contact dermatitis, a chronic skin condition that may include blistering, redness, and discomfort. If a solvent is on your skin, rinse (do not scrub) the affected area for 15 to 20 min. Use a faucet, hose, or other available clean water source. If the solvent is dry, brush it off before beginning to rinse. Remove the contaminated clothing. Do not put the clothing back on until it has been decontaminated. If possible, after rinsing, cover the affected area with a sterile dressing. Do not apply burn ointments or neutralizing solutions.

Inhalation   If a person suddenly breathes in or swallows a solvent, acute symptoms can include headache, nausea, vomiting, sore throat, dizziness, fatigue, giddiness, rapid or irregular heartbeat, and difficulty breathing. Over time, some solvents, when inhaled, can cause liver, kidney, nervous system damage, unconsciousness, or even death. If solvent vapors are inhaled, the symptoms may include headache, dizziness, nausea, vomiting, or difficulty breathing. Seek fresh air immediately. Ask a co-worker to obtain medical help. Artificial respiration may be necessary.

Ingestion   If a solvent is swallowed, ask a co-worker to obtain medical attention immediately and to call the local Poison Control Center (dial 911). Do not eat or drink anything unless so instructed by the solvent label or a medical professional.

Symptoms of solvent exposure may appear immediately, or they may not be noticeable until some time later. Therefore, victims of a solvent emergency should remain under medical observation until the doctor feels it is safe to release them.

4.11.6 Personal Protective Equipment

The MSDS for each solvent will also provide guidance for the appropriate personal protective equipment (PPE). PPE will vary according to the particular job and specific substances involved.

Eyes   PPE should include safety glasses and, in some cases, chemical safety goggles, which have indirect ventilation ports, or face shield.

Skin   The specific glove used for working with solvents is very important. The wrong glove can deteriorate and become spongelike, pulling the solvent next to the skin. Always make sure to wear the proper gloves.
Check the MSDS for the appropriate apron requirements. If clothing becomes contaminated, remove it immediately. Decontaminate, wash, or dispose of it properly.

Inhalation   When a respirator is necessary, it **must** meet company and MSDS standards.

4.11.7 Storage and Disposal

Special precautions necessary to prevent a hazardous situation are noted on the MSDS. These precautions may identify storage procedures, such as storing in a cool, dry place, storing away from sources of heat, and storing solvents separately. In addition, these precautions may include ventilation requirements, as well as special tools and any other pertinent precautions.

Store all solvents in temperature-controlled environments, out of direct sunlight. Dispense solvents from safety-approved nozzles and dispensers only. Store solvents away from oxidizers (any substance that causes fire easily). Check storage containers regularly to make sure the spout, cap, and container are in good working order and do not leak. Immediately replace damaged container parts such as flame arrester screens.

Know the location of spill control stations and materials, eyewash stations, and safety showers.

Know whether contaminated clothing, PPE, rags, and materials should be decontaminated, cleaned, or disposed of, according to company policy. Always dispose of flammable solvents into approved tightly covered safety containers, never into a sewer, storm drain, or the garbage, or onto the ground.

Use common sense when storing or disposing of solvents. Follow company procedures and those listed on the MSDS. If there are questions, ask the supervisor.

4.11.8 Emergency Response

Since it is impossible to predict when an emergency will occur, take the time to learn the emergency procedure before working with solvents. Consult the emergency and first-aid procedure section of the MSDS for the specific substances.

Victims of emergencies with solvents can be seriously hurt. The more rapid the response to a solvent emergency, the less likelihood there will be of serious damage to people and property. Send for medical help as soon as possible, and perform first aid while waiting for emergency personnel. This may include bringing the victim to an eyewash station, safety shower, or fresh air source or removing the victim's clothing. Follow appropriate guidelines in administering first-aid procedures.

Handle an emergency **only** if it is small and if trained to do so. If it is a fire, make sure to use the right kind of extinguisher. Evacuate the area as quickly as possible while alerting others to the emergency. Then, notify the supervisor or the appropriately trained persons immediately. Do not reenter the emergency area without appropriate PPE and training.

4.11.9 Helpful Handling Hints

- Consult the MSDS for the specifics on the solvents handled in the work area. Read the label on the container.
- Keep incompatible chemicals separate. Never mix or store them together.
- Label all containers, even temporary ones. Make sure that seals, screens, caps, and containers are working properly and do not leak.
- Learn first-aid skills, including CPR (cardiopulmonary resuscitation) and the company's emergency plan.
- Stay calm. Someone who is not trained to handle emergencies involving solvents can make the situation worse. An employee who has not received training for this solvent can still help keep possible injury and damage to a minimum by remaining calm.
- Have emergency PPE on hand **before** an emergency arises.

- If PPE is necessary, be sure to wear the appropriate PPE and make sure it fits. Never wear less than the recommended PPE as outlined on the MSDS.
- Know the location of nearest eyewash stations, safety showers, fresh air sources, and fire extinguishers. Use the right kind of extinguisher for the solvent in use — the wrong kind can spread the fire.

4.12 TOLUENE OR XYLENE

4.12.1 TOLUENE OR XYLENE INFORMATION SOURCES

This Hazardous Communication safety information is intended to provide general safety information only. Its intent is to serve as an easy-to-understand reference for the different aspects of handling toluene or xylene. One should always review the product labels and Material Safety Data Sheet (MSDS) for the specific toluene or xylene in use. In some cases, it may even be necessary to consult the toluene or xylene manufacturer before handling the toluene or xylene.

4.12.2 TOLUENE OR XYLENE SYNONYMS, COMMON NAMES, AND SIMILAR PRODUCTS

Toluol, Phenylmethane, Methylbenzene, Thiophene, Solvent.

4.12.3 WHAT ARE TOLUENE AND XYLENE?

Toluene and xylene are used as solvents. They are commonly found in paints, varnishes, inks, adhesives, cleaning fluids, and in manufacturing processes. They are very volatile and vaporize quickly. They are both colorless unless mixed with other compounds such as paints or inks.

4.12.4 COMMON CHARACTERISTICS

Both toluene and xylene are colorless liquids and they have strong recognizable smells even at low levels. The thing to remember is that they can both vaporize at room temperature. That means, they can travel through the air and they are very volatile.

4.12.5 PERSONAL HAZARDS

The following are recommended exposure levels for an 8-hour work shift and should not be exceeded:

Xylene	100 ppm
Toluene	200 ppm

These chemicals can affect the body if they are inhaled, if they contact eyes or skin directly, or if they are swallowed. If toluene or xylene is inhaled in excess of recommended exposure levels, symptoms would probably be eye, nose, and throat irritation, dizziness or confusion, and headache.

These chemicals are flammable. Extreme care must be taken to avoid anything that may trigger a fire or explosion. Keep them away from strong oxidizers.

4.12.6 Personal Protective Equipment

The MSDS for the toluene and xylene will also provide guidance for the appropriate personal protective equipment (PPE). PPE will vary according to the particular job and the specific substances involved.

Emergency-response PPE should be kept in the immediate area in a designated locker or cabinet and be clearly marked.

Eyes   PPE should include safety glasses and, in some cases, chemical safety goggles, which have indirect ventilation ports, or face shields. If liquid splashes in the eyes, remove contacts, which can absorb or trap toluene or xylene in the eye.

Skin   Special gloves are required; check the MSDS for the proper gloves to be worn when handling the specific toluene or xylene. A chemical apron may also be required. If toluene or xylene gets on clothing, keep gloves on while removing the clothes immediately. When the solvent is dried, the clothes should be laundered before they are worn again. If the job deals with continual use of toluene or xylene materials, fire-retardant clothing may be required.

Inhalation When a respirator is necessary, it **must** meet company and MSDS standards. Local ventilation such as an exhaust hood should be used to lower the exposure level when possible. If the exposure level is too high and it cannot be controlled for some reason, use respirators that are approved by NIOSH.

4.12.7 Storage and Disposal

Special precautions necessary to prevent a hazardous situation are also noted on the MSDS. These precautions may identify storage procedures, such as storing in a cool, dry place, storing away from sources of heat, and storing toluene and xylene separately. In addition, these precautions may include ventilation requirements, as well as special tools and any other pertinent precautions.

Toluene and xylene are flammable, so store them separately in the appropriate location in tightly closed approved containers separate from flammable and/or combustible liquids. Dispense them only from approved nozzles and dispensers. Store large drums and containers below eye level whenever possible to avoid splashing the face or eyes. Check storage containers regularly for leaks, and make sure that caps and spouts are in good working order.

Do not smoke around toluene or xylene. Do not carry lighters, matches, or sparking devices when you're handling them.

Know whether contaminated clothing, PPE, rags, and other materials should be decontaminated, cleaned, or disposed of. Dispose of waste as outlined in the appropriate MSDS.

Use common sense when disposing of toluene or xylene. These are some of the most dangerous chemicals. Follow company procedures and those listed on the MSDS. If there are questions, ask the supervisor.

4.12.8 EMERGENCY RESPONSE

Since it is impossible to predict when an emergency will occur, take the time to learn the emergency procedure before working with toluene and xylene. Consult the emergency and first-aid procedure section of the MSDS for the substances.

Do not rush right in and start to clean up. Put on protective clothing and make sure there are no sources of ignition around. Ventilate the area.

Handle a spill, leak, fire or other emergency **only** if it is small and if trained to do so. Otherwise, the situation might become more dangerous. Evacuate the area as quickly as possible and let others know about the emergency.

Eyes	If toluene or xylene gets in the eyes, go directly to the nearest eyewash station. If none is nearby, use any low-pressure clean water source, such as a hose. Remove contact lenses, which can absorb or trap the toluene or xylene. Flush the eyes for 15 to 20 min, letting the water run from the inside to the outside of the eyes. Do not apply neutralizers or ointments, which have the potential to make damage worse. Someone who has toluene or xylene in the eyes may need to be restrained to receive first aid. Immediately after flushing, transport the person to a medical professional for further treatment.
Skin	If toluene or xylene contacts the skin, rinse (do not scrub) the affected area for 15 to 20 min. If there is no safety shower nearby, use a faucet, hose, or any clean water source. Remove any contaminated clothing and, if necessary, dispose of it. Do not put clothing back on until it has been decontaminated or cleaned. Do not apply burn ointments or neutralizing solutions. If possible, after rinsing, cover the affected area with a sterile dressing; then transport the individual to a medical professional immediately.
Inhalation	If toluene or xylene vapors are inhaled, symptoms can range from slight throat irritation to serious difficulty breathing. Seek fresh air immediately, and ask a coworker to obtain medical help right away.
Ingestion	If toluene or xylene is swallowed, ask a co-worker to obtain medical assistance immediately and to call the local Poison Control Center (dial 911). Do not eat or drink anything unless so instructed by the MSDS, the label of the toluene or xylene, or a medical professional.

Toluene and xylene cause much of their damage quickly, but some symptoms may not appear immediately. Victims of an emergency involving toluene or xylene should stay under medical observation until a doctor releases them.

Victims of toluene or xylene emergencies can be seriously hurt. Send for medical help as soon as possible, and perform first aid while waiting for emergency personnel. This may include bringing the victim to an eyewash station, safety shower, or fresh air source or removing the victim's clothing. Follow appropriate guidelines in administering first-aid procedures.

4.12.9 Helpful Handling Hints

- Consult the MSDS for the specifics on the toluene and xylene handled in the work area. Read the label on the container.
- Keep incompatible chemicals separate. Never mix or store them together.
- Label all containers, even temporary ones. Make sure that seals, screens, caps, and containers are working properly and do not leak.
- Learn first-aid skills, including CPR (cardiopulmonary resuscitation) and the company's emergency plan.
- Stay calm. Someone who is not trained to handle emergencies involving toluene and xylene can make the situation worse. An employee who has not received training for this chemical can still help keep possible injury and damage to a minimum by remaining calm.
- Have emergency PPE on hand **before** an emergency arises.
- If PPE is necessary, be sure to wear the appropriate PPE and make sure it fits. Never wear less than the recommended PPE as outlined on the MSDS.
- Know the location of nearest eyewash stations, safety showers, fresh air sources, and fire extinguishers. Use the right kind of extinguisher for the chemical in use — the wrong kind can spread the fire.

4.13 WASTE

4.13.1 Waste Information Sources

This Hazardous Communication safety information is intended to provide general safety information only. Its intent is to serve as an easy-to-understand reference for the different aspects of handling waste. One should always review the product labels and Material Safety Data Sheet (MSDS) for the specific waste in use. In some cases, it may even be necessary to consult the manufacturer of specific wastes before handling them.

4.13.2 Waste Synonyms, Common Names, and Similar Products

All products become waste when they are no longer used for their intended purpose, recycled, or treated as a product. Care must be given to retention of materials because useful materials can transform into waste any time due to deterioration, change in storage conditions or regulatory guidelines.

4.13.3 WHAT IS WASTE?

Waste is generally considered to be any material or product that is no longer used or reused for its intended purpose or as another product.

Hazardous waste is any material not used for its intended purpose or reused that is listed as hazardous by the EPA or states or that exhibits any of the following characteristics:

Ignitability — Has a flash point below 140°F (60°C).
Toxicity — Contains listed substances above specified levels that may pose a hazard to humans or the environment.
Reactivity — Spontaneously reacts in air or water.
Corrosivity — Has a pH $\leq$ 2 or $\geq$ 12.5.

Any container that previously held hazardous waste or material and is not empty as defined in the contaminated container regulation shall be managed as hazardous waste.

Hazardous waste minimization is important because it helps protect the environment, while also reducing the great expense and administrative difficulties associated with disposing of hazardous wastes: *waste minimization does not cost — it pays*. Hazardous waste minimization means limiting inventories and the use of hazardous materials to quantities as small as possible, which improves safety wherever hazardous materials are stored and used.

Waste minimization begins with planning. Choose equipment and techniques that (1) use the least hazardous material, (2) use the smallest quantities, (3) generate the least amount, and (4) are practical and safe for any given task.

4.13.4 COMMON CHARACTERISTICS

Common characteristics are difficult to define because almost anything can become waste. A product that was once inventory can become waste because of expiration, deterioration, damage, specification changes, overproduction, and many other factors.

Waste does not necessarily have to be hazardous to be considered waste. Nonhazardous material can become waste and must be treated as a waste, not as a product or other usable material.

4.13.5 PERSONAL HAZARDS

Personal hazards depend on the material that has become waste; how much waste exists, what kinds of exposures are present, what kind of containers or containment is used and many other variables. To determine what personal hazards exist, one must look at not only what hazards the original product may have presented, but also what hazards are presented in its current condition. The hazards may have increased or decreased. The original MSDS is a good place to start in the assessment of the hazards involved with waste.

4.13.6 PERSONAL PROTECTIVE EQUIPMENT

The MSDS for each waste will also provide guidance for the appropriate personal protective equipment (PPE). PPE will vary according to the particular job and specific substances involved. It will also vary depending on the condition of the waste and of its container. Waste in properly labeled and closed drums will require different PPE than the same waste that has spilled on the ground.

Waste may present special PPE hazards because the waste material is often not marked, not contained, and/or has contaminated other material or equipment. Caution should always be used to wear more PPE than appears needed, not less. Determine what PPE is needed before starting work.

PPE should include safety glasses and, in some cases, chemical safety goggles, which have indirect ventilation ports, or face shields. If gloves are required, check the MSDS for the proper gloves to be worn when handling the specific waste. If the job deals with many waste materials or continual use of waste materials, fire-retardant clothing may be required because even the slightest of unstable conditions can be extremely dangerous. When a respirator is necessary, it **must** meet company and MSDS standards.

4.13.7 STORAGE AND DISPOSAL

Special precautions necessary to prevent a hazardous situation are also noted on the MSDS. These precautions may identify storage procedures, such as storing in a cool, dry place, storing away from sources of heat, and storing waste separately. In addition, these precautions may include ventilation requirements, as well as special tools and any other pertinent precautions.

Always know the material constituting the waste. The identity of items being disposed of must be known prior to disposal; analyzing an unknown material to identify it is usually very expensive and time-consuming. Container labeling, whether HMIS or NFPA, helps quickly communicate the hazards involved with waste, and may save significant cost and burdens of identifying unknown chemicals.

Empty containers that previously held a hazardous material may be subject to specific state or federal regulations. Waste chemicals and spent chemical solutions can pose a threat to the environment if not properly managed.

Under the following criteria, a container is considered empty when (these guidelines pertain to determining emptiness with hazardous material, but also serve as good guidelines for nonhazardous waste):

- *A container that formerly held pourable hazardous material or waste:* No residual material or waste can be poured or drained from that container when the container is held in any orientation (e.g., tilted or inverted).

Federal regulations declare a drum empty, if there is less than 1 in. of liquid left in the drum.

- *A container that formerly held nonpourable hazardous material or waste:* No material or waste remaining in or on the container can be feasibly removed by physical methods commonly employed to remove materials from the containers. A thin, uniform layer of dried material or powder is considered acceptable in the container.
- *A container that formerly held materials or wastes considered to be extremely or acutely hazardous:* The container has been triple-rinsed with a solvent capable of removing the material. Triple rinsing currently requires a treatment permit, conditional exemption, permit, or permit waiver. The rinse may then be considered hazardous.

Detailed information about spills of specific chemicals can be found on the specific MSDS and the Spill Reporting and Emergency Release Plan of the plant. The employee must be trained in advance to handle cleanups. Some spills of waste may trigger reporting requirements to government agencies.

Use common sense when storing or disposing of waste. Follow company procedures and those listed on the MSDS. If there are questions, ask the supervisor.

4.13.8 EMERGENCY RESPONSE

Since it is impossible to predict when an emergency will occur, take the time to learn the emergency procedure before working with the waste or even before it becomes waste. Consult the emergency and first-aid procedure section of the MSDS for the specific substances in use.

Emergency response in dealing with waste material can be as simple as placing waste material in a designated disposal bin or as complicated as regulatory permits, self-contained body suits, and area evacuation. The level of response is naturally dependent on the level of the hazard. The first step in emergency response for waste is to assess the hazards and materials involved. Obtain accurate information on the waste and conditions before attempting any emergency response. When in doubt, evacuate the area until appropriate response information is obtained.

Victims of waste emergencies can be seriously hurt. Send for medical help as soon as possible, and perform first aid while waiting for emergency personnel. This may include bringing the victim to an eyewash station, safety shower, or fresh air source or removing clothing from the victim. Follow appropriate guidelines in administering first-aid procedures.

4.13.9 HELPFUL HANDLING HINTS

- Consult the MSDS for the specifics on the waste handled in the work area. Read the label on the container.
- Keep incompatible chemicals separate. Never mix or store them together.
- Label all containers, even temporary ones. Make sure that seals, screens, caps, and containers are working properly and do not leak.

- Learn first-aid skills, including CPR (cardiopulmonary resuscitation) and the company's emergency plan.
- Share inventories of hazardous chemicals with other users.
- Reuse hazardous material whenever possible.
- Dispose of hazardous materials promptly.
- Have emergency PPE on hand **before** an emergency arises.
- If PPE is necessary, be sure to wear the appropriate PPE and make sure it fits. Never wear less than the recommended PPE as outlined on the MSDS.
- Know the location of the nearest eyewash stations, safety showers, fresh air sources, and fire extinguishers. Use the right kind of extinguisher for the specific waste — the wrong kind can spread the fire.
- Be calm. Someone who is not trained to handle emergencies involving waste can make the situation worse. An employee who has not received training for waste can still help keep possible injury and damage to a minimum by remaining calm.

5 Helpful Audit Checklists

Audits of the various elements of a safety program are important. Long debates ensue over just how important they are and how much value they add to the safety of the workforce. When considering audits, the most important thing to remember is to keep them in perspective. An audit is not an opportunity to find every item that could possibly be out of compliance or even that might cause an accident. The focuses of an audit should be threefold.

1. First, to find any serious areas of violation of law or policy and any item that might cause immediate injury. This should be considered as "minimal requirements." Findings in this area represent fundamental flaws in the safety system. Although these items are important to unearth and bring to light, they are not where the ultimate benefit of audits will be achieved.
2. Second, audits should be used to track progress in areas that have been defined as important to the safety system in a specific organization. This may not have any meaning to any other organization. However, for this specific organization, these are the keys to success. They are items that have been identified as critical to its management and employees.
3. Third, and most important, audits should be used for measures of "constant improvement." Audits can be customized for just about any area that has been determined to be an area of needed improvement, which is the area where the most tremendous changes can be achieved. Audits in this area start to be connected to goals and objectives established by the organization. Used for this purpose, audits cease to be burdensome and dreaded and start to become tools of success and achievement.

In any of these predetermined areas, "20 questions" can be developed that determine whether essential milestones and objectives are being met. The term "20 questions" is used loosely. As will be seen from review of the audit checklists in this area, the number of questions is not as important as the pertinence of the questions. These questions, however many there are, should bring out real problems or point to real solutions. They can be easily designed to steer an organization in specific directions.

The following 20 questions audit checklists should be beneficial in themselves, but also serve as a guide to develop meaningful audit checklists in any area where constant improvement is desired.

INDEX OF "20 QUESTIONS" AUDIT CHECKLISTS

5.1 AUDIT CHECKLIST FOR BAD HABITS AND SHORT-CUTS*

Audit Date			Person Auditing	Overall Condition?	Additional Training Needed?

1.	☐ Yes	☐ No	Are all major tasks covered by Plant Standard Operating Procedures?
2.	☐ Yes	☐ No	Has Job Safety Analysis/Job Hazard Analysis been done on major tasks of each job? (Job Descriptions may be a good place to get a starter list of specific tasks for each job.)
3.	☐ Yes	☐ No	Have all previous accidents involving this job task been reviewed and evaluated for bad habits and shortcuts?
4.	☐ Yes	☐ No	Are employees pressured to maintain a faster-than-normal pace?
5.	☐ Yes	☐ No	Are supervisors' directions to employees for work always combined with safety references?
6.	☐ Yes	☐ No	Are all shortcuts observed addressed immediately?
7.	☐ Yes	☐ No	Are shortcuts considered necessary to get the job done?
8.	☐ Yes	☐ No	Are long-time employees given "exceptions" from safety rules and standard operating procedures because "they know how to do it"?
9.	☐ Yes	☐ No	Is cross-observation among supervisors, employees, departments, task, and teams used to detect hazards, bad habits, and shortcuts?
10.	☐ Yes	☐ No	Is a certain number of "mishaps" accepted as "the best way we know how"?
11.	☐ Yes	☐ No	Are concerns, solutions, and suggestions from employees discussed and considered as ways to eliminate bad habits, shortcuts, and other habits?
12.	☐ Yes	☐ No	Is the right tool *always* used for the job?
13.	☐ Yes	☐ No	Are there jobs that have hazardous operations that have not had detailed Job Safety Analysis, Job Hazard Analysis, or Ergonomic Reviews done on them?
14.	☐ Yes	☐ No	Do employees consider a specific job to be extra hazardous or extra dangerous?
15.	☐ Yes	☐ No	Are employees free to point out shortcuts that compromise safety and not fear repercussions?

*This is not intended to be a comprehensive checklist.

5.2 AUDIT CHECKLIST FOR COMPUTER WORKSTATIONS

Any "no" response indicates a potential problem area, which should receive further investigation.

1.	☐ Yes	☐ No	Does the workstation ensure proper worker posture, such as:
	☐	☐	Horizontal thighs?
	☐	☐	Vertical lower legs?
	☐	☐	Feet flat on floor or a footrest?
	☐	☐	Neutral wrists?
2.	☐ Yes	☐ No	Does the chair
	☐	☐	Adjust easily?
	☐	☐	Have a padded seat with a rounded front?
	☐	☐	Have an adjustable backrest?
	☐	☐	Provide lumbar support?
	☐	☐	Have casters?
3.	☐ Yes	☐ No	Are the height and tilt of the work surface on which the keyboard is located adjustable?
4.	☐ Yes	☐ No	Is the keyboard detachable?
5.	☐ Yes	☐ No	Do keying actions require minimal force?
6.	☐ Yes	☐ No	Is there an adjustable document holder?
7.	☐ Yes	☐ No	Are armrests provided where needed?
8.	☐ Yes	☐ No	Are glare and reflections avoided?
9.	☐ Yes	☐ No	Does the monitor have brightness and contrast controls?
10.	☐ Yes	☐ No	Do the operators judge the distance between eyes and work to be satisfactory for their viewing needs?
11.	☐ Yes	☐ No	Is there sufficient space for knees and feet?
12.	☐ Yes	☐ No	Can the workstation be used for either right- or left-handed activity?
13.	☐ Yes	☐ No	Are adequate rest breaks provided for task demands?
14.	☐ Yes	☐ No	Are high stroke rates avoided by:
	☐	☐	Job rotation?
	☐	☐	Self-pacing?
	☐	☐	Adjusting the job to the skill of the worker?
15.	☐ Yes	☐ No	Are employees trained in:
	☐	☐	Proper postures?
	☐	☐	Proper work methods?
	☐	☐	When and how to adjust their workstations?
	☐	☐	How to seek assistance for their concerns?

5.3   AUDIT CHECKLIST FOR CONFINED SPACE ENTRY*

Audit Date	Person Auditing	Overall Condition?	Additional Training Needed?
_____	_____	_____	_____

1. ☐ Yes ☐ No   Does every confined space have a specific safety checklist for entering that confined space?
2. ☐ Yes ☐ No   Prior to entering the space, is a written permit required?
3. ☐ Yes ☐ No   For "nonpermit confined spaces" is the monitoring for a permit still performed?
4. ☐ Yes ☐ No   Is constant monitoring performed while the space is occupied?
5. ☐ Yes ☐ No   Has emergency equipment been placed on the confined space site prior to entry?
6. ☐ Yes ☐ No   Have attendants been trained on all of their duties?
7. ☐ Yes ☐ No   Is a specific list or accounting system in place to track everyone who goes in and comes out?
8. ☐ Yes ☐ No   Is the confined space clearly posted, such as "Do Not Enter without Authorization"?
9. ☐ Yes ☐ No   Have the individual checklists been updated within the last 12 months?
10. ☐ Yes ☐ No   Does the attendant have means of easy communication with the entrants?
11. ☐ Yes ☐ No   Does the attendant have means of easy communication with emergency response personnel?
12. ☐ Yes ☐ No   Are the operation, production, and maintenance personnel who work on the confined space area in good communication?
13. ☐ Yes ☐ No   Have lockouts, blinds, disconnects, and other safeguards been double-checked?
14. ☐ Yes ☐ No   Has a pre-job safety meeting been held with all those who will be involved?
15. ☐ Yes ☐ No   Is a close watch kept on everything that could possibly change the conditions of the confined space?

*This is not intended to be a comprehensive checklist. See 29 CFR 1910.146 for specific details and more-detailed information.

5.4 AUDIT CHECKLIST FOR EMERGENCY PREPAREDNESS*

Audit Date	Person Auditing	Overall Condition?	Additional Training Needed?

1. ☐ Yes ☐ No Have all employees been trained on the proper evacuation method, paths, and plans?
2. ☐ Yes ☐ No Are evacuation routes posted?
3. ☐ Yes ☐ No Is an emergency evacuation meeting place predetermined?
4. ☐ Yes ☐ No Is a list of employees working in any given area readily available in an emergency?
5. ☐ Yes ☐ No Is there a comprehensive chemical inventory (both internal and external response teams) kept up-to-date and readily available for emergency personnel?
6. ☐ Yes ☐ No Do emergency response personnel have detailed plant maps, chemical inventory, special hazards?
7. ☐ Yes ☐ No Have all exits been clearly identified?
8. ☐ Yes ☐ No Are exit stairways equipped with emergency lighting?
9. ☐ Yes ☐ No Are "nonexit" doorways clearly marked "Not an Exit"?
10. ☐ Yes ☐ No Has the alarm signal been tested recently?
11. ☐ Yes ☐ No Have evacuation drills been conducted recently?
12. ☐ Yes ☐ No Have bad weather take-cover areas been established and are they kept stocked with emergency supplies?
13. ☐ Yes ☐ No Is emergency communication available with entities outside the plant (NOAA weather radio, emergency radio channels, cell phone)?
14. ☐ Yes ☐ No In inclement weather is someone in the facility designated to monitor weather reports?
15. ☐ Yes ☐ No Are first-aid supplies readily available?
16. ☐ Yes ☐ No Are first aid/CPR-qualified personnel available at all times?

*This is not intended to be a comprehensive checklist. See 29 CFR 1910.36,37,38 for specific requirements and more-detailed information.

5.5 AUDIT CHECKLIST FOR FIRE PROTECTION*

Audit Date			Person Auditing	Overall Condition?	Additional Training Needed?

1.	☐ Yes	☐ No	Are flammables and combustibles stored in appropriate cabinets?
2.	☐ Yes	☐ No	Are flammables and combustibles properly identified?
3.	☐ Yes	☐ No	Are "No Smoking" signs posted in all areas where flammables and combustibles are used and stored?
4.	☐ Yes	☐ No	Are oxygen and acetylene stored with required fire barrier between the two?
5.	☐ Yes	☐ No	Are fire extinguishers checked each month?
6.	☐ Yes	☐ No	Are fire extinguishers checked each year in an "annual" inspection?
7.	☐ Yes	☐ No	Have employees been trained on proper use of fire extinguishers within the last 12 months?
8.	☐ Yes	☐ No	Do all employees know the fire signal and how to evacuate?
9.	☐ Yes	☐ No	Are spray cans kept in a designated area?
10.	☐ Yes	☐ No	Are flammables and combustibles materials such as paints stored separately from other flammable materials?
11.	☐ Yes	☐ No	Has the fire water system been tested in the last 12 months?
12.	☐ Yes	☐ No	Is fire response equipment checked on a monthly basis? Is it documented?
13.	☐ Yes	☐ No	Do employees fully understand how they are to respond to a fire?
14.	☐ Yes	☐ No	Are all compressed gases stored outside?
15.	☐ Yes	☐ No	Are Department of Transportation (DOT) testing numbers current on all compressed gas cylinders and canisters?

*This is not intended to be a comprehensive checklist. See 29 CFR 1910.106 for specific details and more-detailed information.

5.6 AUDIT CHECKLIST FOR HAZARDOUS ATMOSPHERES*

			Additional Training
Audit Date	**Person Auditing**	**Overall Condition?**	**Needed?**

1. ☐ Yes ☐ No Have all employees been trained on all known exposures?
2. ☐ Yes ☐ No Are exposure levels monitored and conveyed to all employees?
3. ☐ Yes ☐ No Are methods in place to alert employees when permissible exposure limits are exceeded?
4. ☐ Yes ☐ No Is monitoring conducted when new materials are brought into the workplace?
5. ☐ Yes ☐ No Is an industrial hygienist consulted at the first indication of symptoms of nausea, dizziness, light-headedness, and other telling symptoms?
6. ☐ Yes ☐ No Are employee complaints about strong odors, smells, fumes, other irritations treated as a top priority?
7. ☐ Yes ☐ No Are confined spaces monitored and permitted prior to the entry of any employee?
8. ☐ Yes ☐ No In confined space situations, is the attendant required to stay in constant contact with entrants and has the attendant been trained on emergency procedures?
9. ☐ Yes ☐ No Is welding, cleaning, or other operation that has potential for changing the atmosphere being performed in a confined space?
10. ☐ Yes ☐ No Is an MSDS available for all solvents used in parts washers?
11. ☐ Yes ☐ No Are nonflammable solvents used wherever possible?
12. ☐ Yes ☐ No Are all exhaust fans in operating condition and used when needed?
13. ☐ Yes ☐ No Have ventilation systems been inspected in the last 12 months?
14. ☐ Yes ☐ No Have employees complained that ventilation systems are not adequate?
15. ☐ Yes ☐ No Is all welding done away from solvents and chemical cleaning areas?

*This is not intended to be a comprehensive checklists. See 29 CFR 1910.94 for specific requirements and more-detailed information.

5.7 AUDIT CHECKLIST FOR HAZARDOUS COMMUNICATIONS*

Audit Date			Person Auditing	Overall Condition?	Additional Training Needed?

1. ☐ Yes ☐ No — Is every chemical used in this area listed on the plant chemical inventory?
2. ☐ Yes ☐ No — Has every employee been trained on every chemical or group of chemicals?
3. ☐ Yes ☐ No — Is every container in this area properly labeled?
4. ☐ Yes ☐ No — Do all labels have adequate information?
5. ☐ Yes ☐ No — Are chemicals stored in their proper locations?
6. ☐ Yes ☐ No — Are there chemical spills (even minor leaks) that have not been cleaned up?
7. ☐ Yes ☐ No — Have employees been trained on proper PPE for every chemical in this area?
8. ☐ Yes ☐ No — Can employees readily identify the hazards involved in using a chemical or in a chemical release?
9. ☐ Yes ☐ No — Are chemical stored only with chemicals in the same categories or compatible categories?
10. ☐ Yes ☐ No — Is there a regular program to relabel drums and containers that have damaged or deteriorated labels?
11. ☐ Yes ☐ No — Are used drums not allowed to accumulate around the plant?
12. ☐ Yes ☐ No — Are used drums not allowed to collect water?
13. ☐ Yes ☐ No — Is the chemical inventory updated on a regular basis?
14. ☐ Yes ☐ No — Is the Hazardous Communication program evaluated for effectiveness?
15. ☐ Yes ☐ No — Can every employee explain how the Hazardous Communication program works?

*This is not intended to be a comprehensive checklist. See 29 CFR 1910.1200 for specific details and more-detailed information.

5.8 AUDIT CHECKLIST FOR LADDERS*

Audit Date	Person Auditing	Overall Condition?	Additional Training Needed?
_____	_____	_____	_____

1. ☐ Yes ☐ No Are ladders always inspected before each use?
2. ☐ Yes ☐ No Are ladders only used for the application for which they were designed?
3. ☐ Yes ☐ No Is the height of the ladder adequate for the task at hand?
4. ☐ Yes ☐ No Are ladders ever used as a brace, skid, lever, guy or gin pole, gangway, platform, scaffold, plank, material hoist, or any other use that was not intended?
5. ☐ Yes ☐ No Are stepladders ever used as a single ladder or when partially closed?
6. ☐ Yes ☐ No Are lean-to ladders tied off at the top to a stationary fixed position?
7. ☐ Yes ☐ No Is the 4-to-1 rule for properly setting up extension or straight ladders always used? (The distance from the wall to the base of the ladder should be one fourth the distance from the base of the ladder to where it touches the wall)
8. ☐ Yes ☐ No Are extension ladders extended at least 1 ft, but no more than 3 ft above the point of support at the top?
9. ☐ Yes ☐ No Do employees ever stand on the top of a stepladder, stand above the third rung from the top of an extension ladder, or above the top two steps of a folding stepladder?
10. ☐ Yes ☐ No Do employees climb the back sections of combination ladders when used as stepladders?
11. ☐ Yes ☐ No Are ladders inspected and shoes inspected for grease, mud, or other slippery material prior to using a ladder?
12. ☐ Yes ☐ No Do employees use both hands while climbing up and down ladders and face the ladder?
13. ☐ Yes ☐ No Have the bottoms of rails been insulated with nonskid material?
14. ☐ Yes ☐ No Is there a comprehensive, documented ladder inspection program?
15. ☐ Yes ☐ No Are damaged ladders taken out of service and either repaired or destroyed?

*This is not intended to be a comprehensive checklist.

5.9 AUDIT CHECKLIST FOR MACHINE GUARDING*

Audit Date			Person Auditing	Overall Condition?	Additional Training Needed?

1.	☐ Yes	☐ No	Are all rotating shafts, sprockets, pulleys, belts, and chains completely enclosed?
2.	☐ Yes	☐ No	Are all guards intact, free from wear, tightly secured, and free from binding?
3.	☐ Yes	☐ No	Are guards placed back on equipment prior to starting up the equipment/tool?
4.	☐ Yes	☐ No	Are controls that require both hands to operate utilized on equipment that is applicable?
5.	☐ Yes	☐ No	Are fail-safe guards inspected on a regular basis to ensure equipment will not operate with the guard out of position?
6.	☐ Yes	☐ No	Are remote shutdown switches used for automated equipment where employees could become entangled?
7.	☐ Yes	☐ No	Is there a comprehensive list of guards and guard equipment that can be used for inspection, servicing, and maintenance?
8.	☐ Yes	☐ No	Is all automated equipment equipped with prestart-up warning alarms?
9.	☐ Yes	☐ No	Are there signs clearly notifying all employees that automated equipment will start automatically?
10.	☐ Yes	☐ No	Are "safety bars" that stop the equipment automatically when an employee "trips" them used on all equipment that could entangle an employee?
11.	☐ Yes	☐ No	Are the tail stocks of all shafts, whether smooth or with key way or set screw, covered by a guard?
12.	☐ Yes	☐ No	Is work stopped until guards are replaced on all equipment after repairs?
13.	☐ Yes	☐ No	Are lubrication ports extended through guards to a safe area where possible?
14.	☐ Yes	☐ No	Are bench grinders adjusted for tool rest setting and checked for wheel condition on a regular basis?
15.	☐ Yes	☐ No	Are all foot-petal controls protected from accidental engagement?

*This is not intended to be a comprehensive checklist. See 29 CFR 1910.211-.222 for specific details and more-detailed information.

5.10 AUDIT CHECKLIST FOR NEW HIRE EMPLOYEES*

"New Employee" should be considered any employee who does not have recent experience in the specific work area and specific task being performed.

Audit Date	Person Auditing	Overall Condition?	Additional Training Needed?

1. ☐ Yes ☐ No Has the new employee been trained on the proper evacuation method, paths, and plans?
2. ☐ Yes ☐ No Has the new employee had a personal walk-through of the area of assignment?
3. ☐ Yes ☐ No Has an experienced employee in the work area been assigned to work with the new employee?
4. ☐ Yes ☐ No Have all safety policies been reviewed and competency determined?
5. ☐ Yes ☐ No Have specific standard operating procedures (SOPs) been reviewed and competency determined?
6. ☐ Yes ☐ No Have co-workers in the area been alerted that a new employee will be working in that area?
7. ☐ Yes ☐ No Has the new employee been assigned a "safety buddy" to encourage safety awareness and participation constantly?
8. ☐ Yes ☐ No Has a member of management personally emphasized the importance of safety to the new employee?
9. ☐ Yes ☐ No Have previous accidents in the new employee's work area been reviewed with the new employee?
10. ☐ Yes ☐ No Has someone verified that the techniques and SOPs given the new employee are correct and best-operation practices?
11. ☐ Yes ☐ No Are regularly scheduled follow-up meetings held with the new employee to ensure safety remains a primary focus?
12. ☐ Yes ☐ No Has all OSHA-mandated training been performed for "initial assignment" requirements?
13. ☐ Yes ☐ No Has the supervisor of the new employee spent regular time observing the new employee perform the work?
14. ☐ Yes ☐ No Have hazards and concerns expressed by the new employee been addressed thoroughly?
15. ☐ Yes ☐ No Is the new employee actively involved in safety-related activities, such as safety meetings, safety audits, housekeeping audits, SOP reviews, etc.?

*This is not intended to be a comprehensive checklist.

5.11 AUDIT CHECKLIST FOR NOISE AND HEARING PROTECTION*

Audit Date	Person Auditing	Overall Condition?	Additional Training Needed?

1. ☐ Yes ☐ No Have all employees been trained on hearing conservation within the last 12 months?
2. ☐ Yes ☐ No Have all employees exposed to greater than 85 dBa time-weighted average had an audiogram within the last 12 months?
3. ☐ Yes ☐ No Have all employees with a standard threshold shift of 10 dBa or greater in the 2000, 3000, or 4000 Hz. Range been notified of the threshold shift?
4. ☐ Yes ☐ No Are all areas where people have to raise their voices to be heard designated as "Hearing Protection Required" areas?
5. ☐ Yes ☐ No Are all employees required to wear hearing protection in designated "Hearing Protection Required" areas?
6. ☐ Yes ☐ No Has noise level monitoring been conducted in all areas where new equipment has been added or where equipment has been modified?
7. ☐ Yes ☐ No Are annual sound-level readings available for the area?
8. ☐ Yes ☐ No Have baffles, mufflers, or barriers been installed on noise sources where practical?
9. ☐ Yes ☐ No Are sound levels considered for all new equipment installations?
10. ☐ Yes ☐ No Are ear plugs and/or other hearing protection readily available for all employees?
11. ☐ Yes ☐ No Are "Hearing Protection Required" areas clearly posted?
12. ☐ Yes ☐ No Have all new employees exposed to 85 dBa or greater received a baseline audiogram within 6 months of employment?

*This is not intended to be a comprehensive checklist. See 29 CFR 1910.95 for specific requirements and more-detailed information.

5.12 AUDIT CHECKLIST FOR PERSONAL PROTECTIVE EQUIPMENT*

Audit Date	Person Auditing	Overall Condition?	Additional Training Needed?

1. ☐ Yes ☐ No — Do employees perform tasks or work near someone who generates dust or flying particles?

2. ☐ Yes ☐ No — Do employees handle or work near someone handling hazardous liquids, chemicals, irritants, or other potential hazards?

3. ☐ Yes ☐ No — Are employees exposed to intense light (lasers, flames, molten material)?

4. ☐ Yes ☐ No — Are employees exposed to extreme temperatures or conditions (cold, heat, wind, rain, etc.)?

5. ☐ Yes ☐ No — Do employees work in areas where overhead work is performed, where overhead traffic or movement occurs, or underneath equipment or other activity?

6. ☐ Yes ☐ No — Do employees have to squat or bend to go under anything?

7. ☐ Yes ☐ No — Do employees work near electrical wiring or electrical components?

8. ☐ Yes ☐ No — Do employees come in contact with material that might be sharp, have burrs, or could cause bruises, cuts, scrapes, or scratches?

9. ☐ Yes ☐ No — Do employees handle chemicals or other materials that might irritate the skin or cause contamination?

10. ☐ Yes ☐ No — Are employees in areas where tools, materials, or equipment might drop on their feet, roll over their feet, or hit their feet?

11. ☐ Yes ☐ No — Are employees exposed to splashing of any kind (water, chemicals, hot material, sparks, etc.)?

12. ☐ Yes ☐ No — Are employees exposed to noise loud enough that people have to raise their voices to be heard?

13. ☐ Yes ☐ No — Do employees work in areas that have known levels, or could have levels of material that could asphyxiate, cause illness, explode, or otherwise be hazardous?

14. ☐ Yes ☐ No — Have employees been trained, fit-tested, and certified as able to wear the appropriate respirator for the atmospheres they work in?

15. ☐ Yes ☐ No — Is there a concise list of tasks that require special PPE available for employees to use and have they been trained on it?

*This is not intended to be a comprehensive checklist. See 29 CFR 1910.132 for specific details and more-detailed information.

5.13 AUDIT CHECKLIST FOR POWERED PLATFORMS*

Powered platforms include JLG, JIG lifts, man baskets, scissor lifts, and other powered lifting devices and equipment.

Audit Date	Person Auditing	Overall Condition?	Additional Training Needed?

1. ☐ Yes ☐ No Have all employees been trained on the proper operation of the equipment?
2. ☐ Yes ☐ No Has "hands on training" been conducted and documented?
3. ☐ Yes ☐ No Are daily inspections and maintenance performed and documented?
4. ☐ Yes ☐ No Are the controls marked legibly?
5. ☐ Yes ☐ No Do all controls operate as they should?
6. ☐ Yes ☐ No Is the lift capacity known, clearly marked, and adhered to?
7. ☐ Yes ☐ No Is the swing radius barricaded off prior the beginning of work?
8. ☐ Yes ☐ No Is all work below the powered platform stopped during overhead work?
9. ☐ Yes ☐ No Is the work area of the lift clear of trash, debris, equipment, and other items that could cause the operator to stumble or fall?
10. ☐ Yes ☐ No Is there some method of preventing falling from the work area of the lift (such as bolts, tools, debris, etc.)?
11. ☐ Yes ☐ No Does the operator have good communications with those on the ground and at the elevated work area?
12. ☐ Yes ☐ No When hand signals are required, has a single person been designated to give the hand signals?
13. ☐ Yes ☐ No Will the work be done at less than the maximum capacity of the lift (height, weight, reach)?
14. ☐ Yes ☐ No Has the lift been weight-tested within the last 12 months?
15. ☐ Yes ☐ No Is the lift moved from one location to the next with the personnel lowered?

*This is not intended to be a comprehensive checklist. See 29 CFR 1910.66 and 26 CFR Subparts L and M for specific requirements and more-detailed information.

5.14 AUDIT CHECKLIST FOR SCAFFOLDS*

Audit Date	Person Auditing	Overall Condition?	Additional Training Needed?
_____	_____	_____	_____

1. ☐ Yes ☐ No Have there been recent near misses or accidents involving scaffolding?
2. ☐ Yes ☐ No Has a qualified scaffold builder erected the scaffold?
3. ☐ Yes ☐ No Are all planks or scaffold boards less than 1 in. apart?
4. ☐ Yes ☐ No Are platforms at least 18 in. apart?
5. ☐ Yes ☐ No Are "open-side" platforms less than 14 in. from the work surface?
6. ☐ Yes ☐ No Is fall protection used around "open faces" farther than 14 in. from work surfaces?
7. ☐ Yes ☐ No Are all platforms either "cleated" or extend over support by at least 6 in.?
8. ☐ Yes ☐ No Are scaffold boards used for platforms, not common lumber?
9. ☐ Yes ☐ No Are scaffolds built using components from the same manufacturer and system design?
10. ☐ Yes ☐ No Does the scaffold conform to the 4:1 base-to-height ratio requirement?
11. ☐ Yes ☐ No Has the scaffold been erected on firm footing?
12. ☐ Yes ☐ No Is the scaffold plumb?
13. ☐ Yes ☐ No Have ladders been positioned so as not to tip the scaffold?
14. ☐ Yes ☐ No Are toe boards, mid-rail, and top rail in place on the top level?
15. ☐ Yes ☐ No Are casters pinned into the scaffold?

*This is not intended to be a comprehensive checklist. See 29 CFR 1910.28 for specific requirements and more-detailed information.

5.15 AUDIT CHECKLIST FOR WALKING SURFACES*

Audit Date	Person Auditing	Overall Condition?	Additional Training Needed?

1. ☐ Yes ☐ No Have there been recent near misses or accidents?
2. ☐ Yes ☐ No Are the aisle ways in this area cluttered, poorly maintained, or filled with material?
3. ☐ Yes ☐ No Have all deficiencies from the last audit of this area been corrected? If not, which deficiencies remain unresolved?
4. ☐ Yes ☐ No Are aisle ways clearly marked?
5. ☐ Yes ☐ No Have previous jobs left tools, equipment, debris, etc. in the walkway?
6. ☐ Yes ☐ No Has training been conducted on how the walkways should be maintained? Is additional training or reminders warranted?
7. ☐ Yes ☐ No Are any walkways too narrow for easy passage?
8. ☐ Yes ☐ No Would quick evacuation be hampered by walkways, aisles, or working surfaces?
9. ☐ Yes ☐ No Are platforms clearly posted with load limits? Have employees been trained on these limits?
10. ☐ Yes ☐ No Do aisles, walkways, stairs, work platforms in this area have adequate light?
11. ☐ Yes ☐ No Are all light fixtures in this area operable?
12. ☐ Yes ☐ No Is there oil or grease on any of the walking surfaces?
13. ☐ Yes ☐ No Are there caution cones or caution tape available to barricade unsafe walkways?
14. ☐ Yes ☐ No Are all exits clearly marked?
15. ☐ Yes ☐ No Are all exits maintained in good condition, free from obstacles, and accessible to all employees?

*This is not intended to be a comprehensive checklist. See 29 CFR 1910.22 for specific requirements and more-detailed information.

6 Safety Manuals

Safety manuals come in all shapes, sizes, and styles. A small fledgling organization may have only some copied papers, hardly readable because they have been copied so many times. The "manual" may include only a smattering of safety thoughts or rules and only be thought of or reviewed when new employees are given a copy to start their tenure with the company.

On the other end of the spectrum is the multivolume, multimedia safety manual that is too cumbersome and specific for anyone to ever do more than wipe the dust off and thumb through it in amazement. Just its sheer volume and size will ward off any serious attempts to find much good in it. Many are written from strictly an effort to comply with regulations and often by a person who is not a safety professional or who does not understand much about safety at all.

Somewhere in between lies the perfect safety manual for an organization. A 20-page conglomeration of "safety concerns" in a modern industrial setting will do little but encourage lawsuits and mock the safety system itself. In the same way, a 400-page standard operating procedure for a small, nonhazardous industry is overkill in most cases.

So, how does one determine what type of safety manual is right for an organization? First, the needs of the organization must be determined. Are there hazardous chemicals? Do federal, state, and local regulations apply? What required training is involved? Is there specialized regulation that deals specifically with the company? All these questions, and many others, must be addressed before anyone can really know what will be best for a specific organization.

There are many "canned" safety manuals on the market. For small and nonspecialized situations, some of these may offer a good option or, at least, a place to start. Many safety and health consultants will develop a customized safety manual for an organization that will meet its needs much better than a canned program.

Many organizations choose to develop their own safety manual. Figure 6.1 is a table of contents for the safety manual of a "typical industry." Several things are worth pointing out in this figure. First, this table of contents may include items an organization is not interested in at all, and it may not include items that an organization needs as a priority. Be careful to use it only as a guide to topics a safety manual should include.

Figure 6.1 also shows some "deleted" sections. For example, Standard Operating Procedure 22.100 is shown to have been deleted from the previous revision. This is important to illustrate that once an item is included in a safety manual it may not always remain a valuable part. Notice the revision numbers as well. Revision numbers help to ensure that everyone is working off of the same copy. It also represents that a safety manual is a "living document." The worse thing that can happen is to develop a safety manual and then place it on a shelf for the next 5 or 10 years to gather dust. A good safety manual should reflect changes that arise from policy changes, process changes, and from corrections of errors and shortcomings reported by readers of the manual.

Regardless of how thorough and complete a manual is at its conception, it will need continual updating and revision to keep it a viable tool for all concerned to use and rely on.

Figure 6.2 represents how a typical policy in a safety manual might appear. Again, use this only as a guide to help build ideas.

XYZ COMPANY, INC. — SAFETY STANDARD OPERATING PROCEDURES

TABLE OF CONTENTS

SSOP Number		Date	Revision Number
22.05	Safety Standard Operating Procedures Distribution List	12/31/98	11
22.095	Safety Vision	12/31/98	3
22.098	Developing Safety Standard Operating Procedures and Document Control	12/31/98	4
22.100	~~Strategic Safety Team~~	~~12/3/98~~	~~Deleted~~
22.101	Safety Task Teams	12/31/98	3
22.200	Housekeeping	02/02/99	3
22.201	Lockout/Tagout/Blinding Policy	12/31/98	3
22.202	Personal Protective Equipment	12/31/98	5
22.203	Respiratory Protection	01/15/99	0
22.204	Action Plan	12/31/99	3
22.205	OSHA Vehicle Seat Belt Use and Speed Limit	12/31/98	3
22.206	Outside Contractor Hazard Control Program	12/31/98	3
22.206	Portable Electrical Equipment Inspection Program	12/31/98	3
22.209	Hoisting Equipment Inspection Program	12/31/98	4
22.245	Confined Space Entry	12/31/98	4
22.250	Fire Protection Program	12/31/98	3
22.260	Emergency Action Plan	12/31/98	3
22.265	Hazard Communication Plan	12/31/98	4
22.270	Hazard Material Plan	12/31/98	1
22.280	Noise Control and Hearing Conservation	12/31/98	3
22.300	Asbestos Abatement	12/31/98	3
22.400	First-Aid and Medical Services Training	12/31/98	5
22.450	~~Safety Meeting Program~~	~~10/01/98~~	~~3 Deleted~~
22.470	Material Handling Vehicles	12/31/98	3
22.600	Safety Award Program	02/02/99	4
22.700	Accident Investigation Program	12/31/98	4
22.750	General Duty Policy	12/31/98	3
22.800	Job Safety Analysis and Workplace Ergonomics Review	12/31/98	1
22.900	Safety and Human Factors Checklist	12/31/98	1
22.950	Internal Safety Audit	12/31/98	1

FIGURE 6.1 Table of contents for a typical industrial safety manual.

ABC, INC. —
SAFETY STANDARD OPERATING PROCEDURE 12.0 —
LOCKOUT/TAGOUT

TABLE OF CONTENTS

FIGURE 6.2 Typical policy within a safety manual.

SAFETY STANDARD OPERATING PROCEDURE LOCKOUT/TAGOUT

SUMMARY PAGE

SPIRIT OF SAFETY STATEMENT

Always make sure equipment and systems are locked out before performing service or maintenance activities.

Equipment and systems refer to any piece of machinery or equipment and the supply of any energy source to that machinery or equipment. Such energy sources include electrical, mechanical, hydraulic, pheumatic, chemical, thermal, gravity, or other.

HAZARDS AND DANGERS THAT CAN EXIST

- Equipment starting while persons are exposed to moving parts
- Electrical shock
- Explosion
- Asphyxiation
- Heat or chemical burns
- High-pressure gas or air
- Falling or dropping load, material, or equipment

REMEMBER!!

Every employee will be issued lockout locks if applicable.

- ✓ Each lock has only one key.
- ✓ Anyone working on a piece of machinery or equipment must put his or her lock on the machinery or equipment unless it has been locked with a group lock by an Authorized Employee.
- ✓ The equipment will remain locked out until repairs have been completed, the area has been cleared of all personnel, and the equipment is ready to be restarted.
- ✓ A nonlocation employee or contract employee will perform work only under an ABC Group Authorized Employee's Lock. The nonlocation employee may add his or her lock. ABC Management or its designee may give special authorization for contractors or nonlocation employees to lock equipment out on their own.
- ✓ If there is anything you are not sure about, ask your supervisor.

I. POLICY

It is the policy of ABC to ensure the safety of employees by providing written procedures to control hazardous energy (lockout/tagout) of equipment and systems. Further, it is the policy of ABC to abide by federal laws in the area of lockout/tagout.

II. PURPOSE

A. To establish procedures for ensuring that equipment and systems shall be locked out, and/or tagged out, and/or blinded before performing service or maintenance where the unexpected energization, start-up, or release of stored energy could cause injury or equipment damage.

B. To establish training requirements to ensure employees have the skills and knowledge necessary to comply with these procedures and to meet Occupational Safety and Health Act (OSHA) requirements.

C. To establish a record-keeping system that meets OSHA requirements.

D. To establish Procedures Inspection Requirements that meet OSHA requirements and that ensure established procedures are followed.

III. GENERAL/OBJECTIVES

A. ABC Management or its designee shall implement and audit the Safety Lockout Program.

B. ABC Management or its designee shall investigate methods to convert nonlockable equipment or processes to a condition where they are capable of being locked out.

IV. DEFINITIONS

Affected employee.   An employee whose job requires him or her to service or perform maintenance on a piece of equipment or to work in an area that requires lockout or tagout.

Authorized employee.   An employee who locks out or tags out a machine, equipment, or area, when servicing or performing maintenance.

Group authorized employee.   A person who locks out or tags out a machine or equipment (an enegy source) for a nonstandard lockout situation.

"Capable of being locked out."   An energy-isolating device will be considered to be capable of being locked out if it is designed with a hasp or other attachment or integral device to which, or through which, a lock can be affixed, or if it has a locking mechanism built into it. Other energy-isolating devices will be considered to be capable of being locked out if lockout can be achieved without dismantling, rebuilding, or displacing the device or permanently altering its energy control capability.

Energized.   Connected to an energy source or containing residual or stored energy.

Energy isolating device.   A mechanical device that physically prevents the transmission or release of energy.

Energy source.   Any source of electrical, mechanical, hydraulic, pneumatic, chemical, thermal, or other energy.

Hot tap repair.   Service or maintenance activities that involve welding on a piece of equipment (pipelines, vessels, or tanks) under pressure to install a connection or to add sections of pipeline without interruption of service.

V. PROCEDURES AND RESPONSIBILITIES

A. Procedures for Issuing Lockout Equipment

1. ABC Management or its designee shall:

   a. Issue personal safety lockout locks to all affected employees, and issue multiple lockout locks to individual supervisors or lead persons who require more personal locks.

   b. Establish an identification system for locks and coordinate its implementation. Locks must be durable; of a standard color, shape, or size; and print and format must be standardized (Exhibit 1). Lockout locks may not be used for any use other than lockout procedures.

   c. Issue multiple safety lockout locks and hasps and identify such locks as group lockout equipment

   d. Ensure no more than one key exists for each lockout lock. Establish a system to control the single key for group locks.

   e. Ensure all locks have specific user identification or issue number with lock-holder log identifying the lock owner/user.

B. Procedures for Lockout/Tagout, Standard Situations

Note: ABC Management or its designee shall develop equipment and/or machinery checklists for each piece of equipment, machinery and each process that requires deenergizing prior to performing maintenance or repairs and does not have a single isolatable energy source.

1. Supervisor:

   a. Ensures authorized employee knows type and magnitude of energy that machine or equipment utilizes.

   b. Ensures authorized employee understands the hazards involved.

2. Authorized employee:

   a. Confirms identity of machine or equipment to be worked on.

   b. Makes sure affected employees have been notified that lockout/tagout is in progress and the reason for lockout/tagout.

   c. If machine or equipment is operating, shuts it down by normal stopping procedure (stop button, toggle switch, etc.).

   d. Surveys to identify all isolating devices to be certain which switches, valves, or other energy-isolating devices apply to the equipment to be locked out, tagged out, or blinded. More than one energy sources (electrical, mechanical, or other) may be involved.

   e. Makes sure all energy sources have been locked out, tagged out, or blinded.

3. Affected Employees:

   Install their own personal safety padlock and hasp. The only exception is the nonstandard Lockout Procedure described in Section V.C below.

4. Authorized Employee:

   a. Makes sure all stored energy has been released or restrained by methods such as repositioning, blocking, bleeding down, etc.

b. After making sure no personnel are exposed, operates push-button or normal operating controls to make certain the equipment will not operate.
   Caution: Return operating controls to "neutral" or "off" position after the test.
c. Complete work.
d. Assures all tools are removed and guards replaced.
e. Assures the area is cleared and no one is exposed.
f. Notifies everyone in the area that work is complete.

5. Affected employees: Remove their locks.
6. Authorized employee: removes lockout locks, tags, blinds.

C. Procedures for Nonstandard Lockout Situations
A "Nonstandard" situation arises when the standard "one-person, one-lock, one-key" principle is not practical. Examples of such situations are a major shutdown where several locks shall be required at several locations; where a job shall run for an extended period of time; or several people are involved in the job and applying individual locks is not practical.

1. ABC Management or its designee shall:
   a. Determine who may act as group-authorized employee in a nonstandard lockout situation.
   b. Provide names of group-authorized employees to ABC Management.

2. Group Authorized Employee:
   a. When more than one group or department is involved, acts as employee in charge of lockout.
   b. Assumes responsibility for all employees working under protection of group lockout lock(s).
   c. Ascertains hazard exposure status of individual group members.
   d. Maintains exclusive control of key to multiple lockout locks. Passes key and authorization to next authorized employee at the end of the shift or at any time the authorized employee must surrender control of the lockout situation.
   e. Locks out and/or tags out and/or blinds machine, equipment o process, following Standard Lockout Procedure outlined in Section V.B.

3. Affected employees:
   Confirm that adequate protection has been provided and install their own safety lockout locks if they so desire.

4. Group authorized employee:
   a. Makes sure all stored energy has been released or restrained by methods such as repositioning, blocking, bleeding down, etc.
   b. After making sure no personnel are exposed, operates push-button or normal operating controls to make certain the equipment will not operate.

Caution: Return operating controls to "neutral" or "off" position after the test.

c. Ensures work is completed.

d. Assures all tools are removed and guards replaced.

e. Assures the area is cleared and no one is exposed.

f. Notifies everyone in the area that work is complete.

5. Affected employees: Remove their locks.

6. Group authorized employee: Removes lockout locks, tags, blinds.

D. Procedures for Working on Cord and Plug-Connected Equipment Exposure to the hazards of unexpected energization or start-up of cord and plug-connected equipment can be controlled by unplugging equipment from energy source.

1. Authorized employee keeps cord or plug under his or her exclusive control. If authorized employee leaves area, cannot see plug, or otherwise loses exclusive control of plug, the equipment must be locked out! Tag switch with a "Do Not Operate" tag or use a device specifically made for locking out cord or plug.

E. Procedures for Removing Lockout Devices Prior to Completion of Work:

1. At the end of the employee's portion of the work, or if called off the job for another job, or at the end of the shift, the authorized employee removes personal safety padlock and takes it with him or her, after assuring that the equipment is safe for others to be around.

2. If continued lockout is necessary, the authorized employee makes sure another employee's lockout lock, a group lockout lock, or other device, such as a "Danger: Out-of-Service" tag, is installed.

3. If a "Danger: Out-of-Service" tag is installed, the authorized employee:

   a. Secures the tag with a plastic tie wrap that has a pull rating of 50 lb. No other means of securing the tag is acceptable.

   b. After securing the tag, informs the supervisor or lead person who shall then install his or her lock.

F. Procedures for Restoring Machine or Equipment to Normal Operation:

1. After servicing and/or maintenance is complete and equipment is ready for normal operation, the authorized employee:

   a. Checks the area around the machine or equipment to ensure no one is exposed.

   b. Make sure all tools have been removed from machine or equipment, guards have been reinstalled, and employees are in the clear, then removes all lockout/tagout devices and blinds.

   c. Operates energy-isolating devices to restore energy to the machine or equipment.

G. Procedures for Removing Lockout Locks when Employe Cannot Be Located:

Note 1: If employee forgets to remove lock at the completion of work, the employee shall be called back to work to remove the lock personally. The employee's supervisor shall instruct employee in proper application of safety lockout procedure.

Note 2: A safety lock must never be cut unless a situation exists where the lock cannot be removed without cutting (key lost, key slot plugged, lock damaged, etc.). If it becomes necessary to cut a lock for any reason, the procedures below shall be followed.

Note 3: OSHA requires a reasonable attempt be made to call the employee back to work. The employee must be informed that the lock was removed before the employee resumes work.

1. Supervisor:
   a. Determines the need to remove an employee's lockout lock.
   b. Assures safety of personnel or equipment shall not be jeopardized by removing the lock.
   c. Obtains permission from the policy controller, ABC Management, or management's designee to remove the lock.
2. ABC Management or its designee grants permission to remove the lockout lock.
3. Supervisor:
   a. Removes lock by any means necessary (bolt cutters, etc.). Do not use the employee's key to remove the lock under any circumstance.
   b. Submits a report to ABC Management explaining situation and actions taken.
   c. Immediately upon return of employee to work, and prior to employee doing work on system, notifies employee of lock removal.
   d. Reinstructs employee in proper application of lockout procedures.

      Warning! Unauthorized removal of a safety lockout lock can endanger the lives of others. Anyone who removes another employe's locks without proper authorization shall be disciplined in accordance with the work rules outlined in HR practice and procedures.

H. Multiple Lockout Procedures — Procedures and Responsibilities:
   1. Multiple Lockout is required when a breaker for a piece of equipment is marked with a "Multiple Lockout Required" sign on the front of the breaker box (Exhibit 5).
   2. The checklists shall be filled out by the authorized or group authorized employee, signed, and all safety measures followed before any work is allowed to start.
   3. All contractor work will be handled in the same manner. The group authorized employee will ensure all steps are complete and the checklist is filled out before allowing contractors to proceed with their assigned task.

    4. ABC Management or its designee will ensure that all applicable employees are trained in Multiple Lockout Procedures and retrained as deemed necessary.

I. Special Lockout Procedures:

Some equipment, such as automatic computer-operated equipment, requires special procedures to secure it in a nonoperable state. Special job-safe procedures shall be written and followed for such work.

    1. ABC Management of its designee ensures special job-safe procedures exist for each type of hot work to be performed.

    2. Authorized employees perform work and lockout and/or tagout and/or blind according to special job-safe procedures.

J. Procedures for contractor Lockout Situations:

    1. Group lockout authorized employee:

Installs lockouts in the presence of contractor supervisor and maintains keys to all safety locks.

    2. Affected contract employees may install their own personal lockout locks and hasps.

    3. Group lockout authorized employee:

      a. Instructs contractor supervisor to check locks prior to start of each shift and at each worker turnover to ensure the equipment or process is secured adequately.

      b. Removes lockouts and reactivates equipment at completion of work. Removal and reactivation shall be performed only by the group lockout authorized employee.

      c. Provides lockout protection for outside service people.

K. Procedures for Removing Equipment from Service:

    1. Authorized employee:

Attaches "Danger — Out of Service" tag (Exhibit 3) to main ABConnect or source of power for that piece of equipment and identifies equipment on the tag. Nylon tie wraps should be used to attach the tag. Wire wraps are not permitted.

L. Procedure for ABC Management to Lockout Equipment as Authorized Person:

    1. The ABC Management may lockout equipment as an authorized person when special or otherwise hazardous or potentially hazardous situations arise. Any lockout initiated by ABC Management may be transferred as a group lock would be to a designee of ABC Management (such as the shop manager, safety coordinator, or other specifically designated persons). When ABC Management initiates a lockout, a tag identifying the designated person(s) that can remove the lock shall be attached to the lockout.

VI. TRAINING REQUIREMENTS

A. Initial Training of Employees

    1. ABC Management or its designee:

      a. Provides training to authorized employees in the recognition of applicable hazardous energy sources, the type and magnitude of the energy present at the location, and the methods and means necessary for energy isolation and control.

b. Provides instruction to authorized employees regarding the purpose and use of these procedures.

c. Provides instruction about procedures to all other employees whose work operations are or may be in an area where procedures are being utilized.

d. After lockout and initial test, makes sure affected employees know they are prohibited from attempting to restart or reenergize machines or equipment that are locked out or tagged out.

e. When tagouts are used, makes sure employees are trained in the following limitations of tags:

    i. Tags are warning devices attached to energy-isolating devices and do not provide the physical restraint on those devices that is provided by a lock.

    ii. A tag shall only be removed according to lock removal procedure and it is never to be bypassed, ignored, or otherwise defeated.

    iii. Tags must be legible and understandable by all authorized employees, affected employees, and all other employees whose work operations are or may be in the area.

    iv. Tags must be attached with nylon tie wraps (50-lb pull minimum strength). Wire wraps are not allowed. Tags must be able to withstand the environmental conditions of the location.

    v. Tags may cause a false sense of security, and the use of tags must be understood by all employees.

f. Ensures training and testing of all group authorized employees.

B. Retraining of Employees

  1. ABC Management or its designee:

a. Ensures retraining of all authorized and affected employees whenever there is a change in their job assignments, a change in machines, equipment, or processes, which present a new hazard, or when there is a change in energy control procedures.

b. Ensures retraining whenever periodic inspection (see Section VIII) reveals a need or when there is reason to believe there are deviations form these procedures or inadequacies in the employee's knowledge of energy control procedures.

c. Ensures retraining has reestablished employee proficiency and introduces new or revised control methods and procedures, as necessary.

d. Ensures policy is reviewed at least annually.

VII. RECORD KEEPING REQUIREMENTS

ABC Management or its designee maintains all records associated with this policy.

VIII. AUDIT/INSPECTION REQUIREMENTS
   A. ABC Management of its designee shall:
      1. Conduct periodic (no less than annually) audits of this policy to:
         a. Ensure requirements of this policy are being met.
         b. Evaluate condition of employee training.
         c. Determine the necessity of policy and/or procedure revisions.
         d. Document and maintain the audit in writing.
      2. Implement changes to policy to increase effectiveness and useful-
         ness of policy.
   B. ABC Management or its designee shall ensure an annual audit of this
      policy to ensure compliance and to correct any deviations or inadequa-
      cies observed. The audit checklist (Exhibit 4) may be used for the
      annual audit. It may also be used for periodic audits if desired. Any
      deviations or deficiencies noted during periodic inspections o annual
      audits should be documented in writing.

EXHIBIT 1
EMPLOYEE PERSONAL/GROUP LOCKOUT LOCK EXAMPLE.

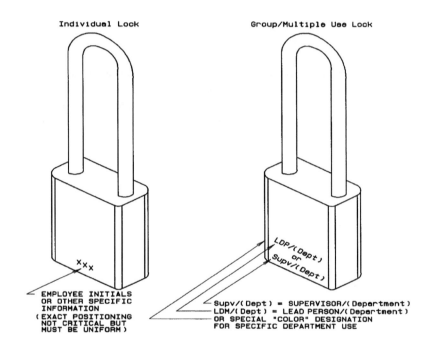

Individual Lock

Group/Multiple Use Lock

x x x

EMPLOYEE INITIALS
OR OTHER SPECIFIC
INFORMATION
(EXACT POSITIONING
NOT CRITICAL BUT
MUST BE UNIFORM)

LDP/(Dept)
or
Supv/(Dept)

Supv/(Dept) = SUPERVISOR/(Department)
LDM/(Dept) = LEAD PERSON/(Department)
OR SPECIAL "COLOR" DESIGNATION
FOR SPECIFIC DEPARTMENT USE

EXHIBIT 2
MULTIPLE LOCKOUT HASP EXAMPLE.

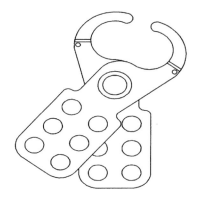

EXHIBIT 3
"DANGER — OUT-OF-SERVICE" TAGS AND TAPE.

DANGER	DANGER
OUT	**DO NOT**
OF	**OPERATE**
SERVICE	REASON:
	INSTALLED BY:
	DATE INSTALLED:
SEE OTHER SIDE	
FRONT	**BACK**

EXHIBIT 4
AUDIT CHECKLIST FOR
LOCKOUT/TAGOUT POLICY

Audit Date	Person Auditing	Overall Condition?	Additional Training Needed?

1. ☐ Yes ☐ No Is the latest revision of this policy in use?
2. ☐ Yes ☐ No If audited, was the audit documented in writing?
3. ☐ Yes ☐ No Have all deficiencies been corrected? If not, which deficiencies remain unresolved?

_____

_____

4. ☐ Yes ☐ No Have all policy suggestions and corrections been forwarded to ABC Management?
5. ☐ Yes ☐ No Has documentation been completed for any violations in compliance to this policy?
6. ☐ Yes ☐ No Is all training documentation accurately entered into the training matrix?
7. ☐ Yes ☐ No Are all employees observed performing their task in compliance with this policy? Based on _____ observations.
8. ☐ Yes ☐ No Have employee complaints been addressed and documented?
9. ☐ Yes ☐ No Have spot checks been done on an adequate sampling of training files to document compliance?
10. ☐ Yes ☐ No Have additional training needs been met?
11. ☐ Yes ☐ No Is training of employees evident?
12. ☐ Yes ☐ No Have old revisions been removed from circulation?
13. ☐ Yes ☐ No Are all locking devices properly marked as required by this policy?
14. ☐ Yes ☐ No Are locking mechanisms adequate to prevent "slide by" while lock is installed?
15. ☐ Yes ☐ No Have any inadequate locking mechanisms been identified and posted?
16. ☐ Yes ☐ No Have any inadequate locking mechanisms been assigned for corrective action?

EXHIBIT 5
MULTIPLE LOCKOUT TAG

	MULTIPLE	
o	LOCKOUT	o
	REQUIRED	

7 Accident Investigations

Accident investigations are invaluable to reduce losses from accidents that have occurred and to prevent future accidents of the same or related nature. Typical accident investigations provide data for statistical analysis of the event. The illustrations in Figures 7.1 and 7.2 show typical ways that accident investigations are categorized for future use in industry. Accident investigation can provide additional information that can ultimately be used to reduce injuries and further losses.

Accident investigation should be done when there is a loss-time accident, when an OSHA-recordable accident occurs, and when there is injury requiring medical or first-aid attention, near misses, or property or equipment damage. Accident investigations should be conducted when:

- Any injury or illness results in loss or restricted work days by an employee.
- There are significant environmental releases.
- There is loss or damage to equipment, property, or product.
- Delay occurs as a result of the accident.
- There are any undesired consequences or incidences.
- There are near misses to employees, equipment, or processes.

When conducting investigations, some of the pitfalls an organization can easily fall into can determine the success and productivity of the investigation. Investigation should not be punitive in nature. The employees must understand the investigation system and know that it will not be used in a punitive way. The investigation cannot be viewed by employees as an effort to place blame or find fault. The investigation can never be "altered" to protect management or a supervisor's turf. In other words, if the investigation is viewed as trying to cover up something or hide the errors of a department or manager, the system will lose its credibility. If the investigation fails to identify accurately the true cause of the accident, the investigation will be of no value or benefit. When more serious or underlying causes are ignored in favor of surface findings, the accident investigation is diminished in value. It is important not to stop the process before a thorough investigation has been completed.

Another area that often diminishes the long-term value of accident investigations is failure to abate fully the problem identified in the investigation. When the real problem or cause is identified but no corrective action is taken, future investigations are disabled from the start. When positive steps are taken based on information obtained through an accident investigation, it lends credibility to the whole process. Aggressive and persistent investigation of all serious and potentially serious incidences is critical to improve safety performance consistently.

	Industry	ABC
LTA	8.8	0.53
REC	7.8	12.63
Total	16.5	13.16

FIGURE 7.1   SIC Code 3331: Primary Nonferrous Metals.

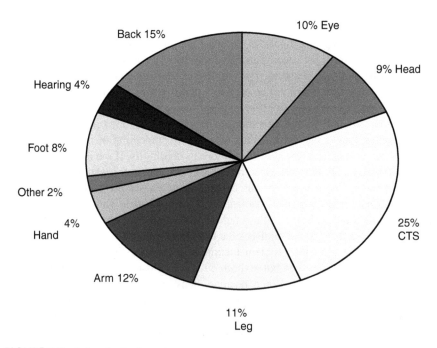

FIGURE 7.2   Injury by body part.

Accident investigation should begin with clear goals understood by everyone involved. Some goals that should be considered before establishing an investigation effort consist of the following items.

- Share the findings with all concerned.
- Identify related problems, not just the surface causes.
- Focus on preventing future injury and property damage.
- Limit business interruptions and losses.
- Limit unfavorable impact on the public.
- Identify weaknesses in the system (production, maintenance, safety, etc.).

Accident investigation should work to develop a climate of openness, trust, and respect. A follow-up system on abatement items, system failures, and other items that are uncovered can add extra value to the accident investigation. These are simple and straightforward, but if left to fall through the cracks, they can take the value out of the investigation time, effort, and expense.

Communication of results back to employees is one of the most important aspects of a first-rate investigation. Without the communication of results, employees feel left out or, even worse, that something is being hidden. The communication must be presented to employees in a timely and fair way. Short of this, employees will not buy into any solution.

The atmosphere of the accident investigation interview is very important. The atmosphere must be nonconfrontational and must put the person being interviewed at ease. Remember that it is not an interrogation. Do not make accusations. It is better if the interview is conducted with a third person present. It always helps if there is someone to substantiate the story. When there are at least three people in the meeting, all involved are less likely to change their story. Investigators should remember that they are seeking *help* from the people they are interviewing; thus, it is important to be friendly. Most importantly, do not lead the witness. Let the witness tell the story, even if the story already is well known.

One might ask, " Why do all the work that a thorough investigation demands? I already know what caused the accident." This is a common fallacy. Even in situations where the suspected causes are known, there is much more to gain by a thorough investigation. Quality can be improved, losses can be limited, future incidents can be avoided. More importantly, the investigation can be used to improve lines of communications, improve follow-up on critical items, and, most of all, heighten the awareness level of every accident.

Some items that an organization might need to consider prior to performing accident investigations include:

- What training will be needed?
- Who will be involved in the investigations?
- How many investigations will likely be done. Will several teams be needed, rather than just a small group?
- Will there be a 24-hour response time? If so, how will it be implemented?

- Will the investigators have the authority to do what is necessary to prevent recurrence?
- Will reports be submitted in a timely manner? Who will prepare the reports?
- Will help need to be provided to the investigators to produce the investigation report?

Some minimal physical tools are needed to conduct a typical accident investigation. That does not necessarily mean that this equipment is mandatory, but its availability does make it easier and helps with the overall professionalism and expertise available.

Clipboard, paper, pencils, and pens — Basics, but they need to already be assembled when the need for them arises.

Chalk — A valuable tool for marking items for short-term references and for pictures.

Barrier tape — Essential for securing the area against additional accidents as well as protecting the scene from contamination.

Latex examination gloves — Often used to prevent contaminating any samples taken.

Cassette recorder — Much easier to use for note-taking, and to make certain no points are missed.

Tape measure — Provides identification of the exact location of evidence.

Identification tags — Essential because samples often get confused once back in the office or when they have been moved.

Plastic bags, bottles, microscope slides — This may seem a little deeper than the investigation team is intended to go, but they are inexpensive and can easily be added to the investigation kit. If needed, they are invaluable for isolating and protecting small samples that are easily lost or damaged.

Compass — Provides exact references for later reference.

Adhesive tape — Good for collecting samples and marking material.

Digital camera — Has evolved to become an invaluable part of the investigation kit. The investment will be well worth the results. A picture really is worth a thousand words.

Laser pointer — A useful tool for pointing out high or out-of-reach items. The pointer can pin-point the exact position referred to.

Video camera — Another valuable tool. Be careful in performing any "reinactments" of the accident. More investigators have been injured in this way than one would think.

This kit may seem excessive, but remember two important points. First, this is all in aid of digging to the bottom of a situation and preventing it from ever happening again. Second, the information collected may very well end up in court. In today's tort-friendly environment, facts with professional documentation are always welcome in the witness box.

Some time should be given to discussing the difference between an eyewitness and a participant. While they may sound like the same thing, they are not. Keep in mind that an eyewitness is someone who actually saw the event happen — someone who saw not the aftermath of the event or who wishes to have seen the event, but someone who was actually there and looking. Participants actually played a part in the event. Granted they may have seen it take place as well, but if they participated, they should be kept in the participant category. An eyewitness may be less biased than the person who was actually involved.

Another subtle difference that should be considered is the difference between a witness and a resource. Resources should be used because they know what the witness is relating from the standpoint of experience. Witnesses do not make good resources because their view may be tainted by what they saw; whereas the resource is uninvolved and can give an unbiased viewpoint. A resource may be a person in a similar job, an outside expert, or someone present but not involved as a participant or an eyewitness.

A word about "hostile witnesses." A hostile witness does not necessarily mean the person is angry or guilty. It merely means that the person may have reason not to provide all the information wanted or needed. Examples of hostile witnesses include someone hiding something, such as drinking or drugs; someone involved in horseplay; someone with at least a portion of fault; someone who has made an operational error; someone who has missed some job function. All of these can be hostile witnesses.

Dealing with hostile witnesses is not difficult, it is just necessary to know where they are coming from. Ask leading but not intimidating questions. Verify the information given. Focus on the goal; preventing future accidents. The old adage, "why, why, why, why," will be very helpful in accident investigation.

Corrective action may be the single most important link between the accident and building support for future accident investigations. There absolutely must be a solid corrective action plan as well as feedback to the employees informing them what was done to correct the situation. It builds credibility for future occurrences. Although corrective action is an essential part, it must also be remembered that the decision to *not* abate the hazard is sometimes a valid decision. Even though valid, if not successfully communicated to employees, again, credibility for the entire program is lost.

Accident *reports* are cumbersome and often are the proverbial bottleneck. They do not have to be that way. Employees must be given some basic tools to work with. They do not have to be English majors to write a successful accident report. Bulleted points are often sufficient and are easily used to convey information. The reports are important. They give management something to "sell," and they give employees something to "buy." They offer the regulatory agencies something to "accept." But most of all, they offer a road map to constant improvement.

8 Reference Materials

8.1 ESSENTIAL REFERENCE MATERIAL FOR ENVIRONMENTAL, HEALTH, AND SAFETY PRACTITIONERS

Accident Facts, National Safety Council, 2000
   (Order Information: National Safety Council 630.285.1121).
Accident Prevention Manual for Business and Industry, Administration and Programs,
   11th ed., National Safety Council, 1997
   (Order Information: National Safety Council 630.285.1121).
Accident Prevention Manual for Business and Industry, Engineering and Technology, 11th ed.,
   National Safety Council, 1997
   (Order Information: National Safety Council 630.285.1121).
Code of Federal Regulation 29 — General Industry, Government Institutes, 2000
   (Order Information: NFPA 301.921.2323).
Code of Federal Regulation 26 — Construction Industry, Government Institutes, 2000
   (Order Information: NFPA 301.921.2323).
Fire Protection Handbook, 17th ed., National Fire Protection Association (NFPA)
   (Order Information: NFPA 800.344.3555).
Fundamentals of Industrial Hygiene, 4th ed., National Safety Council
   (Order Information: National Safety Council 630.285.1121).
National Electric Code — 1996, National Fire Protection Association (NFPA)
   (Order Information: NFPA 800.344.3555).
Taber's Cyclopedic Medical Dictionary, 16th ed., F.A. Davis, Philadelphia, 1985
   (Order Information: F.A. Davis 215.568.2270).
The Occupational Environment — Its Evaluation and Control, DiNardi, S.R., Ed., American
   Industrial Hygiene Association, Fairfax, Virginia, 1997
   (Order Information: American Industrial Hygiene Association 703.849.8888).
1999 TLVs and BEIs, Threshold Limit Values for Chemical Substances and Physical Agents
   and Biological Exposure Indices, ACGIH
   (Order Information: ACGIH 513.742.2020).
2000 Emergency Response Guidebook, J.J. Keller & Associates, Neenah, Wisconsin
   (Order Information: 800-327-6868).

8.2 ESSENTIAL REFERENCE MATERIAL FOR THE CERTIFIED INDUSTRIAL HYGIENIST

The Certified Industrial Hygienist (CIH) is a designation offered by the American Board of Industrial Hygiene (ABIH). It is a rigorous examination by any standard. Applicants must meet educational requirements and work experience requirements before they can be approved to sit for the examination. The following reference material has been suggested as useful information in preparing for the examination and will reflect the type of information with which a CIH candidate should be familiar. These references are listed here as a quick reference source of good information dealing with a wide array of industrial hygiene-related subject matter.

A reader choosing to pursue the CIH designation should contact the ABIH and request updated information on prerequisites and other qualification criteria.

Alden, J.L. and Kane, J.M., *Design of Industrial Ventilation Systems*, 5th ed., Industrial Press, New York, 1982.

Bioerosols: Assessment and Control, ACGIH, Cincinnati, OH, 1999.

Burgess, W.A., *Recognition of Health Hazards in Industry: A Review of Materials and Processes,* 2nd ed., Wiley-Interscience, New York, 1995.

Cember, H., *Introduction to Health Physics*, 2nd ed., Pergamon Press, Elmsford, NY, 1983.

Chaffin, D.B. and Anderson, G.B., *Occupational Biomechanics*, 2nd ed., John Wiley & Sons, New York, 1991.

Checkoway, H., Pearce, N.E., and Crawford-Brown, D.J., *Research Methods in Occupational Epidemiology*, Oxford University Press, New York, 1989.

Clayton, G.D. and Clayton, F.E., *Patty's Industrial Hygiene and Toxicology*, Eds., John Wiley & Sons, New York, 1991–1994.

Cralley, L.V. and Cralley, L.J., Eds., *In-Plant Practices for Job-Related Health Hazards Control,* John Wiley & Sons, New York, Vol. I, *Production Prices,* 1989. Vol. II, *Engineering Aspects,* 1989.

DiNardi, S.R., Ed., *The Occupational Environment: Its Evaluation and Control*, AIHA Press, Fairfax, VA, 1997.

Eastman Kodak Company Staff, *Ergonomic Design for People at Work,* Van Nostrand Reinhold; New York, Vol. I, 1983, Vol. II (Subtitled *The Design of Jobs*), 1986.

Finkel, A.J., Ed. *Hamilton and Hardy's Industrial Toxicology*, 4th revised ed., PSG Publishing, Littleton, MA, 1991.

Gammage, R.B. and Berven, B.A., *Indoor Air and Human Health*, 2nd ed., CRC/Lewis Publishers, Boca Raton, FL, 1996.

Garrett, J.T., Cralley, L.J., and Cralley, L.V., eds., *Industrial Hygiene Management*, John Wiley & Sons, New York, 1988.

Hathaway, G., Proctor, N.H., and Hughes, J.P., *Proctor and Hughes' Chemical Hazards of the Workplace*, 4th ed., Van Nostrand Reinhold, New York, 1996.

Hemeon, W.C.L. and Burton, D.J., *Plant and Process Ventilation*, 3rd ed., Lewis Publishers, Boca Raton, FL, 1998.

Klaassen, C.D., Amdur, M.O., and Doul, J., eds., *Casarette and Doull's Toxicology: The Basic Science of Poisons*, 5th ed., McGraw-Hill, New York, 1995.

Konz, S.A., *Work Design: Industrial Ergonomics*, 4th ed., Holcomb Hathaway, Scottsdale, AZ, 1995.

Levine, S.P. and Martin, W.F., Eds., *Protecting Personnel at Hazardous Waste Sites,* 2nd ed., Butterworth, Stoneham, MA, 1994.

Lillienfeld, D.E. and Lillienfeld, A.M., *Foundations of Epidemiology*, 3rd ed., Oxford University Press, New York, 1994.

McDermott, H.J., *Handbook of Ventilation for Contaminant Control*, 2nd ed., Butterworth, Stoneham, MA, 1985.

NIOSH, OSHA, U.S. Coast Guard, and EPA, *Occupational Safety and Health Guidance Manual for Hazardous Waste Site Activities*, NIOSH, Cincinnati, OH, 1985.

Perkins, J.L., *Modern Industrial Hygiene,* Vol. I, *Recognition and Evaluation of Chemical Agents*, Van Nostrand Reinhold, New York, 1997.

Plog, B.A., Niland, J., and Quinlan, P.J., Eds., *Fundamentals of Industrial Hygiene*, 4th ed., National Safety Council, Chicago, 1996.

Practitioner's Approach to Indoor Air Quality Investigations, AIHA Press Fairfax, VA, 1989.

Shapiro, J., *Radiation Protection: A Guide for Scientists and Physicians,* 3rd ed., Harvard University Press, Cambridge, MA, 1989.

Sliney, D.H. and Wolbarsht, M.L., *Safety with Lasers and Other Optical Sources: A Comprehensive Handbook*, Plenum Publishing Corp., New York, 1980.

Toca, F.M. and Woodhull, D., *Management of People and Programs in Industrial Hygiene.* AIHA Press, Fairfax, VA, 1996.

Wadden, R.A. and Scheff, P.A., *Engineering Control of Workplace Hazards*, McGraw-Hill, New York, 1987.

Walsh, P.J., Dudney, C.S., and Copenhaver, E.L., *Indoor Air Quality*, 2nd ed., CRC Press, Boca Raton, FL, 1993.

8.3 ESSENTIAL REFERENCE MATERIAL FOR THE CERTIFIED SAFETY PROFESSIONAL

The Certified Safety Professional (CSP) is a designation offered by the Board of Certified Safety Professionals (BCSP). It is a rigorous examination by any standard. Applicants must meet educational requirements and work experience requirements before they can be approved to sit for the examination. The following reference material has been suggested as useful information in preparing for the examination and reflects the type of information with which a CSP candidate should be familiar. These references are listed here as a quick reference source of good information dealing with a wide array of safety-related subject information.

A reader choosing to pursue the CSP designation should contact the BCSP and request updated information on prerequisites and other qualification criteria.

Brauer, R.L., *Safety and Health for Engineers,* Van Nostrand Reinhold, New York, 1990.

Cote, A.E., Ed., *Fire Protection Handbook,* 17th ed., National Fire Protection Association, Quincy, MA, 1991.

Crowl, D.A. and Louvar, J.F., *Chemical Process Safety: Fundamentals with Applications,* Prentice-Hall, Englewood Cliffs, NJ, 1990.

Ellis, J.N., *Introduction to Fall Protection,* 2nd ed., American Society of Engineers, Des Plaines, IL, 1998.

Frein, J.P., Ed., *Handbook of Construction Management and Organization,* 2nd ed., Van Nostrand Reinhold, New York, 1980.

Griffin, R.D., *Principles of Hazardous Materials Management,* Lewis Publishers, Chelsea, MI, 1988.

Grimaldi, J.V. and Simonds, R.H., *Safety Management*, 2nd ed., American Society of Engineers, Des Plaines, IL, 1989.

Hammerr, W., *Product Safety Management and Engineering*, 2nd ed., Society of Engineers, Des Plaines, IL, 1993.

National Safety Council Accident Prevention Manual for Business and Industry, Administration and Programs, 10th ed., National Safety Council, Itasca, IL, 1992.

National Safety Council Accident Prevention Manual for Business and Industry, Engineering and Technology, 10th ed., National Safety Council, Itasca, IL, 1992.

Plog, B.A., Ed., *Fundamentals of Industrial Hygiene,* 3rd ed., National Safety Council, Itasca, IL, 1988.

Pulat, M. and Alexander, D.C., *Industrial Ergonomics,* Institute of Industrial Engineers, Norcross, GA, 1991.

Roland, H.E. and Moriarty, B., *Systems Safety Engineering and Management,* 2nd ed., John Wiley & Sons, New York, 1990.

Rossnagel, W.E., Higgings, L.R., and MacDonald, J.A., *Handbook of Rigging,* 4th ed., McGraw-Hill, New York, 1988.

Sanders, M.S. and McCormick, E.J., *Human Factors in Engineering and Design,* 4th ed., McGraw-Hill, New York, 1992.

Slote, L., Ed., *Handbook of Occupational Safety and Health,* 2nd ed., John Wiley & Sons, New York, 1987.

Society of Fire Protection Engineers, *Handbook of Fire Protection Engineering,* National Fire Protection Association, Quincy, MA, 1988.

Spiegel, M.R., *Statistic,* Schaum's Outline Series, 4th ed., McGraw-Hill, New York, 1988.

Tapley, B., Ed., *Esshbach's Handbook of Engineering Fundamentals,* 4th ed., John Wiley & Sons, New York, 1990.

Thaamhain, H.J., *Engineering Management,* John Wiley & Sons, New York, 1992.

8.4   METRIC SYSTEM

Mass and Weight

Unit	Abreviation or Symbol	Number of Grams	Approx. U.S. Equivalent
metric ton	t	1,000,000.0000	1.102 short tons
kilogram	kg	1,000.0000	2.2046 pounds
hectogram	hg	100.0000	3.527 ounces
dekagram	dag	10.0000	0.353 ounce
gram	g	1.0000	0.035 ounce
decigram	dg	0.1000	1.543 grains
centigram	cg	0.0100	0.154 grain
milligram	mg	0.0010	0.015 grain

Length

Unit	Abreviation or Symbol	Number of Meters	Approx. U.S. Equivalent
kilometer	km	1,000.0000	0.62 mile
hectometer	hm	100.0000	109.36 yards
dekameter	dam	10.0000	32.81 feet
meter	m	1.0000	39.37 inches
decimeter	dm	0.1000	3.94 inches
centimeter	cm	0.0100	0.39 inch
millimeter	mm	0.0010	0.039 inch

Area

Unit	Abreviation or Symbol	Number of Square Meters	Approx. U.S. Equivalent
square kilometer	km^2	1,000,000.0000	0.3861 square mile
hectare	ha	10,000.0000	2.47 square yards
are	a	100.0000	119.60 square yards
square centimeter	cm^2	0.0001	0.155 square inch

Volume

Unit	Abreviation or Symbol	Number of Cubic Meters	Approx. U.S. Equivalent
cubic centimeter	cm^3, cc	0.0000	0.061 cubic inch
cubic decimeter	dm^3	0.0010	61.023 cubic inches
cubic meter	m^3	1.0000	1.307 cubic yards

Capacity

Unit	Abreviation or Symbol	Number of Liters	Cubic	Dry	Liquid
kiloliter	kl	1,000.0000	1.31 cubic yards		
hectoliter	hl	100.0000	3.53 cubic yards	2.84 bushels	

8.4   (CONT.)

Capacity

Unit	Abreviation or Symbol	Number of Liters	Approx. U.S. Equivalent		
			Cubic	Dry	Liquid
dekaliter	dal	10.0000	0.35 cubic yards	1.14 pecks	2.64 gallons
liter	l	1.0000	61.02 cubic inches	0.908 quarts	1.057 quarts
cubic decimeter	dm³	1.0000	61.02 cubic inches	0.908 quarts	1.057 quarts
deciliter	dl	0.1000	6.1 cubic inches	0.18 pint	0.21 pint
centiliter	cl	0.0100	0.61 cubic inches		0.338 fluid ounce
milliliter	ml	0.0010	0.061 cubic inches		0.27 fluid dram

8.5   WEIGHTS AND MEASURES

Unit	Abreviation or Symbol	Equivalent in Other Units of Same System	Metric Equivalent
		Weight	
Avoirdupois			
ton (short)		20 short hundredweight, 2000 pounds	0.907 metric ton
ton (long)		20 long hundredweight, 2240 pounds	1.016 tons
hundredweight	cwt		
hundredweight (short)		100 pounds, 0.05 short ton	45.359 kilograms
hundredweight (long)		112 pounds, 0.05 long ton	50.802 kilograms
pound	lb or lb avdp or #	16 ounces, 7000 grains	0.454 kilogram
ounce	oz or oz avdp	16 drams, 437.5 grains	28.350 grams
dram	dr or dr avdp	27.344 grains, 0.0625 ounce	1.772 grams
grain	gr	0.037 dram, 0.02286 ounce	0.0648 gram
Troy			
pound	lb t	12 ouces, 240 pennyweight, 5760 grains	0.373 kilogram
ounce	oz t	20 pennyweight, 480 grains	31.103 grams
pennyweight	dwt or pwt	24 grains, 0.05 ounce	1.555 grams
grain	gr	0.042 pennyweight, 0.002083 ounce	0.0648 gram
Apothecaries'			
pound	lb ap	12 ounces, 5760 grains	0.373 kilogram
ounce	oz ap	8 drams, 480 grains	31.103 grams
dram	dr ap	3 scruples, 60 grams	3.888 grams
scruple	s ap	20 grains, 0.33 dram	1.296 grams
grain	gr	0.05 scruple, 0.002083 ounce, 0.0166 dram	0.0648 gram
		Capacity	
U.S. Liquid Measurement			
gallon	gal	4 quarts (231 cubic inches)	3.785 liters
quart	qt	2 pints (57.75 cubic inches)	0.946 liter
pint	pt	4 gills (28.875 cubic inches)	0.473 liter

8.5   (CONT.)

gill	gi	4 fluid ounces (7.219 cubic inches)	118.294 milliliters
fluid ounce	fl oz	8 fluid drams (1.805 cubic inches)	29.573 milliliters
fluid dram	fl dr	60 minims (0.226 cubic inches)	3.697 milliliters
minim	min	1/10 fluid dram (0.003760 cubic inch)	0.061610 milliliter

U.S. Dry Measure

bushel	bu	4 pecks (2150.42 cubic inches)	35.239 liters
peck	pk	8 quarts (537.605 cubic inches)	8.810 liters
quart	qt	2 pints (67.201 cubic inches)	1.101 liters
pint	pt	1/2 quart (33.600 cubic inches)	0.551 liter

British Imperial Liquid and Dry Measure

bushel	bu	4 pecks (2219.36 cubic inches)	0.036 cubic meter
peck	pk	2 gallons (554.84 cubic inches)	0.0091 cubic meter
gallon	gal	4 quarts (277.420 cubic inches)	4.546 liters
quart	qt	2 pints (69.355 cubic inches)	1.136 liters
pint	pt	4 gills (34.678 cubic inches)	568.26 cubic centimeters
gill	gi	5 fluid drams (8.669 cubic inches)	142.066 cubic centimeters
fluid ounce	fl oz	8 fluid drams (1.7339 cubic inches)	28.4112 cubic centimeters
fluid dram	fl dr	60 minims (0.216734 cubic inches)	3.5516 cubic centimeters
minim	min	1/60 fluid dram (0.003612 cubic inch)	0.059194 cubic centimeter

Length

mile	mi	5280 feet, 320 rods, 1760 yards	1,609 kilometers
rod	rd	5.50 yards, 16.5 feet	5.029 meters
yard	yd	3 feet, 36 inches	0.9144 meter
foot	ft or '	12 inches, 0.333 yard	30.48 centimeters
inch	in or "	0.083 foot, 0.028 yard	2.54 centimeters

8.5   (CONT.)

Area

square mile	mi^2	640 acres, 102,400 square rods	2.590 square kilometers
acre		484 square yards, 43,560 square feet	0.405 hectare, 4047 square meters
square rod	rd^2	30.25 square yards, 0.00625 acre	25.293 square meters
square yard	yd^2	1296 square inches, 9 square feet	0.836 square meter
square foot	ft^2	144 square inches, 0.11 sqare yard	0.093 square meter
square inch	in^2	0.0069 square foot, 0.00077 square foot	6.452 square centimeters

Volume

cubic yard	yd^3	27 cubic feet, 46,656 cubic inches	0.765 cubic meter
cubic foot	ft^3	1728 cubic inches, 0.0370 cubic yard	0.028 cubic meter
cubic inch	in^3	0.00058 cubic foot, 0.000021 cubic yard	16.387 cubic centimeters

8.6 APPROXIMATE SPECIFIC GRAVITIES AND DENSITIES

Substance	Specific Gravity	Avg. density, lb/ft³	Substance	Specific Gravity	Avg. density, lb/ft³
Metals, Alloys, Ores			**Timber, air-dried**		
Aluminum, cast-hammered	2.55–2.80	165	Apple	0.66–0.74	44
Aluminum bronze	7.7	481	Ash, black	0.55	34
Brass, cast-rolled	8.4–8.7	534	Ash, white	0.64–0.71	42
Bronze, 7.9 to 14% Sn	7.4–8.9	509	Birch, sweet, yellow	0.71–0.72	44
Bronze, phosphor	8.88	554	Cedar, white, red	0.35	22
Copper, cast-rolled	8.8–8.95	556	Cherry, wild red	0.43	27
Copper ore, pyrites	4.1–4.3	262	Chestnut	0.48	30
German silver	8.58	536	Cypress	0.45–0.48	29
Gold, cast-hammered	19.25–19.35	1205	Fir, Douglas	0.48–0.55	32
Gold coin (U.S.)	17.18–17.2	1073	Fir, balsam	0.40	25
Iridium	21.78–22.42	1383	Elm, white	0.56	35
Iron, gray cast	7.03–7.13	442	Hemlock	0.45–0.50	29
Iron, cast, pig	7.2	450	Hickory	0.74–0.80	48
Iron, wrought	7.6–7.9	485	Locust	0.67–0.77	45
Iron, spiegel-eisen	7.5	468	Mahogany	0.56–0.85	44
Iron, ferrosilicon	6.7–7.3	437	Maple, sugar	0.68	43
Iron ore, hematite	5.2	325	Maple, white	0.53	33
Iron ore, limonite	3.6–4.0	237	Oak, chestnut	0.74	46
Iron ore, magnetite	4.9–5.2	315	Oak, live	0.87	54
Iron slag	2.5–3.0	172	Oak, red, black	0.64–0.71	42
Lead	11.34	710	Oak, white	0.77	48
Lead ore, galena	7.3–7.6	465	Pine, Oregon	0.51	32
Manganese	7.42	475	Pine, red	0.48	30
Manganese ore, pyrolusite	3.7–4.6	259	Pine, white	0.43	27
Mercury	13.546	847	Pine, Southern	0.61–0.67	38–42
Monel metal, rolled	8.97	555	Pine, Norway	0.55	34
Nickel	8.9	537	Poplar	0.43	27
Platinum, cast-hammered	21.5	1330	Redwood, California	0.42	26
Silver, cast-hammered	10.4–10.6	656	Spruce, white, red	0.45	28
Steel, cold-drawn	7.83	489	Teak, African	0.99	62
Steel, machine	7.80	487	Teak, Indian	0.66–0.88	48
Steel, tool	7.70–7.73	481	Walnut, black	0.59	37
Tin, cast-hammered	7.2–7.5	459	Willow	0.42–0.50	28
Tin ore, cassiterite	6.4–7.0	418			

8.6 (CONT.)

Substance	Specific Gravity	Avg. density, lb/ft³	Substance	Specific Gravity	Avg. density, lb/ft₃
Tungsten	19.22	1200			
			Various Liquids		
Zinc, cast-rolled	6.9–7.2	440	Alcohol, ethyl (100%)	0.789	49
Zinc, ore, blende	3.9–4.2	253	Alcohol, methyl (100%)	0.796	50
			Acid, muriatic (HCl), 40%	1.20	75
Various			Acid, nitric, 91%	1.50	94
Cereals, oats, bulk	0.41	26	Acid, sulfuric, 87%	1.80	112
Cereals, barley, bulk	0.62	39	Chloroform	1.500	95
Cereals, corn, rye, bulk	0.73	45	Ether	0.736	46
Cereals, wheat, bulk	0.77	48	Lye, soda, 66%	1.70	106
Cork	0.22–0.26	15	Oils, vegetable	0.91–0.94	58
Cotton, flax, hemp	1.47–1.50	93	Oils, mineral, lubricants	0.88–0.94	57
Fats	0.90–0.97	58	Turpentine	0.861–0.867	54
Flour, loose	0.40–0.50	28	Water, 4°C, max. density	1.0	62.428
Flour, pressed	0.70–0.80	47	Water, 100°C	0.9584	59.83
Glass, common	2.40–2.80	162	Water, ice	0.88–0.92	56
Glass, plate or crown	2.45–2.72	161	Water, snow, fresh fallen	0.125	8
Glass, crystal	2.90–3.00	184	Water, seawater	1.02–1.03	64
Glass, flint	3.2–4.7	247			
Hay and straw, bales	0.32	20	**Ashlar Masonry**		
Leather	0.86–1.02	59	Granite, syenite, gneiss	2.4–2.7	159
Paper	0.70–1.15	58	Limestone	2.1–2.8	153
Potatoes, piled	0.67	44	Marble	2.4–2.8	162
Rubber, Caoutchouc	0.92–0.96	59	Sandstone	2.0–2.6	143
Rubber goods	1.0–2.0	94	Bluestone	2.3–2.6	153
Salt, granulated, piled	0.77	48			
Saltpeter	2.11	132	**Rubble Masonry**		
Starch	1.53	96	Granite, syenite, gneiss	2.3–2.6	153
Sulfur	1.93–2.07	125			
Wool	1.32	82			

Note: At room temperature with reference to water at 39.

8.7 C.E. LAPPLE'S PARTICLE SIZE DISTRIBUTION CHART

 CHARACTERISTICS OF PARTICLES AND PARTICLE DISPERSOIDS

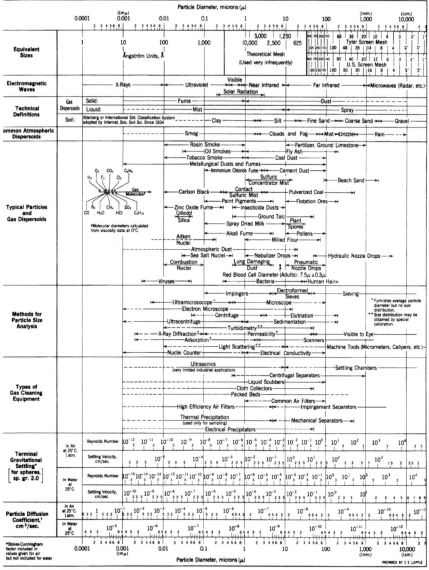

Printed from Stanford Research Institute Journal. Third Quarter.

SRI International   333 Ravenswood Ave. • Menlo Park, California 94025 • (415) 326-6200

(From C.E. Lapple, *Stanford Res. Inst. J.,* 5, 94, 1961. With permission).

8.8 PERIODIC TABLE OF ELEMENTS

Group	1	2	3	4	5	6	7	8	9	10	11	12	13	14	15	16	17	18
	1A	2A	3B	4B	5B	6B	7B	8B	8B	8B	1B	2B	3A	4A	5A	6A	7A	8A
Period																		
1	1 H																	2 He
2	3 Li	4 Be											5 B	6 C	7 N	8 O	9 F	10 Ne
3	11 Na	12 Mg											13 Al	14 Si	15 P	16 S	17 Cl	18 Ar
4	19 K	20 Ca	21 Sc	22 Ti	23 V	24 Cr	25 Mn	26 Fe	27 Co	28 Ni	29 Cu	30 Zn	31 Ga	32 Ge	33 As	34 Se	35 Br	36 Kr
5	37 Rb	38 Sr	39 Y	40 Zr	41 Nb	42 Mo	43 Tc	44 Ru	45 Rh	46 Pd	47 Ag	48 Cd	49 In	50 Sn	51 Sb	52 Te	53 I	54 Xe
6	55 Cs	56 Ba	71 Lu	72 Hf	73 Ta	74 W	75 Re	76 Os	77 Ir	78 Pt	79 Au	80 Hg	81 Tl	82 Pb	83 Bi	84 Po	85 At	86 Rn
7	87 Fr	88 Ra	103 Lr	104 Rf	105 Db	106 Sg	107 Bh	108 Hs	109 Mt	110 Uun	111 Uuu	112 Uub	113 Uut	114 Uuq	115 Uup	116 Uuh	117 Uus	118 Uuo

lanthanides	57 La	58 Ce	59 Pr	60 Nd	61 Pm	62 Sm	63 Eu	64 Gd	65 Tb	66 Dy	67 Ho	68 Er	69 Tm	70 Yb
actinides	89 Ac	90 Th	91 Pa	92 U	93 Np	94 Pu	95 Am	96 Cm	97 Bk	98 Cf	99 Es	100 Fm	101 Md	102 No

8.9 ALPHABETICAL LISTING OF ELEMENTS AND ATOMIC NUMBERS

Symbol	Element	Atomic Number
Ac	Actinium	89
Al	Aluminum	13
Am	Americium	95
Sb	Antimony	51
Ar	Argon	18
As	Arsenic	33
At	Astatine	85
Ba	Barium	56
Bk	Berkelium	97
Be	Beryllium	4
Bi	Bismuth	83
B	Boron	5
Br	Bromine	35
Cd	Cadmium	48
Ca	Calcium	20
Cf	Californium	98
C	Carbon	6
Ce	Cerium	58
Cs	Cesium	55
Cl	Cholorine	17
Cr	Chromium	24
Co	Cobalt	27
Cu	Copper	29
Cm	Curium	96
Dy	Dysprosium	66
Es	Einsteinium	99
Er	Erbium	8
Eu	Europium	63
Fm	Fermum	100
F	Fluorine	9
Fr	Francium	87
Gd	Gadolinium	64
Ga	Gallium	31
Ge	Germanium	32
Au	Gold	79
Hf	Hafnium	72
Ha	Hahnium	105
Hs	Hassium	108
Hi	Helium	2
Ho	Holmium	67
H	Hydrogen	1
In	Indium	49
I	Iodine	53
Ir	Iridium	77
Fe	Iron	26
Kr	Krpton	36

8.9 (CONT.)

Symbol	Element	Atomic Number
La	Lanthanum	57
Lr	Lawrencium	103
Pb	Lead	82
Li	Lithium	3
Lu	Lutetium	71
Mg	Magnesium	12
Mn	Manganesse	25
Mt	Meitnerium	109
Md	Mendelevium	101
Hg	Mercury	80
Mo	Molybdenum	42
Ns	Neilsborium	107
Nd	Neodymium	60
Ne	Neon	10
Np	Neptunium	93
Ni	Nickel	28
Nb	Niobium	41
N	Nitrogen	7
No	Nobelium	102
Os	Osmian	76
O	Oxygen	8
Pd	Palladium	46
P	Phosphorus	15
Pt	Platinum	78
Pu	Plutonium	94
Po	Polonium	84
K	Potassium	19
Pr	Praseodymium	59
Pm	Promethium	61
Pa	Protactinium	91
Ra	Radium	88
Rn	Radon	86
Re	Rhenium	75
Rh	Rhodium	45
Rb	Rubidium	37
Ru	Ruthenium	44
Rf	Rutherfordium	104
Sm	Samarium	62
Sc	Scandium	21
Sg	Seaborgium	106
Se	Selenium	34
Si	Silicon	14
Ag	Silver	47
Na	Sodium	11
Sr	Strontium	38
S	Sulfur	16
Ta	Tantalum	73

8.9 (CONT.)

Symbol	Element	Atomic Number
Tc	Technetium	43
Te	Tellurium	52
Tb	Terbium	65
Tl	Thalium	81
Th	Thorium	90
Tm	Thulium	69
Sn	Tin	50
Ti	Titanium	22
W	Tungsten	74
U	Uranium	92
V	Vadum	23
Xi	Xenon	54
Yb	Ytterbium	70
Y	Yttrium	39
Zn	Zinc	30
Zr	Zirconium	40

8.10 WIRE ROPE DEFORMATION / INSPECTION

All wire ropes will wear out eventually and gradually lose work capability throughout their service life. That is the reason periodic inspections are critical. Applicable industry standards such as ASME B30.2 for overhead and gantry cranes or federal regulations such as OSHA refer to specific inspection criteria for varied applications.

Regular inspection of wire rope and equipment should be performed. There are three purposes for inspection:

1. It reveals the condition of the rope and indicates the need for replacement.
2. It can indicate if the most suitable type of rope is being used.
3. It makes possible the discovery and correction of faults in equipment or operation that can cause costly accelerated rope wear.

All wire ropes should be thoroughly inspected at regular intervals. The longer the rope has been in service or the more severe the service, the more thoroughly and frequently it should be inspected. Be sure to maintain records of each inspection.

Inspections should be carried out by a person who has learned through special training or practical experience what to look for and who knows how to judge the importance of any abnormal conditions that may be discovered. It is the inspector's responsibility to obtain and follow the proper inspection criteria for each application inspected.

Figure 8.1(a) illustrates what happens when a wire breaks under tensile load exceeding its strength. It is typically recognized by "the cup and cone" appearance at the point of failure. The necking down of the wire at the point of failure to form the cup and cone indicates failure has occurred while the wire retained its ductility.

Figure 8.1(b) illustrates a wire with a distinct *fracture break*. It is recognized by the square *end* perpendicular to the wire. This break was produced by a torsion machine that is used to measure the ductility. This break is similar to wire failures in the field caused by fatigue.

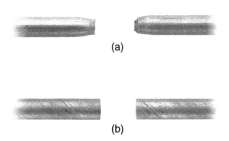

(a)

(b)

FIGURE 8.1 (a) "Cup and cone" break. (b) Distinct fracture break. (Courtesy of the Wire Rope Corporation of America, St. Joseph, MO.)

FIGURE 8.2 Fracture breaks in individual wires. (Courtesy of the Wire Rope Corporation of America, St. Joseph, MO.)

FIGURE 8.3 Fatigue failure. (Courtesy of the Wire Rope Corporation of America, St. Joseph, MO.)

FIGURE 8.4 Strand nicking. (Courtesy of the Wire Rope Corporation of America, St. Joseph, MO.)

Figure 8.2 illustrates wire rope that has been subjected to repeated bending over sheaves under normal loads. This results in *fatigue breaks* in individual wires — these breaks are square and usually in the crown of the strands.

An example of *fatigue failure* of a wire rope subjected to heavy loads over small sheaves is shown in Figure 8.3. The breaks in the valleys of the strands are caused by "strand nicking." There may be crown breaks as well.

Figure 8.4 illustrates a single strand removed from a wire rope subjected to *strand nicking.* This condition is a result of adjacent strands rubbing against one another. Although this is normal in the operation of a rope, the nicking can be accentuated by high loads, small sheaves, or loss of core support. The ultimate result is individual wire breaks in the valleys of the strands.

A *birdcage* is caused by sudden release of tension and the resulting rebound of rope. These strands and wires will not return to their original positions. The rope should be replaced immediately (see Figure 8.5).

Figure 8.6 shows a typical failure of a rotary drill line with a poor cutoff practice. These wires have been subjected to continued *peening,* causing fatigue-type failures. A predetermined, regularly scheduled cutoff practice can help eliminate this type of problem.

FIGURE 8.5 (a) Birdcage. (Courtesy of the Wire Rope Corporation of America, St. Joseph, MO.)

FIGURE 8.6 Peening. (Courtesy of the Wire Rope Corporation of America, St. Joseph, MO.)

FIGURE 8.7 Localized wear. (Courtesy of the Wire Rope Corporation of America, St. Joseph, MO.)

FIGURE 8.8 High strand. (Courtesy of the Wire Rope Corporation of America, St. Joseph, MO.)

Figure 8.7 illustrates *localized wear* over an equalized sheave. The danger here is that the wear is invisible during the operation of the rope, and that is the reason it is necessary to inspect this portion of an operating rope regularly. The rope should be pulled off the sheave during inspection and bent to check for broken wires.

Shown in Figure 8.8 is a wire rope with a *high strand* — a condition in which one or more strands are worn before adjoining strands. This is caused by improper socketing or seizing, kinks, or dog-legs. On top, is shown a close-up view of the concentration of wear. The bottom image shows how it recurs every sixth strand in a rope.

FIGURE 8.9 Kinked wire rope. (Courtesy of the Wire Rope Corporation of America, St. Joseph, MO.)

FIGURE 8.10 Curled rope. (Courtesy of the Wire Rope Corporation of America, St. Joseph, MO.)

FIGURE 8.11 Drum crushing. (Courtesy of the Wire Rope Corporation of America, St. Joseph, MO.)

A *kinked wire rope* is shown in Figure 8.9. The damage is caused by pulling down a loop in a slack line during handling, installation, or operation. Note the distortion of the strands and individual wires. This rope must be replaced.

Figure 8.10 illustrates a wire rope that has jumped a sheave. The rope *curled* as it went over the edge of the sheave. Two types of breaks are evident here: tensile "cup and cone" breaks and shear breaks that appear to have been cut on an angle.

Drum crushing, caused by small drums, high loads, and multiple winding conditions, is shown in Figure 8.11.

8.11 INTERNET SITES

Hundreds, if not thousands, of safety-related Internet sites exist. It is futile to try to list them individually here. However, there are two specific areas of information that are expecially useful. The first is an outstanding book dedicated to these Internet addresses: *Safety and Health on the Internet*, by Ralph B. Stuart, III, which discusses Internet use in detail and provides many addresses.

   The second area of information of particular usefulness is the addresses of the major regulatory agencies. These should prove beneficial for speed of research.

http://www.open.gov.uk	— Health and Safety Executive (United Kingdom)
http://turva.me.tut.fi/cis	— International Occupational Safety and Health Information Center
http://www.msha.gov	— Mine Safety and Health Administration
http://www.cdc.gov/niosh	— National Institute of Occupational Health and Safety
http://www.osha.gov	— Occupational Safety and Health Administration
http://www.epa.gov	— Environmental Protection Agency

9 Glossary: Terms of Interest to Environmental, Safety, Health, and Industrial Hygiene Practitioners

Every environmental, safety, health, or industrial hygiene practitioner should have a vocabulary of words, symbols, and acronyms to communicate with other professionals in similar or related fields. Many of these terms may have different meanings in different settings or in different fields of work. The definitions provided here are intended to provide a quick, sometimes abbreviated meaning, which would most generally help the safety practitioner understand the meaning and context of a term.

The following is a comprehensive list of words, symbols, and acronyms the typical environmental, safety, health, or industrial hygiene practitioner should typically know.

A-, an- (prefix). Absent, lacking, deficient, without.

AAIH. American Academy of Industrial Hygiene (an association of certified industrial hygienists, CIH).

AAOHN. American Association of Occupational Health Nurses.

Abrasive blasting. A process for cleaning surfaces by means of such materials as sand, alumina, or steel grit in a stream of high-pressure air.

Absorption. The condensation of gases, liquids, or dissolved substances on the surfaces of solids.

Absorption coefficient. *See* Sound absorption coefficient.

AC. *See* Alternating current.

Accelerator. A device for imparting very high velocity to charged particles such as electrons or protons. Also, a chemical additive that increases the speed of a chemical reaction.

Access door. A sliding or hinged door in a duct or fan housing used for clean-out, inspection, and/or maintenance.

Accident. An unplanned, undesired event, not necessarily resulting in injury, but damaging to property and/or interrupting the activity in process.

Accident causes.   Hazards and those factors that, individually or in combination, directly cause accidents.

Accident prevention.   The application of countermeasures designed to reduce accidents.

Accident rate.   Accident experience in relation to a base unit of measurement (e.g., accidents per hours worked, accidents per days worked, accidents per 100 employees).

Acclimation, Acclimatization.   The physiological process of becoming accustomed to environmental conditions (e.g., a hot environment).

Accommodation.   The ability of the eye to adjust focus for various distances.

Accuracy.   The agreement of a reading or measurement with the true value. For example, suppose the actual sound pressure level in a space is 80 dB and there are three measurements of 81, 81, and 81. For these measurements, the accuracy is *pretty* good (only 1 db off), but the precision is *very* good (no variation between measurements).

Accuracy (instrument).   Often used incorrectly as precision. Accuracy is the agreement of a reading or observation obtained from an instrument or a technique with a true value.

acfm.   Actual cubic feet per minute of gas or air flowing at existing temperature and pressure (*See also* scfm).

ACH, AC/H,N.   Air changes per hour. The number of times air is theoretically replaced in a space during an hour.

ACGIH.   The American Conference of Governmental Industrial Hygienists. An association whose membership is open to anyone who is engaged in the practice of industrial hygiene or occupational and environmental health and safety. The association supports and produces the TLV List, the Industrial Ventilation Manual, bioaerosol documents, and other activities.

Acid.   A proton donor.

Acid pickling.   A bath treatment to remove scale and other impurities from metal surfaces before plating or other surface treatment. Sulfuric acid is commonly used.

ACM.   Asbestos-containing materials.

Acne.   *See* Oil dermatitis.

Acoustic, acoustical.   Containing, producing, arising from, actuated by, related to, or associated with sound.

Acoustics.   The study, science, or application of principles associated with sound.

Acoustic trauma.   Hearing loss caused by sudden loud noise in one ear or by a sudden blow to the head. In most cases, hearing loss is temporary, although there may be some permanent loss.

Acro (prefix).   Topmost; outer end. An extremity of the body. Acro-osteolysis is degeneration of the terminal or distal end of bone tissue.

Acrylic.   A family of synthetic resins made by polymerizing esters of acrylic acids.

Action level. A term used by OSHA and NIOSHA to express the level of intoxicant that requires medical surveillance, usually one half of the permissible exposure limit.

Activated charcoal. An amorphous form of carbon formed by burning wood, nutshells, animal bones, and other carbonaceous materials. Charcoal becomes activated by heating it with steam to 800 to 900°C. During this treatment, an aporous, submicroscopic internal structure is formed that gives it an extensive internal surface area. Activated charcoal is commonly used as a gas or vapor absorbent in air-purifying respirators and as a solid sorbent in air sampling.

Activation. Making a substance artificially radioactive in an accelerator or by bombarding it with protons or neutrons in a reactor.

Activity. Often used as a shortened form of radioactivity; refers to the radiating power of a radioactive substance. Activity may be given in terms of atoms disintegrating per second.

Acuity. Pertaining to the sensitivity of receptors used in hearing or vision.

Acute. Health effects that show up a short time after exposure. An acute exposure runs a comparatively short course.

ADA. Americans with Disabilities Act. A 1991 federal law prohibiting discrimination against people with disabilities in most public activities, including the workplace.

Additives. An inclusive name for a wide range of chemical substances that are added in low percentage to stabilize certain end products, such as antioxidants in rubber.

Aden- (prefix). Pertaining to a gland. Adenoma is a tumor of glandlike tissue.

Adenoma. An epithelial tumor, usually benign, with a glandlike structure (the cells lining glandlike depressions or cavities in the stroma).

Adhesion. The ability of one substance to stick to another. There are two types of adhesion: mechanical, which depends on the penetration of the surface, and molecular or polar adhesion, in which adhesion to a smooth surface is obtained because of polar groups such as carboxyl groups.

Administrative controls, administrative method. A method of controlling emissions and / or exposures by controlling some aspect of the basic job procedure. (e.g., job rotation, training, work procedures, time periods worked, work cycle).

Adsorption. Condensation of gases and vapors on the surface of solids.

AEC. Atomic Energy Commission. Now called the Nuclear Regulatory Commission in the U.S. Department of Energy.

Aerobe. Microorganisms that require the presence of oxygen.

Aerodynamic diameter. The diameter of a unit density sphere having the same settling velocity as the particle in question of a given shape and density. Compares the settling velocity of a small particle to the equivalent diameter of a sphere with a density of $62.4 lb/ft^3$ (unit density).

Aerodynamic forces. The forces exerted on a particle in suspension by either the movement of air or gases around the particle or the resistance of the gas or air to movement of the particle through the medium.

Aerosols. Liquid droplets of solid particles dispersed in air that are of fine-enough particle size (0.01 to 100 μm) to remain so dispersed for a period of time.

Affirmative action. Positive action taken to assure nondiscriminatory treatment of all groups in employment regardless of sex, religion, age, handicap, or national origin.

Agency, agent. The principal object, such as a tool, machine, or material, involved in an accident that inflicts injury, illness, or property damage.

Agglomeration. Term that implies consolidation of solid particles into larger shapes by means of agitation alone, that is, without application of mechanical pressure in molds, between rolls, or through dies. Industrial agglomeration is usually implemented in balling devices such as rotating disks, drums, or cones, but it can occur in a simple mixer. Agglomeration has also been used to describe the entire field of particulate consolidation.

Agricultural Hazards. Noise, ammonia exposure, vibration, pesticide exposure, trauma histoplasmosis (from bird droppings), poisonous plants, sunlight, fatigue.

AHU. Air-handling unit. Refers to ventilation equipment in HVAC systems.

AIDS. Acquired Immune Deficiency Syndrome.

AIHA. American Industrial Hygiene Association.

Air. The mixture of gases that surrounds Earth; its major components are as follows: 78.08% nitrogen, 20.95% oxygen, 0.03% carbon dioxide, and 0.93% argon. Water vapor (humidity) varies. *See also* standard air.

Air bone gap. The difference in decibels between the hearing levels for a particular frequency as determined by air conduction and bone conduction.

Airborne microorganisms. Biologically active contaminants suspended in air either as free-floating particles surrounded by a film of organic or inorganic material, or attached to the surface of other suspended particulates.

Air cleaner. A device designed to remove atmospheric airborne impurities, such as dusts, gases, vapors, fumes, and smokes. Examples include filters, scrubbers, electrostatic precipitators, cyclones, and afterburners.

Air conditioning. The process of treating air to control its temperature, humidity, cleanliness, and distribution to meet requirements of the conditioned space.

Air conduction. The process by which sound is conducted to the inner ear through air in the outer ear canal.

Air density. Also weight density. The weight of air in pounds per cubic foot. Dry, standard air at temperature = 70°F, barometric pressure = 29.92 in. Hg has a density of 0.075 lb/ft³.

Air filter. An air-cleaning device to remove light particulate matter from normal atmospheric air.

Air hammer. A percussion-type pneumatic tool fitted with a handle at one end of the shank and a tool chuck at the other, into which a variety of tools may be inserted.

Air horsepower (ahp).   The theoretical horsepower required to drive a fan if there are no losses in the fan, that is, if it is 100% efficient.

Air line regulator.   A respirator that is connected to a compressed breathing air source by hose.

Air monitoring.   Sampling for the measurement of pollutants in the atmosphere.

Air mover.   Any device that is capable of causing air to be moved from one space to another. Such devices are generally used to exhaust, force, or draw gases through specific assemblies.

Air-powered tools.   Tools that use air under pressure to drive various rotating, percussion attachments. Pneumatic.

Air-purifying respirator.   Device that uses filters or sorbents to remove harmful substances from the air.

Air quality criteria.   The amounts of pollution and lengths of exposure at which specific adverse effects to health and welfare take place.

Air-regulating valve.   An adjustable valve used to regulate airflow to the face piece, helmet, or hood of an air line respirator.

Air, standard.   *See* Standard air.

Air-supplied respirator.   Device that provides a supply of breathable air from a clean source outside the work area.

Albumin.   A protein material found in animal and vegetable fluids, characterized by solubility in water.

Albuminuria.   The presence of albumin or other protein substance, such as serum globulin, in the urine.

-algia (suffix).   Pain. A prefix such as a neur- indicates where the pain is located (neuralgia = pain in a nerve, for example).

Algorithm.   A precisely stated procedure or set of instructions that can be applied stepwise to solve a problem.

Aliphatic.   Derived from the Greek word for oil. Pertaining to an open change carbon compound. Usually applied to petroleum products derived from a paraffin base and having a straight or branched chain, saturated or unsaturated molecular structure. Substances such as methane or ethane, are typical aliphatic hydrocarbons. *See also* Aromatic.

Alkali.   A compound that has the ability to neutralize an acid and form salt. Sodium hydroxide, known as caustic soda or lye, is an example. Used in soap manufacture and other applications. Turns litmus paper blue. *See also* Base.

Alkaline earths.   Usually considered to be the oxides of alkaline earth metals: barium, calcium, strontium, beryllium, and radium. Some authorities also include magnesium oxide.

Alkyd.   A synthetic resin that is the condensed product of a polybasic acid, such as phthalic, apolyhydric alcohol, glycerin, or an oil fatty acid.

Alkylation.   The process of introducing one or more alkyl radicals by addition or substitution into an organic compound.

Allergy.   An abnormal response of a hypersensitive person to chemical or physical stimuli. Allergic manifestations of major importance occur in about 10% of the population.

Alloy. A mixture of metals (and sometimes a nonmetal), as in brass, tin, or copper.

Alpha-emitter. A radioactive substance that gives off alpha particles (two neutrons and two protons).

Alpha particle. A small, positively charged particle made up of two (alpha-ray, alpha-radiation) neutrons and two protons and of very high velocity, thrown off by many radioactive materials, including uranium and radium.

Alternating current (AC). Electric current that reverses direction. Ordinary house current in the United States reverses direction 60 times per second.

Aluminosis. A form of pneumoconiosis due to the presence of aluminum-bearing dust in the lungs, especially that of alum, bauxite, or clay.

Alveoli. Tiny air sacs of the lungs, formed at the ends of bronchioles; through the thin walls of the alveoli, the blood takes in oxygen and gives up carbon dioxide in respiration.

Alveolus. A general term used in anatomical nomenclature to designate a small sac-like dilation.

Amalgamation. The process of alloying metals with mercury. This is the process used in extracting gold and silver from their ores.

Ambient. The existing, unaltered environment.

Ambient air. Air found naturally in the environment under consideration.

Ambient noise. The all-encompassing noise associated with a given environment; usually a composite of sounds from many sources.

AMCA. Air Moving and Control Association. A fan-manufacturing association that sets certification criteria and testing methods for fan performance.

American Conference of Governmental Industrial Hygienists. *See* ACGIH.

Amorphous. Noncrystalline.

Ampere. The standard unit for measuring the strength of an electrical current.

Anaerobe. A microorganism that grows without oxygen. Facultative anaerobes are able to grow without oxygen; obligate anaerobes grow *only* in the absence of oxygen.

Anaerobic bacteria. Any bacteria that can survive in a partial or complete absence of air, in a nonoxygen atmosphere.

Anaphylaxis. Hypersensitivity resulting from sensitization following prior contact with a chemical or compound.

Andro- (prefix). Man, male. An androgen is an agent that produces masculinizing effects.

Anechoic room. Room whose boundaries effectively absorb all the sound (free-field room) incident therein, thereby affording essentially free-field conditions.

Anemia. Deficiency in the hemoglobin and erythrocyte content of the blood. Term refers to a number of pathological states that may be attributed to a large variety of causes and appear in many different forms.

Anemometer. A device that measures velocity of air. Common types include the swing vane, and the hot wire anemometer.

Anesthesia. Loss of sensation; in particular, the temporary loss of feeling induced by certain chemical agents.

Angi-, angio- (prefix). Blood or lymph vessel. Angitis is the inflammation of a blood vessel.

Angle of abduction. Angle between the longitudinal axis of a limb and a sagittal plane.

Angstrom. A unit of measure of wavelength equal to 10^{-10} m or 0.1 nanometer.

Anneal. To treat by heat with subsequent cooling for drawing the temper of metals, that is, to soften and render them less brittle. *See also* Temper.

Anode. The positive electrode.

Anorexia. Lack or loss of the appetite for food.

ANSI. American National Standards. A voluntary membership organization (run with private funding) that develops consensus standards nationally for a wide variety of devices and procedures.

Antagonist. A muscle opposing the action of another muscle. An active antagonist is essential for control and stability of action by a prime mover.

Antagonistic interaction. Interaction of two chemicals in which the resultant toxic effect is lower than the individual action of the chemicals.

Anthracosilicosis. A complex form of pneumoconiosis; a chronic disease caused by breathing air containing dust that has free silica as one of its components and that is generated in various processes of mining and preparing anthracite (hard) coal and, to a lesser degree, bituminous (soft) coal.

Anthracosis. A disease in the lungs caused by prolonged inhalation of dust that contains particles of carbon and coal.

Anthrax. A highly infectious bacterial infection communicated from infected animals and animal products.

Anthropometric evaluation. A study of human body and its modes of action to improve design of tools and machines to correspond with human capabilities.

Anthropometry. The branch of anthropology having to do with measurement of the human body to determine differences in individuals or groups of individuals.

Anti- (prefix). Against. An antibiotic is "against life" in the sense that it is a drug directed against the life of disease-causing germs.

Antibody. Any of the body globulins that combine specifically with antigens to neutralize toxins, agglutinate bacteria or cells, and precipitate soluble antigens. It is found naturally in the body or produced by the body in response to the introduction into its tissues of a foreign substance.

Antigen. A substance that when introduced into the body stimulates antibody production.

Antioxidant. A compound that retards deterioration by oxidation. Antioxidants for human food and animal feeds, sometimes referred to as freshness preservers, retard rancidity of fats and lessen loss of fat-soluble vitamins (A, D, E, K). Antioxidants are also added to rubber, motor lubricants, and other materials to inhibit deterioration.

Antiparticle. A particle that interacts with its counterpart of the same mass but opposite electric charge and magnetic properties (e.g., proton and antiproton), with complete annihilation of both and production of an equivalent amount of radiation energy. The positron and its antiparticle, the electron, annihilate each other upon interaction and produce gamma (γ) rays.

Antiseptic.   A substance that prevents or inhibits the growth of microorganisms; a substance used to kill microorganisms on biological surfaces, such as skin.

Aplastic anemia.   A condition in which the bone marrow fails to produce an adequate number of red blood corpuscles.. .

Approved.   Tested and listed as satisfactory by an authority having jurisdiction, such as U.S. Department of Health and Human Services (HHS), NIOSH-MSHA, or the U.S. Department of Agriculture.

Aqueous humor.   Fluid in the anterior chamber of the eye.

Arc welding.   A form of electrical welding using either uncoated or coated rods.

Arc-welding electrode.   A component of the welding circuit through which current is conducted between the electrode holder and the arc.

Argyria.   A slate-gray or bluish discoloration of the skin and deep tissues caused by the deposit of insoluble albuminate of silver, occurring after the medicinal administration for a long period of a soluble silver salt; formerly fairly common after the use of insufflations of silver-containing materials into the nose and sinuses. Also seen with occupational exposure to silver-containing chemicals.

Aromatic.   Term applied to a group of hydrocarbons and their derivatives characterized by the presence of the benzene nucleus (molecular ring structure). *See also* Aliphatic.

Aromatic Compound.   A ringed hydrocarbon (e.g., benzene-derived compounds).

Arrestance.   The ability of a filter to remove a coarse dust particle.

Arthr- (prefix).   Joint. Arthropathy is a disease affecting a joint.

Artificial abrasive.   Materials such as carborundum or emery substituted for a natural abrasive such as sandstone.

Artificial radioactivity.   That produced by bombardment of a target element with nuclear particles. Iodine-131 is an artificially produced radioactive substance.

Asbestos.   A hydrated magnesium silicate in fibrous form.

Asbestosis.   A disease of the lung caused by inhalation of fine airborne asbestos fibers. Typically has a very long latency period.

Asepsis.   Clean and free of microorganisms.

Aseptic technique.   A procedure or operation that prevents the introduction of septic material.

ASHRAE.   American Society of Heating, Refrigeration, and Air Conditioning Engineers.

Aspect ratio.   Length-to-width ratio.

Asphyxia.   Suffocation from lack of oxygen. Chemical asphyxia is produced by a substance such as carbon monoxide that combines with hemoglobin to reduce the capacity of the blood to transport oxygen. Simple asphyxia is the result of exposure to a substance, such as methane, that displaces oxygen.

Asphyxiant. A gas whose primary or most acute health effect is asphyxiation. There are two classes of asphyxiant: simple asphyxiant, such as nitrogen or methane, which acts by replacing oxygen; and chemical asphyxiant, such as carbon monoxide, which causes asphyxiation by preventing oxygen uptake at the cellular level.

Assigned Protection. The level of respiratory protection expected from a factor (APF) respirator that is properly functioning, has been properly fitted, and is worn by a worker trained in its use. APFs can be used to help provide an estimate of the maximum concentrations of a contaminant in which a particular respirator can be used.

ASTM. American Society for Testing and Materials. Voluntary membership organization with members from a broad spectrum of individuals, agencies, and industries concerned with materials.

Asthma. Constriction of the bronchial tubes in response to irritation, allergy, or other stimulus.

Ataxia. Lack of muscular coordination caused by any of several nervous system diseases.

Atmosphere-supplied respirator. A respirator that provides breathing air from a source independent of the surrounding atmosphere.

Atmospheric pressure. The pressure exerted in all directions by the atmosphere. At sea level, mean atmospheric pressure is 29.92 in. Hg; 14.7 psi, or 407 in. wg.

Atom. All materials are made of atoms. The elements, such as iron, lead, and sulfur, differ from each other because their atomic structures are different. The word *atom* comes from the Greek word meaning indivisible. It is now known that an atom can be split and consists of an inner core (nucleus) surrounded by electrons that rotate around the nucleus. As a chemical unit, it remains unchanged during any chemical reaction, yet may undergo nuclear transmutations to other atoms, as in atomic fission.

Atomic energy. Energy released in nuclear reactions. Of particular interest is the energy released when a neutron splits the nucleus of an atom into smaller pieces (fission) or when two nuclei are joined together under millions of degrees of heat (fusion). Atomic energy is a popular misnomer; it is more correctly called nuclear energy.

Atomic hydrogen welding. A shielded gas-electric welding process using hydrogen as the reducing atmosphere.

Atomic number. The number of protons found in the nucleus of an atom. All elements have different atomic numbers. The atomic number of hydrogen is 1, that of oxygen 8, iron 26, lead 82, uranium 92. The atomic number is also called charge number and is usually denoted by Z.

Atomic power. The name given to the production of thermal power in a nuclear reactor or power facility.

Atomic waste.   The radioactive ash produced by the splitting of uranium fuel, as in a nuclear reactor. It may include products that have been made radioactive in such a device.

Atomic weight.   The atomic weight is approximately the sum of the number of protons and neutrons found in the nucleus of an atom. This sum is also called mass number. The atomic weight of oxygen is approximately 16, with most oxygen atoms containing 8 neutrons and 8 protons. Aluminum is 27; it contains 14 neutrons and 13 protons.

Atom smasher.   Accelerator that speeds up atomic and subatomic particles so that they can be used as projectiles literally to blast apart the nuclei of other atoms.

Atrophy.   Arrested development or wasting away of cells and tissue.

Attenuate.   To reduce in amount. Usually refers to noise or ionizing radiation.

Attenuation.   The reduction of intensity at a designated first location as compared with the intensity at a second location, which is farther from the source.

Attenuation block.   A block or stack, having dimensions 20 by 20 by 3.8 cm, of Type 1100 aluminum alloy or aluminum alloy having equivalent attenuation.

Audible range.   The frequency range across which normal ears hear: at approximately 20,000 Hz, the term ultrasonic is used; below 20 Hz, the term subsonic is used.

Audible sound.   Sound containing frequency components lying between 20 and 20,000 Hz.

Audiogram.   A record of hearing loss or hearing level measured at several different frequencies — usually 500 to 6000 Hz. The audiogram may be presented graphically or numerically. Hearing level is shown as a function of frequency.

Audiologist.   A person with graduate training in the specialized problems of hearing and deafness.

Audiometer.   A signal generator or instrument for measuring objectively the sensitivity of hearing. Pure-tone audiometers are standard instruments for industrial use for audiometric testing.

Audiometric technician.   A person who is trained and qualified to administer audiometric examinations.

Audiometric zero.   The threshold of hearing: 0.0002 microbars of sound pressure. *See also* decibel.

Auditory.   Pertaining to or involving the sense or organs of hearing.

Aural insert.   Usually called earplugs or inserts. A pliable material is inserted into the ear canal to reduce the amount of noise reaching the inner ear.

Auricle.   Part of the ear that projects from the head; medically, the pinna. Also, one of the two upper chambers of the heart.

Autoclave.   An apparatus using pressurized steam for sterilization.

Autoignition temperature.   The lowest temperature at which a flammable gas–air or vapor–air mixture ignites from its own heat source or a contacted heated surface without necessity of spark or flame. Vapor and gases spontaneously ignite at a lower temperature in oxygen that in air, and their autoignition temperature may be influenced by the presence of catalytic substances.

Avogadro's number. The number of molecules in a gram-mole of a material; 6.022×1023. One gram mole of a gas or vapor occupies about 24.1 liters at 70°F. and 29.92 in. Hg; one pound mole occupies 387 ft^3 at the same conditions.

Axial-flow fan. A propeller-type fan useful for moving large volumes of air against little resistance.

Axis of rotation. The true line about which angular motion takes place at any instant. Not necessarily identical with anatomical axis of symmetry of a limb, or necessarily fixed. Thus, the forearm rotates about an axis that extends obliquely from the lateral side of the elbow to a point between the little finger and ring finger. The elbow joint has a fixed axis maintained by circular joint surfaces, but the knee has a moving axis as its cam-shaped surfaces articulate. Axis of rotation of tools should be aligned with true limb axis of rotation. System of rotation of tools should be aligned with true limb axis of rotation. Systems of predetermined motion times often specify such an axis incorrectly.

Axis of thrust. The line along which thrust can be transmitted safely. In the forearm, it coincides with the longitudinal axis of the radius. Tools should be designed to align with this axis.

Babbitt. An alloy of tin, antimony, copper, and lead used as a bearing metal.

Babbitting. The process of applying babbitt to a bearing.

Bacillus. A rod-shaped bacterium.

Background noise. Noise coming from sources other than the particular noise source being monitored.

Background radiation. The radiation coming from sources other than the radioactive material to be measured. This background is primarily a result of cosmic rays that constantly bombard Earth from outer space.

Bacteria. Microscopic organisms living in soil, water, organic matter, or the bodies of plants and animals characterized by lack of a distinct nucleus and lack of ability to photosynthesize. Singular: bacterium.

Bactericide. Any agent that destroys bacteria.

Bacteriophage. Viruses that infect bacteria and lyse the bacterial cell.

Bacteriostat. An agent that stops the growth and multiplication of bacteria but does not necessarily kill them. Usually growth resumes when the bacteriostat is removed.

Bagasse. Sugarcane pulp residues.

Bagassosis. Respiratory disorder believed to be caused by breathing fungi found in bagasse.

Bag house. Many different trade meanings. Commonly connotes the housing containing bag filters for recovery of fumes of arsenic, lead, sulfur, and other materials from the flues of smelters.

Balancing. Creating appropriate flows in ductwork by means of careful design, or through the use of dampers.

Balancing by dampers. Method for designing local exhaust system ducts using adjustable dampers to distribute airflow after installation.

Balancing by static. Method for designing local exhaust system ducts by selecting the duct diameters that generate static pressure to distribute airflow without dampers.

Ball mill. A grinding device using balls usually made of steel or stone in a revolving container.

Banbury mixer. A mixing machine that permits control over the temperature of the batch; commonly used in the rubber industry.

Band-pass filter. A wave filter that has a single transmission band extending from a lower cutoff frequency greater than zero to a finite upper cutoff frequency.

Band-pressure level. Level of a sound for a specified frequency band that is the sound-pressure level for the sound contained within the restricted band. The reference pressure must be specified.

Bandwidth. When applied to a band-pass filter, bandwidth is determined by the interval of transmitted waves between the low and high cutoff frequencies.

Baritosis. An inert pneumoconiosis produced by the inhalation of insoluble barium compounds.

Barotrauma. An injury to the ear caused by a sudden alteration in barometric (atmospheric) pressure; aerotitis.

Barrier cream. A lotion-type material that is often used to provide some minimal protection from irritant materials.

Barrier guard. Physical protection for operators and other individuals from hazard points on machinery and equipment (e.g., Fixed barrier guard, interlocked barrier guard, adjustable barrier guard).

Basal metabolism. A measure of the amount of energy required by the body at rest.

Base. A compound that reacts with an acid to form a salt; another term for alkali. It turns litmus paper blue.

Basilar. Of, relating to, or situated at the base.

Bauxite. Impure mixture of aluminum oxides and hydroxides; the principal source of aluminum.

Bauxite pneumoconiosis. Shaver's disease. Found in workers exposed to fumes containing aluminum oxide and minute silica particles arising from smelting bauxite in the manufacture of corundum.

Beam axis. A line from the source through the centers of the X-ray fields.

Beam divergence. Angle of beam spread measured in mrad (1 mrad = 3.4 min of arc).

Beam-limiting device. A device that provides a means to restrict the dimensions of an X-ray field.

Beat elbow. Bursitis of the elbow; occurs from use of heavy vibrating tools.

Beat knee. Bursitis of the knee joints caused by friction or vibration; common in mining.

Becquerel (Bq).   One disintegration per second; a measure of the rate of radio-active disintegration. There are 37 billion Bqs per curie.

Beehive kiln.   An oven shaped like a large beehive usually used for calcining ceramics.

BEI.   *See* Biological exposure indices.

Bel.   A unit of sound level based on a logarithmic scale. A decibel is 1/10th of a Bel.

Belding–Hatch index.   Estimate of the body heat stress of a standard man for various degrees of activity; also relates to sweating capacity. *See also* Heat stress index.

Benign.   Not malignant. A benign tumor is one that does not metastasize or invade tissue. A benign tumor may still be lethal because of pressure on vital organs.

Benzene, CH.   A major organic intermediate and solvent derived from coal or petroleum. The simplest member of the aromatic series of hydrocarbons.

Beryl.   A silicate of beryllium and aluminum.

Berylliosis.   Chronic beryllium intoxication.

Beta decay.   The process whereby some radioactive emitters give off a beta particle (electron). Also called beta disintegration.

Beta particle.   A small electrically charged particle thrown off by many (beta-radiation) radioactive materials; identical to the electron. Beta particles emerge from radioactive material at high speeds.

Betatron.   A large, doughnut-shaped accelerator in which electrons (beta particles) are whirled through a changing magnetic field, gaining speed with each trip and emerging with high energies. Energies of the order of 100 million electron volts have been achieved. The betatron produces artificial beta radiation.

bhp.   The actual horsepower required to move air through a ventilation system against a fixed total pressure plus the losses in the fan. Bhp = ahp × 1/eff, where eff is the fan mechanical efficiency.

Biceps brachii muscle.   The large muscle in the front of the upper arm. Supinates the forearm.

Bicipital tuberosity.   A protuberance on the medial surface of the radius to which the biceps brachii attaches.

Billet.   A piece of semifinished iron or steel, nearly square in section, made by rolling and cutting an ingot.

Binder.   The nonvolatile portion of a coating vehicle that is the film-forming ingredient used to bind paint pigment particles together.

Binding energy.   The energy that holds the neutrons and protons of an atomic nucleus together. Represents the difference between the mass of an atom and the sum of the masses of protons and neutrons that make up its nucleus.

Bioaerosol.   Suspensions in air of viruses, bacteria, fungi, pollen, and their nonviable components.

Biodegradable.   Capable of being broken down into innocuous products by the action of living things.

Bioengineering. Designing equipment, machines, and other structures to fit the characteristics of people.

Biohazard. "Biological hazard." Hazards associated with exposure to microorganisms or their products (e.g., bacteria, rickettsia, virus, fungi). Most work with these organisms is performed in biosafety cabinets.

Biohazard area. Any area (a complete operating complex, a single facility, a room within a facility, and so on) in which work has been or is being performed with biohazardous agents or materials.

Biohazard control. Any set of equipment and procedures used to prevent or minimize the exposure of humans and their environment to biohazardous agents or materials.

Biological Exposure Indices (BEI). Advisory biological limit values adopted by the ACGIH for some substances. Indices are based on urine, blood, or expired air samples. A BEI may be a value for the substance itself or it may refer to a level of metabolite. BEIs represent the value of the biological determinant that is most likely to be the value of the determinant obtained from a worker exposed at the 8-hour TLV-TWA for the substance in question.

Biological half-life. The time required to reduce the amount of an exogenous substances in the body by half.

Biological oxygen demand (BOD). Quantity of oxygen required for the biological and chemical oxidation of waterborne substances under test conditions.

Biomechanics. The study of a human body as a system operating under two sets of laws: the laws of Newtonian mechanics and the biological laws of life.

Biopsy. Careful removal of small sections of tissue from the body for further examination, usually under the microscope.

Black light. Ultraviolet (UV) light (0.3 to 0.4 μm) responsible for pigmentation of the skin following exposure to the UV light.

Black liquor. A liquor composed of alkaline and organic matter resulting from digestion of wood pulp and cooking acid during the manufacture of paper.

Blast gate. A sliding, sheet-metal valve used in ducts to create additional pressure loss in the duct and to restrict flow.

Bleaching bath. Chemical solution used to bleach colors from a garment preparatory to dyeing it; a solution of chlorine or sodium hypochlorite is commonly used.

Bleph- (prefix). Pertaining to the eyelid.

Blind spot. Normal defect in the visual field due to the position at which the optic nerve enters the eye.

Bloodborne pathogen program. A 1992 OSHA standard mandates exposure control plans and the use of universal precautions for places of employment where there is risk of employee exposure to blood or other potentially infectious material. Hepatitis B and HIV are the most-often-discussed pathogens, but the program is not limited to these two areas.

Blood count. A count of the number of corpuscles per cubic millimeter of blood. Separate counts may be made for red and white corpuscles.

BLS. Bureau of Labor Statistics.

Body burden. The amount of noxious material in the body at a given time. The body burden of a radionuclide that, if maintained at a maximum permissible constant level, would produce the maximum permissible dose equivalent in the critical organ.

Boiler codes. Standards prescribing requirements for the design, construction, testing, and installation of boilers and unfired pressure vessels.

Boiling point. The temperature at which the vapor pressure of a liquid equals atmospheric pressure.

Bombardment. Shooting neutrons, alpha particles, and other high-energy particles at atomic nuclei, usually in an attempt to split the nucleus or to form a new element.

Bone conduction test. A special test conducted by placing an oscillator on the mastoid process to determine the nerve-carrying capacity of the cochlea and the eighth cranial (auditory) nerve.

Bone marrow. A soft tissue that constitutes the central filling of many bones and that produces blood corpuscles.

Bone seeker. Any element or radioactive species that lodges in the bone when introduced into the body.

Brachialis. Short, strong muscles originating at the lower end of the humerus and inserting into the ulna. Powerful flexor of forearm; employed when lifting.

Brady- (prefix). Slow. Bradycardia is slow heartbeat.

Bradycardia. Abnormal slowness of the heartbeat, as evidenced by slowing of the pulse rate to 50 beats per minute or less.

Brake horsepower. The horsepower required to drive a unit; it includes the energy losses in the unit and can be determined only by actual test. It does not include drive losses between the motor and unit.

Branch. In a junction of two ducts, the branch is the duct with the lowest volume flow rate. The branch usually enters the main at some angle less than 90°.

Branch of greatest resistance. The path from hood to fan that causes the largest potential static pressure loss; usually used when design is performed with blast gates or dampers as the balancing approach.

Brass. An alloy of copper and zinc that may contain a small amount of lead.

Brattice. A partition constructed in underground passageways to control ventilation in mines.

Braze. To solder with any relatively infusible alloy.

Brazing furnace. Used for heating metals to be joined by brazing. Requires a high temperature.

Breathing tube. A tube through which air or oxygen flows to a face piece, helmet, or hood.

Breathing zone (BZ). Imaginary globe of 2- to 4-ft radius surrounding the head within which a person breathes.

Breathing zone sample. An air sample collected in the breathing zone of workers to assess their exposure to airborne contaminants.

Bremsstrahlung. Secondary X radiation produced when a beta particle is slowed down or stopped by a high-density surface.

Briquette. Coal or ore dust pressed into oval or brick-shaped blocks.

Broach. A cutting tool for cutting nonround holes.

Bronch-, broncho- (prefix). Pertaining to the air tubes of the lung.

Bronchial tubes. Branches or subdivisions of the trachea (windpipe). A bronchiole is a branch of a bronchus, which is a branch of the windpipe.

Bronchiectasis. A chronic dilation of the bronchi or bronchioles marked by fetid breath and paroxysmal coughing; with the expectoration of mucopurulent matter. It may affect the tube uniformly, or may occur in irregular pockets, or the dilated tubes may have terminal bulbous enlargements.

Bronchiole. The slenderest of the many tubes that carry air into and out of the lungs. Branch of the bronchus that connects to the trachea.

Bronchiolitis. *See* Bronchopneumonia.

Bronchitis. Inflammation of the bronchi or bronchial tubes.

Bronchoalveolitis. Bronchopneumonia.

Bronchopneumonia. A name given to an inflammation of the lungs that usually begins in the terminal bronchioles. These become clogged with a mucopurulent exudate forming consolidated patches in adjacent lobules. The disease is essentially secondary in character, following infections of the upper respiratory tract, specific infectious fevers, and debilitating diseases.

Bronzing. Act or art of imparting a bronze appearance with powders, painting, or chemical processes.

Brownian motion. The irregular (random) movement or particles suspended in a fluid as a result of bombardment by atoms and molecules.

Brucella. A genus of short, rod-shaped to coccoid, encapsulated, Gram-negative, parasitic, pathogenic bacteria.

Brucellosis. A group of diseases caused by an organism of the *Brucella* genus. Undulant fever. One source is unpasteurized milk from cows suffering from Bang's disease (infectious abortion).

Bubble chamber. A chamber containing a liquefied gas such as liquid hydrogen, under conditions such that a charged particle passing through the liquid forms bubbles that make its path visible.

Bubble tube. A device used to calibrate air-sampling pumps.

Buffer. Any substance in a fluid that tends to resist the change in pH when acid or alkali is added.

Building code. An assembly of regulations that set forth the standards to which buildings must be constructed.

Bulk density. Mass of powdered or granulated solid material per unit of volume.

Bulk facility. That portion of a property where flammable or combustible liquids are received by tank vessel, pipeline, tank car, or tank vehicle, and are sorted or blended in bulk for the purpose of distributing such liquids.

Bump cap. A hard-shell cap, without interior suspension systems, designed to protect the wearer's head in situations where the employee might bump into something.

Burns. Result of the application of too much heat to the skin. First-degree burns show redness of the unbroken skin; second degree, skin blisters and some breaking of the skin; third degree, skin blisters and destruction of the skin and underlying tissues, which can include charring and blackening.

Burn-up. The extent to which the nuclear fuel in a fuel element has been consumed by fission, as in a nuclear reactor.

Burr. The thin, rough edges of a machined piece of metal.

Bursa. A synovial lined sac that facilitates the motion of tendons; usually near a joint.

Bursitis. Inflammation of a bursa.

Byssinosis. Disease occurring in those who experience prolonged exposure to heavy air concentrations of cotton or flax dust.

Calcination. The heat treatment of solid material to bring about thermal decomposition, to lose moisture or other volatile material, or to oxidize or reduce.

Calendar. An assembly of rollers for producing a desired finish on paper, rubber, artificial leather, plastic, or other sheet material.

Calorimeter. A device for measuring the total amount of energy absorbed from a source of electromagnetic radiation.

Cancer. A cellular tumor the natural course of which is fatal and that is usually associated with formation of secondary tumors.

Capitulum of humerus. A smooth, hemispherical protuberance at the distal end of the humerus articulating with the head of the radius. Irritation caused by pressure between the capitulum and head of the radius may be a cause of tennis elbow.

Capture velocity. Air velocity at any point in front of the hood necessary to overcome opposing air currents and to capture the contaminated air by causing it to flow into the exhaust hood.

Carbohydrate. An abundant class of organic compounds, serving as food reserves or structural elements for plants and animals. Compounded primarily of carbon, hydrogen, and oxygen, they constitute about two thirds of the average daily adult caloric intake. Sugar, starches, and plant components (cellulose) are all carbohydrates.

Carbon black. Essentially a pure carbon, produced around the world under controlled conditions. There are several processes: furnace black, thermal black, acetylene, and others. Widely used in tires and plastics.

Carbon monoxide. A colorless, odorless toxic gas produced by any process that involves the incomplete combustion of carbon-containing substances. It is emitted through the exhaust of gasoline-powered vehicles.

Carbonizing. The immersion in sulfuric acid of semiprocessed felt to remove any vegetable matter present.

Carborundum. A trade name for silicon carbide widely used as an abrasive.

Carboy. A large glass bottle, usually protected by a crate.

Carboxyhemoglobin. The reversible combination of carbon monoxide with hemoglobin.

Carcinogen. A substance or agent that can cause a growth of abnormal tissue or tumors in humans or animals.

Carcinogenic. Cancer producing.

Carcinoma. Malignant tumors derived from epithelial tissues, that is, the outer skin, the membranes lining the body cavities, and certain glands.

Cardi-, cardio- (prefix). Denoting the heart.

Cardiac. Pertaining to (1) the heart; (2) cordial or restorative medicine; (3) a person with a heart disorder.

Carding. The process of combing or untangling wool, cotton, and so on.

Carding machine. A textile industry machine that prepares wool, cotton, or their fibers for spinning.

Cardiovascular. Relating to the heart and to the blood vessels or circulation.

Carp- (prefix). The wrist.

Carpal tunnel. A passage in the wrist through which the median nerve and many tendons pass to the hand from the forearm.

Carpal tunnel syndrome. A common affliction caused by the compression of the median nerve in the carpal tunnel. Often associated with tingling, pain, or numbness in the thumb and first three fingers — may be job related.

Carrier. A person in apparent good health who harbors a pathogenic microorganism.

Carrier gas. A mixture of gases that contains and moves a contaminant material. Components of the carrier gas are not considered to cause air pollution or react with the contaminant material.

Case-hardening. A process of surface-hardening metals by raising the carbon or nitrogen content of the outer surface.

Cask (or coffin). A thick-walled container (usually lead) used for transporting radioactive materials.

CAS number. Identifies a particular chemical by the Chemical Abstract Service, a service of the American Chemical Society that indexes and compiles abstracts of worldwide chemical literature called *Chemical Abstracts.*

Casting. Pouring a molten material into a mold and permitting it to solidify to a desired shape.

Catalyst. A substance that changes the speed of a chemical reaction but that undergoes no permanent change itself. In respirator use, a substance that converts a toxic gas (or vapor) into a less toxic gas (or vapor). Usually catalysts greatly increase the reaction rate, as in conversion of petroleum to gasoline by cracking. In paint manufacture, catalysts that hasten film-forming sometimes become part of the final product. In most uses, however, they do not, and can often be reused.

Cataract. Opacity in the lens of the eye that may obscure vision.

Cathode. The negative electrode.

Catwalk. A narrow suspended footway usually used for inspection or maintenance purposes.

Caulking.   The process or material used to fill seams of boats, cracks in tile, etc.

Causal factor (of an accident).   One or a combination of simultaneous or sequential circumstances directly or indirectly contributing to an accident. Modified to identify several kinds of causes such as direct, early, mediated, proximate, distal, etc.

Caustic.   Something that strongly irritates, burns, corrodes, or destroys living tissue. *See also* Alkali.

Ceiling limit (C).   An airborne concentration of a toxic substance in the work environment that should never be exceeded.

-cele (suffix).   Swelling or herniation of a part, as in rectocele (prolapse of the rectum).

Cell.   The structural unit of tissues. There are many types: nerve cells, muscle cells, blood cells, connective tissues cells, fat cells, and others. Each has a special form to serve a particular function.

Cellulose.   A carbohydrate that makes up the structural material of vegetable tissues and fibers. Its purest forms are chemical cotton and chemical pulp; it is the basis of rayon, acetate, and cellophane.

Celsius.   The Celsius temperature scale is a designation of the scale previously known as the centigrade scale.

Cement, Portland.   Portland cement commonly consists of hydraulic calcium silicate to which the addition of certain material in limited amounts is permitted. Ordinarily, the mixture consists of calcareous materials such as limestone, chalk, shells, marl, clay, shale, blast furnace slag, and so on. In some specifications, iron ore and limestone are added. The mixture is fused by calcining at temperatures usually up to 1000°C.

Centrifugal fan.   Wheel-type fan useful where static pressure is medium to high.

Centrifuge.   An apparatus that uses centrifugal force to separate or remove particulate matter suspended in a liquid.

Cephal- (prefix).   Pertaining to the head. Encephal-, "within the head," pertains to the brain.

Ceramic.   A term applied to pottery, brick, and tile products molded from clay and subsequently calcined.

CERCLA.   Comprehensive Environmental Response, Compensation and Liability Act.

Cerumen.   Earwax.

Cervi- (prefix).   Neck.

CEU.   Continuing education units, needed by individuals for some educational and certification programs such as Certified Safety Professional and Certified Industrial Hygienist.

CFR.   Code of Federal Regulations (e.g., 29 CFR is General Industry, 26 CFR is Construction Industry). The rules promulgated under U.S. law, published in the *Federal Register*, and actually enforced are incorporated in this code at the end of a calendar year.

Chain reaction.   When a fissionable nucleus is split by a neutron it releases energy and one or more neutrons. These neutrons split other fissionable nuclei releasing more energy and more neutrons, making the reaction self-sustaining for as long as there are sufficient fissionable nuclei present.

Charged particles. A particle that possesses at least a unit electrical charge and that does not disintegrate upon a loss of charge. Charged particles are characterized by particle size, number, and. sign of unit charges and mobility. *See also* Ion.

CHCM. Certified Hazard Control Manager, issued by the Board of Certified Hazard Control Management.

Chelating agent or chelate. Derived from the Greek *kelos,* for claw. Any compound that inactivates a metallic ion with the formation of an inner ring structure in the molecule, the metal ion becoming a member of the ring. The original ion, thus chelated, is effectively out of action.

Chemical burns. Generally similar to those caused by heat. After emergency first aid, their treatment is the same as that for thermal burns. In certain instances, such as with hydrofluoric acid, special treatment is required.

Chemical cartridge. The type of absorption unit used with a respirator for removal of low concentrations of specific vapors and gases.

Chemical engineering. That branch of engineering concerned with the development and application of manufacturing processes in which chemical or certain physical changes of materials are involved. These processes may usually be resolved into a coordinated series of unit physical operations and unit chemical processes. The work of the chemical engineer is concerned primarily with the design, construction, and operation of equipment and facilities in which these unit operations and processes are applied.

Chemical hygiene plan. Required by OSHA to protect laboratory employees from hazardous chemicals.

Chemical reaction. A change in the arrangement of atoms or molecules to yield substances of different composition and properties. Common types of reactions are combination, decomposition, double decomposition, replacement, and double replacement.

Chemotherapy. Use of chemicals of particular molecular structure in the treatment of specific disorders on the assumption that known structures exhibit an affinity for certain parts of malignant cells or infectious organisms, and thereby tend to destroy or inactivate them.

CHEMTREC. Chemical Transportation Emergency Center. Operates a 24-hour emergency help line linking chemical emergency experts to emergency sites.

Chert. A microcrystaline form of silica. An impure form of flint used in abrasives.

Cheyne–Stokes. The peculiar kind of breathing usually observed with the respiration of unconscious or sleeping individuals who seen to stop breathing altogether for 540 seconds, then start up again with gradually increasing intensity, stop breathing once more, and then repeat the performance. Common in healthy infants.

Chimney effect.   Also called stack affect. Hot air has a lower density than cooler ambient air, resulting in the buoyant rising of hot air in a stack or chimney, and/or the development of negative static pressure at the base of the stack.

Chloracne.   Skin dermatitis caused by chlorinated naphthalenes and polyphenyls acting on sebaceous glands in the skin.

Chol-, chole- (prefix).   Relating to bile. Cholesterol is a substance found in bile.

Chon-, chondro- (prefix).   Cartilage.

Chromatograph.   An instrument that separates and analyzes mixtures of chemical substances.

Chromosome.   Important rod-shaped constituent of all cells. Chromosomes contain the genes and are made up of deoxyribonucleic acids (DNA).

Chronic.   Long acting, persistent, prolonged; as opposed to acute. Safety is usually concerned with acute effects, health with chronic effects.

CIH.   Certified Industrial Hygienist.

Cilia.   Tiny, hair-like whips in the bronchi and other respiratory passages that aid in the removal of dust trapped on these moist surfaces.

Ciliary.   Pertaining to the cilium (pl. cilia), a minute, vibratile hairlike process attached to the free surface of a cell.

Circuit.   A complete path over which electrical current may flow.

Circuit breaker.   A device that automatically interrupts the flow of an electrical current when the current exceeds a specified level (i.e., 20 amps).

Citation.   A written charge issued by regulatory representatives alleging specific conditions or actions that violate maritime, construction, environmental, mining, or general industry laws and standards.

Clays.   A great variety of aluminum silicate–bearing rocks that are plastic when wet and hard when dry. Used in pottery, stoneware, tile, bricks, cements, fillers, and abrasives. Kaolin is one type of clay. Some clay deposits may include appreciable quartz. Commercial grades of clays may contain up to 20% quartz.

Clean Air Act.   U.S. law enacted to regulate/reduce air pollution. Administered by the Environmental Protection Agency.

Clean room.   A room in which parts or all of the space are maintained at low particulate loadings.

Clean Water Act.   U.S. law enacted to regulate/reduce water pollution. Administered by the Environmental Protection Agency.

Clo.   A unit of clothing; used in heat and cold stress assessments (e.g., one Clo is a typical business suit, long underwear adds 0.7 Clo, etc.).

Clostridium botulinum.   Human pathogenic bacteria that produce an exotoxin, botulinin, which causes botulism.

Cloud chamber.   A glass-domed chamber filled with moist vapor. When certain types of atomic particles pass through the chamber, they leave a cloudlike track much like the vapor trail of a jet plane. This permits scientists to see these particles and study their motion. The cloud chamber and bubble chamber serve the same purpose.

CNS.   Central nervous system.

The Safety Officer's Concise Desk Reference

Coagulase. An enzyme produced by pathogenic staphylococci; causes coagulation of blood plasma.

Coalesce. To unite into a whole; to fuse; to grow together.

Coated welding rods (electrodes). The coatings of welding rods vary. For the welding of iron and most steel, the rods contain manganese, titanium, and a silicate.

Coccidioidomycosis. A fungal disease (also known as valley fever or San Joaquin Valley fever) that can affect agricultural, horticultural, construction workers, and any workers who disturb soil containing spores. Although most often a respiratory disease, in rare cases it can be systemic and fatal. It is transmitted by inhalation of dust containing spores of *Coccidioides immitis.*

Coccus. A spherical bacterium. Plural cocci.

Cochlea. The auditory part of the internal ear, shaped like a snail shell. It contains the basilar membrane on which the end organs of the auditory nerve are distributed.

Code of Federal Regulations. The rules promulgated under U.S. law, published in the *Federal Register*, and actually enforced are incorporated in this code (CFR) at the end of a calendar year.

Coefficient of discharge. A factor used in figuring flow through an orifice. The coefficient takes into account that a fluid flowing through an orifice contracts to a cross-sectional area that is smaller than that of the orifice, and there is some dissipation of energy caused by turbulence.

Coefficient of entry (Ce). The actual rate of flow caused by a given hood static pressure compared with the theoretical flow that would result if the static pressure could be converted to velocity pressure with 100% efficiency; it is the ratio of actual to theoretical flow.

Coefficient of variation. The ratio of the standard deviation to the mean value of a population of observations.

Coffin. A thick-walled container (usually lead) used for transporting radioactive materials.

Cohesion. Molecular forces of attraction between particles of like compositions.

Coils. Heating and cooling heat exchanger that either adds heat or takes heat from the air in an HVAC system.

Colic. A severe cramping, gripping pain in or around the abdomen.

Collagen. An albuminoid, the main supportive protein of skin, tendon, bone, cartilage, and connective tissue.

Collection efficiency. The percentage of a specific substance removed and retained from air by an air-cleaning or sampling device. A measure of cleaner or sampler performance.

Collimated beam. A beam of light with parallel waves.

Colloid. Generally a liquid mixture or suspension in which the particles of suspended liquid or solid are very finely divided. Colloids do not appreciably settle out of suspension.

Colloid mill. A machine that grinds materials into a very fine state of suspension, often simultaneously placing this suspension in a liquid.

Colorimetry (colorimetric). The term applied to all chemical analysis techniques involving reactions in which a color is developed when a particular contaminant is present in the sample and reacts with the collection medium. The resultant color intensity is measured to determine the contaminant concentration.

Coma. A level of unconsciousness from which a patient cannot be aroused.

Combustible. Able to catch fire and burn.

Combustible gas indicator (CGI). A device that measures flammable gases or vapors at concentrations of 10 to 1000 times the TLV. The CGI quantifies combustible gases. Operation consists of drawing air across a platinum filament. The combustible gas or vapor oxidizes or burns on the filament, which raises the temperature and changes its electrical resistance. A Wheatstone bridge is used to measure gas concentrations.

Combustible liquids. Those liquids having a flash point at or above 100°F (37.8°C).

Comedones. Blackheads. Blackened, oily masses of dead epithelial matter clogging the openings of oil glands and hair follicles.

Comfort ventilation. Airflow intended to maintain the comfort of room occupants (heat, humidity, and odor).

Comfort zone. The range of effective temperatures over which the majority of adults feels comfortable.

Common name. Any designation or identification such as code name, code number, trade name, brand name, or a generic name used to identify something other than by its proper name.

Communicable. Pertaining to disease whose causative agent is readily transferred from one person to another.

Compaction. The consolidation of solid particles between rolls or by tamp, piston, screw, or other means of applying mechanical pressure.

Competent person. One who is capable of identifying existing and predictable hazards in the surroundings or working conditions that are unsanitary, hazardous, or dangerous to employees, and who has the authorization to take prompt corrective measures to eliminate them (29 CFR 1926.32).

Complex sounds. Music, speech, and almost all noise consists of a collection of frequencies, i.e., is complex. Speech frequencies are usually found between 250 and 3000 Hz. Hearing loss in this range is considered more critical because of the resulting disability.

Compound. A substance composed of two or more elements joined according to the laws of chemical combination. Each compound had its own characteristic properties different from those of its constituent elements.

Compressed gas cylinder. A cylinder containing vapor or gas under higher-than-atmospheric pressure, sometimes to the point where it is liquefied.

Compressible flow. Flow of high-pressure gas or air that undergoes a pressure drop resulting in significant reduction of its density.

Compton effect.   The glancing collision of a gamma ray with an electron. The gamma ray gives up part of its energy to the electron.

Concentration.   The amount of a given substance in a stated unit of measure. Common methods of stating concentration are percent by weight or by volume, weight per unit volume, normality, and so on.

Conchae.   *See* Turbinates.

Condensate.   The liquid resulting from the process of condensation. In sampling the term is generally applied to the material that is removed from a gas sample by means of cooling.

Condensation.   Act or process of reducing from one form to another reducing form such as steam to water.

Condensoid.   A dispersoid consisting of liquid or solid particles formed by the process of condensation. The dispersoid is commonly referred to as a condensation aerosol.

Conductive hearing loss.   Type of hearing loss; not caused by noise exposure, but by any disorder in the middle or external ear that prevents sound from reaching the inner ear.

Confined space.   Any enclosed area not designed for human occupancy that has a limited means of entry and egress and in which existing ventilation is not sufficient to ensure that the space is free of hazardous atmosphere, oxygen deficiency, or other known or potential hazard. Examples are storage tanks, boilers, sewers, and tank cars. A permit-required confined space, as defined by OSHA standard, is one that requires a permit process and implementation of a comprehensive confined space entry program prior to entry.

Congenital.   Pertaining to a problem that originates before birth.

Conjunctiva.   The delicate mucous membrane that lines the eyelids and covers the exposed surface of the eyeball.

Conjunctivitis.   Inflammation of the conjunctiva, the sensitive mucous membrane lining the eyelid.

Consensus standard.   A standard developed through a consensus process or general opinion among representatives of various interested or affected organizations and individuals.

Contact dermatitis.   Dermatitis caused by contact with a substance — gaseous, liquid, or solid. May be caused by primary irritation or allergy.

Controlled areas.   A specified area in which exposure to radiation or radioactive material is controlled and that is under the supervision of a person who knows appropriate radiation protection practices, including pertinent regulations, and who is responsible for applying them.

Control rod.   A rod (containing an element such as boron) used to control the power of a nuclear power reactor. The control rod absorbs neutrons that would normally split the fuel nuclei. Pushing the rod in reduces the release of atomic power; pulling out the rod increases the release.

Convection.   Motion in fluids resulting from differences in density and the action of gravity.

Converter. A nuclear reactor that uses one kind of fuel and produces another. For example, a reactor charged with uranium isotopes might consume uranium-235 and produce plutonium from uranium-238. A breeder reactor produces more atomic fuel than it consumes; a converter does not.

Coolants. Transfer agents used in a flow system to convey heat from its source.

Copolymers. Mixed polymers or heteropolymers. Products of the polymerization of two or more substances at the same time.

Core. (1) The heart of a nuclear reactor where the nuclei of the fuel fission (split) and release energy. The core is usually surrounded by a reflecting material that bounces stray neutrons back to the fuel. It usually consists of fuel elements and a moderator. (2) A shaped, hard-baked cake of sand with suitable compounds that is placed within a mold, forming a cavity in the casting when it solidifies. (3) The vital centers of the body — heart, viscera, brain — as opposed to the shell, the limbs and integument.

Corium. The deeper skin layer containing the fine endings of the nerves and the finest divisions of the blood vessels, capillaries. Also called the derma.

Cornea. Transparent membrane covering the anterior portion of the eye.

Corpuscle. A red or white blood cell.

Corrected effective temperature (CET). An index of thermal stress similar to the effective temperature index except that globe temperature is used instead of dry-bulb temperature.

Corrective lens. A lens ground to the wearer's individual prescription to improve vision.

Corrosion. Physical change, usually deterioration or destruction, brought about through chemical or electrochemical action, as contrasted to erosion, caused by mechanical action.

Corrosive. A substance that causes visible destruction or permanent changes in human skin tissue at the site of contact.

Corundum. An impure form of aluminum oxide.

Cosmic rays. High-energy rays that bombard the earth from outer space. Some penetrate to the earth's surface and others may go deep into the ground. Although each ray is energetic, the number bombarding the planet is so small that the total energy reaching the earth is about the same as that from starlight.

Costo- (prefix). Pertaining to the ribs.

Cottrell. An electrostatic precipitator used to separate small particles from an exhaust airstream.

Cottrell precipitator. A device for dust collection using high-voltage electrodes.

Coulometry. Measurement of the number of electrons that are transferred across an electrode solution interface when a reaction in the solution is created and carried to completion. The reaction is usually caused by a contaminant in a sample gas that is drawn through or onto the surface of the solution. The number of electrons transferred in terms of coulombs is an indication of the contaminant concentrations.

Count. A click in a Geiger counter or the numerical value for the activity of a radioactive specimen.

Counter. A device for counting. *See* Geiger counter; Scintillation counter.

Count median size. The size of the particle in a sample of particulate matter containing equal numbers of particles larger and smaller than the stated size.

Covered electrode. A composite filler metal electrode consisting of a core of bare electrode or metal-cored electrode to which a covering (sufficient to precede a slag layer on the weld metal) has been applied. The covering may contain materials providing such functions as shielding from the atmosphere, deoxidation, and arc stabilization and can serve as a source of metallic additions to the weld.

CPR. Cardiopulmonary resuscitation.

Cps. Cycles per second, now called hertz.

CPSC. Consumer Product Safety Commission. U.S. agency with responsibility for regulating hazardous materials when they appear in consumer goods.

Cracking. Process used almost exclusively in the petroleum industry. Thermal or catalytic decomposition of organic compounds, usually for the manufacture of gasoline. Petroleum constituents are also cracked for the purpose of manufacturing chemicals.

Cramps. Painful muscular contractions that may affect almost any voluntary or involuntary muscle.

Cranio- (prefix). Skull. As in craniotomy, incision through a skull bone.

Cristobalite. A crystalline form of free silica, extremely hard and inert chemically, and very resistant to heat. Quartz in refractory bricks and amorphous silica in diatomaceous earth are altered to cristobalite when exposed to high temperatures (calcined).

Critical pressure. The pressure under which a substance may exist as a gas in equilibrium with the liquid at the critical temperature.

Critical temperature. The temperature above which a gas cannot be liquefied by pressure alone.

Crucible. A heat-resistant barrel-shaped pot used to hold metal during melting in a furnace or in other applications.

Crude petroleum. Hydrocarbon mixtures that have a flash point below 150°F (65.6°C) and that have not been processed in a refinery.

Cry-, cryo- (prefix). Very cold.

Cryogenics. The field of science dealing with the behavior of matter at very low temperatures.

CSP. Certified Safety Professional, a designation from the Board of Certified Safety Professionals.

CTD. *See* Cumulative trauma disorder.

Cubic centimeter (cm³). A volumetric measurement that is equal to 1 milliliter (ml).

Cubic meter (m³). A measure of volume in the metric system.

Culture (biology). A population of microorganisms or tissue cells grown in a medium.

Culture medium. Any substance or preparation suitable for the growth of cultures and cultivation of microorganisms. Selective medium, a medium composed of nutrients designed to allow growth of a particular type of microorganism; broth medium, a liquid medium; agar medium, solid culture medium.

Cumulative trauma disorder (CTD). A disorder of a musculoskeletal or nervous system component caused or aggravated by repeated and/or forceful movements of the same system.

Curie. A measure of the rate at which a radioactive material decays. The radioactivity of 1 gram of radium is a curie. It is named for Pierre and Marie Curie, pioneers in radioactivity and discoverers of the elements radium, radon, and plutonium; 1 curie corresponds to 37 billion disintegrations per second.

Current. Flow of electrons in an electrical circuit measured in amperes (amps).

Cutaneous. Pertaining to or affecting the skin.

Cuticle. The superficial scarfskin or upper strata of skin.

Cutie-pie. A portable instrument equipped with a direct-reading meter used to determine the level of ionizing radiation in an area.

Cutting fluids (oils). As used in industry today, an oil or an oil–water emulsion used to cool and lubricate a cutting tool. Cutting oils are usually light or heavy petroleum fractions.

CW laser. Continuous wave laser.

Cyan- (prefix). Blue.

Cyanosis. Blue appearance of the skin, especially on the face and extremities, indicating a lack of sufficient oxygen in the arterial blood.

Cyclone. A centrifugal force device used to separate particles from an airstream; a sampling instrument used to collect respirable particles.

Cyclone separator. A dust-collecting device that has the ability to separate particles by size. Typically used to collect respirable dust samples.

Cyclotron. A particle accelerator. In this atomic "merry-go-round," atomic particles are whirled around in a spiral between the ends of a huge magnet, gaining speed with each rotation in preparation for their assault on the target material.

Cyst- (prefix). Pertaining to a bladder or sac, normal or abnormal, filled with gas, liquid, or semisolid material. The term appears in many words concerning the urinary bladder (cystocele, cystitis).

Cyto- (prefix). Cell.

Cytoplasm. Cell plasma (protoplasm) that does not include the nucleus of the cell.

Cytotoxin. A substance developed in the blood serum, having a toxic effect on cells.

Damage risk criterion. The suggested baseline of noise tolerance, which, if not exceeded, should result in no hearing loss due to noise. A damage risk criterion may include in its statement a specification of such factors as time of exposure, noise level, frequency, amount of hearing loss considered significant, percentage of the population to be protected, and method of measuring the noise.

Damp.   A harmful gas or mixture of gases occurring in coal mining.

Dampers.   A form of gate or vale in a duct.

Dangerous to life or health, immediately (IDLH).   Used to describe very hazardous atmospheres where employee exposure can cause serious injury or death within a short time or serious delayed effects.

Daughter.   As used in radioactivity, this refers to the product nucleus or atom resulting from decay of the precursor or parent.

dBA.   Sound level in decibels read on the A scale of a sound-level meter. The A scale discriminates against very low frequencies (as does the human ear) and is therefore better for measuring general sound levels. *See also* Decibel.

dBC.   Sound level in decibels read on the C scale of a sound-level meter. The C scale discriminates very little against very low frequencies. *See also* Decibel.

DC.   *See* Direct current.

Decay.   When a radioactive atom disintegrates, it is said to decay. What remains is a different element. An atom of polonium decays to form lead, ejecting an alpha particle in the process.

Decibel (dB).   A unit used to express sound power level, L_o. Sound power is the total acoustic output of a sound source in watts. The decibel is defined as a ratio: $L_o = 10 \log W/W_o$, where W is the sound power of the source and W_o is a small reference sound power.

Decipol.   (From "pollution.") A unit of perceived outdoor air quality. For example, air on mountains or at sea has a decipol = 0.01. In cities with moderate air pollution, air quality in decipols = 0.05 to 0.3. Acceptable air quality indoors = 1.4 decipols (for 80% satisfaction.) *See also* olf.

Decomposition.   The breakdown of a chemical or substance into different parts or simpler compounds. Decomposition can occur because of heat, chemical reaction, decay, etc.

Decontaminate.   To make safe by eliminating poisonous or otherwise harmful substances, such as noxious chemicals or radioactive material.

Degree-days.   A term used to estimate the required heating or cooling for a location; one heating degree-day is the difference between the actual temperature and the target temperature. For example, if the outside temperature is 40° and the inside temperature is desired to be 70°, then there are 30 degree-days. The number of degree-days are added over an entire heating season (e.g., the total heating degree-days for Chicago is 7468 when using daily mean outdoor winter temperatures and an indoor temperature of 70°F).

Deltoid muscle.   The muscle of the shoulder responsible for abducting the arm sideways and for swinging the arm at the shoulder. Overuse of the deltoid muscle may cause fatigue and pain in the shoulder.

Density. The ratio of mass to volume. Also, weight density. The ratio of the mass of material to its volume (e.g., air at standard conditions has a weight density of 0.075 lb/ft^3, and water has a density of 62.4 lb/ft^3, or 8.32 lb/gal).

Density correction (d). A factor to "correct" air density at any temperature and pressure to equivalent conditions at standard conditions (e.g., for ventilation: 70°F, 29.92 in. Hg, dry air), and vice versa. For example,

$$\text{Actual air density} = (0.075 \text{ lb/ft}^3) \cdot \text{d}.$$

where
$$d = \frac{530}{460 + T} \cdot \frac{BP}{29.92 \text{ in. HG}}$$

T = temperature in °F, BP = barometric pressure in in. Hg.

Dent-, dento- (prefix). Pertaining to a tooth or teeth, from Latin.

Derma- (prefix). The corium or true skin.

Dermatitis. Inflammation of the skin from any cause.

Dermatology. Branch of medicine concerned with the diagnosis and treatment, including surgery and prevention, of the skin, hair, and nails.

Dermatophytosis. Athlete's foot.

Dermatosis. A broader term than dermatitis, it includes any cutaneous abnormality. Thus it encompasses folliculitis, acne, pigmentary changes, and nodule sand tumors.

Desiccant. Material that absorbs moisture.

Deuterium. Heavy hydrogen. It is called heavy hydrogen because it weighs twice as much as ordinary hydrogen.

Deuteron. The nucleus of an atom of heavy hydrogen containing one proton and one neutron. Deuterons are often used for the bombardment of other nuclei.

Diagnostic X-ray system. An X-ray system designed for irradiation of any part of the human body for the purpose of diagnosis or visualization.

Diaphragm. (1) The musculomembranous partition separating the abdominal and thoracic cavities. (2) Any separating membrane or structure. (3) A disk with one or more openings, or with an adjustable opening, mounted in relation to a lens, by which part of the light may be excluded from the area.

Diatomaceous earth. A soft, gritty amorphous silica composed of minute siliceous skeletons of small aquatic plants. Used in filtration and decolorization of liquids, insulation, filler in dynamite, wax, textiles, plastics, paint, and rubber. Calcined and flux-calcined diatomaceous earth contains appreciable amounts of cristobalite, and dust levels should be controlled the same as for cristobalite.

Die. A hard metal or plastic form used to shape material to a particular contour or section.

Differential pressure. The difference in static pressure between two locations.

Diffuser. An outlet designed to distribute air in varying directions and planes.

Diffuse sound field.   One in which the time average of the mean-square sound pressure is everywhere the same and the flow of energy in all directions is equally probable.

Diffusion, molecular.   A process of spontaneous intermixing of different substances attributable to molecular motion and tending to produce uniformity of concentration.

Diffusion rate.   A measure of the tendency of one gas or vapor to disperse into or mix with another gas or vapor. This rate depends on the density of the vapor or gas as compared with that of air.

Dike.   A barrier constructed to control or confine solid or liquid substances and prevent or control their movement.

Dilution.   The process of increasing the proportion of solvent or diluent (liquid) to solute or particulate matter (solid).

Dilution ventilation.   *See* General ventilation.

Diopters.   A measure of the power of a lens or prism, equal to the reciprocal of its focal length in meters.

Direct current (DC).   Electricity flowing in one direction only.

Direct-reading instrumentation.   Instruments that give an immediate indication of the concentration of aerosols, gases, or vapors or the magnitude of physical hazard by some means, such as a dial or meter.

Disease.   A departure from a state of health, usually recognized by a sequence of signs and symptoms.

Disinfectant.   An agent that frees from infection by killing the vegetative cells of microorganisms.

Disintegration.   A nuclear transformation or decay process that results in the release of energy in the form of radiation.

Dispersion.   The general term describing systems consisting of particulate matter suspended in air or other fluid; also, the mixing and dilution of contaminant in the ambient environment.

Distal.   Away from the central axis of the body.

Distal phalanx.   The last bony segment of a toe or finger.

Distillery.   A facility or that portion of a facility where flammable or combustible liquids produced by fermentation are concentrated and where the concentrated products may also be mixed, stored, or packaged.

DNA.   Deoxyribonucleic acid. The genetic material within the cell.

DOL.   U.S. Department of Labor; includes the Occupational Safety and Health Administration (OSHA) and Mine Safety Health Administration (MSHA).

DOP.   Dioctyl phthalate, a powered chemical that can be aerosolized to an extremely uniform size, i.e., 0.3 μm for a major portion of any sample.

Dose.   (1) Used to express the amount of a chemical or of ionizing radiation energy absorbed in a unit volume or an organ or individual. Dose rate is the dose delivered per unit of time. (*See also* Roentgen, Rad, Rem.) (2) Used to express the amount of exposure to a chemical substance.

Dose, absorbed.   The energy imparted to matter in a volume element by ionizing radiation divided by the mass of irradiated material in that volume element.

Dose equivalent.   The product of absorbed dose, quality factor, and other modifying factors necessary to express on a common scale, for all ionizing radiations, the irradiation incurred by exposed persons.

Dose equivalent, maximum permissible dose (MPD).   The largest dose equivalent received within a specified maximum permissible period that is permitted by a regulatory agency or other authoritative group on the assumption that receipt of such a dose equivalent creates no appreciable somatic or genetic injury. Different levels of MPD may be set for different groups within a population. (In popular usage, "dose, maximum permissible" is an accepted synonym.).

Dose–response relationship.   Correlation between the amount of exposure to an agent or toxic chemical and the resulting effect on the body.

Dosimeter (dose meter).   An instrument used to determine the full-shift exposure a person has received to a physical hazard.

DOT.   Department of Transportation.

Double insulation.   A method of encasing electric components of tools so that the operator cannot touch parts that could become energized during normal operation or in the event of tool failure.

Drier.   Any catalytic material that, when added to drying oil, accelerates drying or hardening of the film.

Drop forge.   To forge between dies using a drop hammer or drop press.

Droplet.   A liquid particle suspended in a gas. The liquid particle is generally of such size and density that it settles rapidly and remains airborne for an appreciable length of a time only in a turbulent atmosphere.

Dross.   The scum that forms on the surface of molten metals, consisting largely of oxides and impurities.

Dry-bulb temperature.   The temperature of air measured with a dry-bulb thermometer.

Dry-bulb thermometer.   An ordinary thermometer, especially one with an unmoistened bulb, not dependent on atmospheric humidity. The reading is the dry-bulb temperature.

Dry chemical.   A powered fire-extinguishing agent usually composed of sodium bicarbonate, monoammonium phosphate, potassium bicarbonate, etc.

Duct.   A conduit used for conveying air at low pressures.

Ductile.   Capable of being molded or worked, as metals.

Duct velocity.   Air velocity through the duct cross-section. When solid particulate material is present in the airstream, the duct velocity must exceed the minimum transport velocity.

Dust.   Solid particles generated by handling, crushing, grinding, rapid impact, detonation, and decrepitation of organic of inorganic materials, such as rock, ore, metal, coal, wood, and gain. Dusts do not tend to flocculate, except under electrostatic forces; they do not diffuse in air but settle under the influence of gravity.

Dust collector.   An air-cleaning device to remove heavy particulate loadings from exhaust systems before discharge to the outdoors; usual range is loadings of 0.003 g/ft^3 (0.007 mg/m^3) and higher.

Dynometer. Apparatus for measuring force or work output external to a subject. Often used to compare external output with associated physiological phenomena to assess physiological work efficiency.

Dys- (prefix). Difficult, bad. This prefix occurs in a large number of medical words because it is attachable to a term for any organ of process that is not functioning as well as it should.

Dysfunction. Disturbance, impairment, or abnormality of the functioning of an organ.

Dyspnea. Shortness of breath, difficult or labored breathing. More strictly, the sensation of shortness of breath.

Dysuria. Difficulty or pain in urination.

EAP. Employee Assistance Program.

Ear. The entire hearing apparatus, consisting of three parts: external ear (or pinna), the middle ear or tympanic cavity, and the inner ear or labyrinth. Sometimes the pinna is called the ear.

Ecology. The science of the relationships between living organisms and their environments.

Economizer cycle. The use of up to 100% outside air (OA) during favorable weather conditions, e.g., temperatures 55 to 70°F; may cause problems when ambient air has high humidity.

-ectomy (suffix). A cutting out; surgical removal. Denotes any operation in which all or part of a named organ is cut out of the body.

Eczema. A skin disorder. Dermatitis.

Edema. A swelling of body tissues as a result of waterlogging with fluid.

Effective temperature (ET). An arbitrary index that combines into a single value the effects of temperature, humidity, and air movement on the sensation of warmth and cold on the human body.

Effective temperature index. An empirically determined index on the degree of warmth perceived on exposure to different combinations of temperature, humidity, and air movement. The determination of effective temperature requires simultaneous determinations of dry-bulb and wet-bulb temperatures.

Efficiency, fractional. The percentage of particles of a specified size that are removed and retained by a particular type of collector or sampler. A plot of fractional efficiency values vs. the respective sized particles yields a fractional efficiency curve that may be related to the total collecting efficiency of air-cleaning or air-sampling equipment.

Efflorescence. A phenomenon where by a whitish crust of fine crystals forms on a surface. These are usually sodium salts that diffuse from the substrate.

Effluent. Generally something that flows out or forth, like a stream flowing out into a lake. In terms of pollution, an outflow of a sewer, storage tank, canal, or other channel.

Ejector. An air mover consisting of a two-flow system wherein a primary source of compressed gas is passed through a Venturi and the vacuum developed at the throat of the Venturi is used to create a secondary flow of fluid. In the case of air movers for sampling applications, the secondary flow is the sample gas.

Elastomer. In a chemical industry sense, a synthetic polymer with rubberlike characteristics; a synthetic or natural rubber or a soft, rubbery plastic with some degree of elasticity at room temperature.

Electrical current. The flow of electricity measured in amperes.

Electrical precipitators. A device that removes particles from an airstream by charging the particles and collecting the charged particles on a suitable surface.

Electrolysis. The process of conduction of an electric current by means of a chemical solution.

Electromagnetic field. The propagation of varying magnetic and electrical fields through space; often associated with electrical equipment and magnets (e.g., video terminals, electric motors, transformers).

Electromagnetic radiation. The propagation of varying electric and magnetic fields through space at the speed of light, exhibiting the characteristics of wave motion.

Electron. A minute atomic particle possessing a negative electric charge. In an atom the electrons rotate around a nucleus. The weight of an electron is so infinitesimal that it would take 500 octillion (500 followed by 27 zeros) of them to make a pound. It is only about a two-thousandth of the mass of a proton or neutron.

Electron volt (eV). A small unit of energy. An electron gains this much energy when it is acted upon by a volt. Energies of radioactive materials may be millions of electron volts (MeV), whereas particle accelerators generate energies of billions of electron volts (BeV).

Electroplate. To cover with a metal coating (plate) by means of electrolysis.

Electrostatic precipitator (ESP). A collector that removes aerosols from an airstream by charging the particle (usually with a wire) and collecting the charged particles on an oppositely charged collector (usually a flat plate).

Element. Solid, liquid, or gaseous matter that cannot be further decomposed into simpler substances by chemical means. The atoms of an element may differ physically but do not differ chemically. All atoms of an element contain a definite number of protons and thus have the same atomic number.

ELF. Extremely low frequency electromagnetic field.

Elutriator. A device used to separate particles according to mass and aerodynamic size by maintaining a laminar flow system at a rate that permits the particles of greatest mass to settle rapidly while the smaller particles are kept airborne by the resistance force of the flowing air for longer times and distances. The various times and distances of deposit may be used to determine representative fractions of particle mass and size.

Embryo. The name for the early stage of development of an organism. In humans, the period from conception to the end of the second month.

Emergency plan. A plan of action for an anticipated, unwanted, occurrence/disaster. (e.g., evacuation plans, emergency equipment, employee first aid, chemical showers, etc.).

Emergent beam system. Diameter of a laser beam at the exit aperture of the system.

Emery. Aluminum oxide, natural and synthetic abrasive.

Emission. A release of some by-product or product from an operation or process.

Emission factor. Statistical average of he amount of a specific pollutant emitted from each type of polluting source in relation to a unit quality of material handled, processed, or burned.

Emission inventory. A list of primary air pollutants emitted into a given community's atmosphere, in amounts per day, by type of source.

Emission standards. The maximum amount of pollutant permitted to be discharged from a single polluting source.

Emmetropia. A state of perfect vision.

Emphysema. A lung disease in which the walls of the air sacs (alveoli) have been stretched too thin and have broken down.

Emulsifier or emulsifying agent. A chemical that holds one insoluble liquid in suspension in another. Casein, for example, is a natural emulsifier in milk, keeping butterfat droplets dispersed.

Emulsion. A suspension, each in the other, of two or more unlike liquids that usually do not dissolve in each other.

Enamel. A paintlike oily substance that produces a glossy finish to a surface to which it is applied, often containing various synthetic resins. It is lead free, in contrast to ceramic enamel, that is, porcelain enamel, which contains lead.

Encapsulate. To cover or coat over with another substance.

Endemic. (1) Present in a community or among a group of people; usually refers to a disease prevailing continually in a region. (2) The continuing prevalence of a disease, as distinguished from an epidemic.

Endo- (prefix). Within, inside of, internal. The endometrium is the lining membrane of the uterus.

Endocrine. Secreting without the means of a duct or tube. The term is applied to certain glands that produce secretions that enter the bloodstream or the lymph directly and are then carried to the particular gland or a tissue whose function they regulate.

Endothermic. Characterized by or formed with absorption of heat.

Endotoxin. A toxin that is part of the wall of a microorganism and is released when that organism dies.

Energy-control program. A program consisting of an energy-control procedure and employee training to ensure that a machine or equipment is isolated and inoperative before servicing or maintenance, thus protecting the employee from unexpected machine start-up or energizing. *See also* Lockout/tagout.

Energy density. The intensity of electromagnetic radiation per unit area per pulse expressed in joules per square centimeter.

Energy isolation. A control procedure consisting of physical locks, blinds, restraints, or other methods to prevent stored or accessible energy sources from coming in contact with the employee.

Energy-isolation device. A mechanical device that physically prevents the release or transmission of energy. Some examples of energy-isolation devices include a manually operated circuit breaker, a disconnect switch, a line valve, a block or blank, a blind, or other similar device. The following are not energy-isolation devices; push-buttons, on/off switches, selector switches.

Engineering controls. Methods of controlling employee exposures by modifying the source or reducing the quantity of contaminants released into the work environment.

Enteric. Intestinal.

Entero- (prefix). Pertaining to the intestines.

Enterotoxin. A toxin specific for cells of the intestine; gives rise to symptoms of food poisoning.

Entrainment velocity. The gas flow velocity, which tends to keep particles suspended and to cause deposited particles to become airborne.

Entrance loss. The loss in static pressure of a fluid that flows from an area into and through a hood or duct opening. The loss in static pressure is caused by friction and turbulence resulting from the increased gas velocity and configuration of the entrance area.

Entry loss. The loss in pressure caused by air flowing into a duct or hood.

Environmental toxicity. Information obtained as a result of conducting environmental testing designed to study the effects on aquatic and plant life.

Enzymes. Chemical substances, mostly proteins, that enter into and bring about chemical reactions in living organisms.

EOE. Equal Opportunity Employer.

EPA. Environmental Protection Agency.

EPA number. The number assigned to chemicals regulated by the Environmental Protection Agency.

Epicondylitis. Inflammation of certain bony prominences in the area of the elbow, for example, tennis elbow.

Epidemiology. The study of disease in human populations.

Epidermis. The superficial scarfskin or upper (outer) layer of skin.

Epilation. Temporary or permanent loss of body hair.

Epithelioma. Carcinoma of the epithelial cells of the skin and other epithelial surfaces.

Epithelium. The purely cellular, avascular layer covering all the free surfaces — cutaneous, mucous, and serous including the glands and other structures derived therefrom; for example, the epidermis.

Equivalent chill temperature (ECT). Also known as wind-chill index. A temperature index used to account for heat loss from skin exposed to the combined effects of cold temperatures and air speed.

erg.   The force of 1 dyne acting through a distance of 1 centimeter. It would be equivalent to the work done by a June bug climbing over a stone 0.5 in. (1 cm) high, or the energy required to ionize about 20 billion molecules of air.

Ergonomics.   A multidisciplinary activity dealing with interactions between humans and their total working environment plus stresses related to such environmental elements as atmosphere, heat, light, and sound as well as all tools and equipment of the workplace.

Erysipeloid.   A bacterial infection affecting slaughterhouse workers and fish handlers.

Eryth-, erythro- (prefix).   Redness. Erythema is indicated by redness of the skin (including a deep blush). An erythrocyte is a red blood cell.

Erythema.   Reddening of the skin.

Erythemal region.   Ultraviolet light radiation between 2800 and 3200 angstroms (280 to 320 millimicrons); it is absorbed by the cornea of the eye.

Erythrocyte.   A type of red blood corpuscle.

Eschar.   The crust formed after injury by a caustic chemical or heat.

ESP.   *See* Electrostatic precipitator.

Essential oil.   Any of a class of volatile, odoriferous oils found in plants and imparting to the plants odor and often other characteristic properties. Used in perfumery, for example.

Esters.   Organic compounds that may be formed by interaction between an alcohol and an acid, or by other means. Esters are nonionic compounds, including solvents and natural fats.

Etch.   To cut or eat away material with acid or another corrosive substance.

Ethylene oxide.   A carcinogenic hospital sterilant regulated by OSHA. Ethylene oxide is also a reproductive hazard.

Etiologic agent.   Refers to organisms, substances, or objects associated with the cause of disease or injury.

Etiology.   The study or knowledge of the causes of disease.

Eu- (prefix).   Well and good. A euthyroid person has a thyroid gland that could not be working better. A euphoric person has a tremendous sense of well-being.

Eukaryote.   An organism whose cells contain mitochondria and a nuclear membrane. Describes organisms from yeast to humans.

Eustachian tube.   A structure about 2.5 in. (6 cm) long leading from the back of the nasal cavity to the middle ear. It equalizes the pressure of air in the middle ear with that outside the eardrum.

Evaporation.   The process by which a liquid is changed to the vapor state.

Evaporation rate.   The ratio of the time required to evaporate a measured volume of a liquid to the time required to evaporate the same volume of a reference liquid (ethyl ether) under ideal test conditions. The higher the ratio, the slower the evaporation rate.

Evase.   A cone-shaped exhaust stack that recaptures static pressure from velocity pressure. Pronounced (eh-va-say).

Exhalation valve.   A device that allows exhaled air to leave a respirator and prevents outside air from entering through the valve.

Exhaust ventilation.   The removal of air, usually by mechanical means, from any space. The flow of air between two points is due to a pressure difference. The pressure difference causes air to flow from the high-pressure to the low-pressure zone.

Exogenous.   Growing or developing outside the body; originating outside the body.

Exothermic, exothermal.   Characterized by or formed with evolution of heat.

Exotoxin.   A toxin excreted by a microorganism into the surrounding medium.

Explosion.   A reaction that causes a sudden, almost instantaneous release of pressure, gas, and heat.

Explosive limit.   *See* Flammable limit.

Exposure.   Contact with a chemical, biological, or physical hazard.

Exposure routes.   Usually by inhalation, ingestion, or contact with the skin.

Extension.   Movement whereby the angle between the bones connected by a joint is increased. Motions of this type are produced by contraction of extensor muscles.

Extensor muscles.   A muscle that, when active, increases the angle between limb segments, for example, the muscles that straighten the knee or elbow, open the hand, or straighten the back.

Extensor tendon.   Connecting structure between an extensor muscle and the bone into which it inserts. Examples are the hard, longitudinal tendons found on the back of the hand when the fingers are fully extended.

External mechanical environment.   The synthetic physical environment, for example, environment equipment, tools, machine controls, clothing. Antonym: internal (bio)mechanical environment.

Extinguishing media.   The fire-fighting substance used to stop combustion. It is usually referred to by its generic name, such as CO_2, foam, water, dry chemical, etc.

Extravasate.   To exude a substance from the body's vessels into tissues.

Extrusion.   The forcing of raw material through a die or a form in either a hot or cold state, in a solid state, or in partial solution. Long used with metals and clays, it is now extensively used in the plastics industry.

Eyepiece.   Gas-tight, transparent window in a full face piece through which the wearer may see.

Eye protection.   "Safety" glasses, goggles, faces shield, etc., used to protect against physical, chemical, and nonionizing radiation hazards.

Face piece.   The portion of a respirator that covers the wearer's nose and mouth (in a half-mask face piece), or the nose, mouth, and eyes (in a full face piece). It is designed to make a gas-tight or dust-tight fit with the face and includes the headbands, exhalation valves, and connections for an air-purifying device or respirable gas source, or both.

Face velocity.   Average air velocity into the exhaust system measured at the opening into the hood or booth.

Facilitator. A person who makes learning easier, assists interactions and the execution of tasks, and clarifies goals and processes.

Facing. In foundry work, the final touch-up work of the mold surface to come in contact with metal is called the facing operation, and the fine powdered material used is called the facing.

Factor of safety (safety factor). The ratio of ultimate strength of a material or structure to the specified stress available.

Fainting. Technically called syncope, a temporary loss of consciousness as a result of a diminished supply of blood to the brain.

Fallout. Dust particles that contain radioactive fission products resulting from a nuclear explosion. The wind can carry fallout particles many miles.

Fan. A mechanical device that physically moves air and creates static pressure.

Fan coil unit. A single-room ventilation system, consisting of a fan, coils, and dampers. Usually mounted under windows, they often assist HVAC systems in rooms located on the periphery of a building.

Fan laws. Statements and equations that describe the relationship among fan volume, pressure, brake horsepower, size, and rotating speed.

Fan laws (rpm). Equations that describe the relationships between pressure, flow, horsepower, and fan rotations per minute (rpm).

Fan rating curve or table. Data that describe the volumetric output of a fan at different static pressures.

Fan static pressure. The pressure added to the system by the fan. It equals the sum of pressure losses in the system minus the velocity pressure in the air at the fan inlet.

Fan total pressure (FTP). The total pressure difference between the inlet and outlet of a fan; $FTP = SP_{out} - SP_{in}$ when $VP_{in} = VP_{out}$.

Fan types. Two families: axial and centrifugal. Three types of centrifugal: radial, backward inclined, and forward inclined.

Far field (free field). In noise measurement, this refers to the distance from the noise source where the sound-pressure level decreases 6 dBA for each doubling of distance (inverse square law).

Farmer's lung. Fungus infection and ensuing hypersensitivity from grain dust.

FDA. The U.S. Food and Drug Administration. The FDA establishes requirements for the labeling of foods and drugs to protect consumers from improperly labeled, unwholesome, ineffective, and hazardous products. FDA also regulates materials for food contact service and the condition under which such materials are approved.

Federal Register. Publication that compiles U.S. government documents officially promulgated under the law, documents whose validity depends upon such publication. It is published on each day following the end of a government working day. It is, in effect, the daily supplement to the Code of Federal Regulations (CFR).

Feral animal. A wild animal, or a domestic animal that has reverted to the wild state.

Fertilizer. Plant food usually sold in a mixed formula containing basic plant nutrients: compounds of nitrogen, potassium, phosphorus, sulfur, and sometimes other minerals.

Fetus. The term used to describe the developing organism (human) from the third month after conception to birth.

FEV. Forced expiratory volume.

Fever. A condition in which the body temperature is above its regular or normal level.

Fibrillation. Very rapid irregular contractions of the muscle fibers of the heart resulting in a lack of synchronism of the heartbeat.

Fibrosis. A condition marked by an increase of interstitial fibrous tissue. Exposures to contaminants via inhalation can lead to fibrosis or scarring of the lung, a particular concern in industrial hygiene.

Film badge. A piece of masked photographic film worn by nuclear workers. It is darkened by nuclear radiation, and radiation exposure can be checked by inspecting the film.

Filter. (1) A device for separating components of a signal on the basis of its frequency. It allows components in one or more frequency bands to pass relatively unattenuated, and it greatly attenuates components in other frequency bands. (2) A fibrous medium used in respirators to remove solid or liquid particles from the airstream entering the respirators. (3) A sheet of material that is interposed between patient and the source of X rays. (4) A fibrous or membrane medium used to collect dust, fumes, or mist air samples.

Filter efficiency. The efficiency of various filters can be established on the basis of entrapped particles (that is, collection efficiency), or on the basis of particles passed through the filter (that is, penetration efficiency).

Filter, HEPA. Hgh-efficiency particulate air filter, one that is at least 99.97% efficient in removing thermally generated monodisperse dioctyl phthalate smoke particles with a diameter of 0.0003 mm.

Firebrick. A special clay that is capable of resisting high temperatures without melting or crumbling.

Fire Brigade. An organized group trained in fire-fighting operations.

Fire damp. In mining, the accumulation of an explosive gas, chiefly methane gas. Miners call all dangerous underground gases "damps."

Fire point. The lowest temperature at which a material can evolve vapors to support continuous combustion.

Fire resistant. Material that is difficult to burn, has difficulty supporting combustion, and offers some protection from fire. Commonly misunderstood as "unable to burn" or "fireproof."

Fireproof. Material incapable of burning without sustained heat and energy source. The term "fireproof" is false. No material is immune to the effects of fire possessing sufficient intensity and duration. The term is commonly, although erroneously, used synonymously with fire resistant.

Fission. The splitting of an atomic nucleus into two parts accompanied by the release of a large amount of radioactivity and heat. Fission reactions occur only with heavy isotopes, such as uranium-233, uranium-235, and plutonium-239.

Fissionable. A nucleus that undergoes fission under the influence of neutrons, even very slow neutrons.

Fission product. The highly radioactive nuclei into which a fissionable nucleus splits (fissions) under the influence of neutron bombardment.

Flagellum. A flexible, whiplike appendage on cells used as an organ of locomotion.

Flame ionization detector (FID). A direct-reading monitoring device that ionizes gases and vapors with an oxyhydrogen flame and measures the differing electrical currents generated.

Flame-proofing material. Chemicals that catalytically control the decomposition of cellulose material at flaming temperature. Substances used as fire retardants are borax–boric acid, borax–boric acid diammonium phosphate, ammonium bromide, stannic acid, antimony oxide, and combinations containing formaldehyde.

Flame propagation. *See* Propagation of flame.

Flammable. Any substance that is easily ignited, burns intensely, or has a rapid rate of flame spread. Flammable and inflammable are identical in meaning; however, because the prefix "in" indicates the negative in many words and can cause confusion, flammable is the preferred term.

Flammable aerosol. An aerosol that is required to be labeled *Flammable* under the Federal Hazardous Substances Labeling Act (15 USC 1261).

Flammable limits. Also, upper and lower explosive limits: UEL, LEL. The concentration (in percent) of a gas or vapor in air in which flames will propagate. When the concentration is below the LEL, the mixture is too lean; when above the UEL, the mixture is too rich. The concentrations between the LEL and the UEL are called the "explosive range."

Flammable liquid. A liquid with a flash point below 100°F.

Flammable range. The difference between the lower and upper flammable limits, expressed in terms of percentage of vapor or gas in air by volume, also often called the explosive range.

Flange. A rim of edge added to a hood to reduce the quantity of air entering from behind the hood.

Flashback. Occurs when flame from a torch burns back into the tip, the torch, or the hose in a oxygen-acetylene cutting rig.

Flash blindness. Temporary visual disturbance resulting from viewing an intense light source.

Flash point. The lowest temperature at which a liquid gives off enough vapor to form an ignitable mixture with air and produce a flame when a source of ignition is present. Two tests are used: open cup and closed cup.

Flask. In foundry work, the assembly of the cope and the drag constitutes the flask. It is the wooden or iron frame containing sand into which molten metal is poured. Some flasks may have three or four parts.

Flexion. Movement where by the angle between to bones connected by a joint is reduced. Motions of this type are produced by contraction of flexor muscles.

Flexor muscles. A muscle that, when contracting, decreases the angle between limb segments. The principal flexor of the elbow is the brachialis is muscle. Flexors of the fingers and the wrist are the large muscles of the forearm originating at the elbow. *See also* extensor muscles.

Flocculation. The process of forming a very fluffy mass of material held together by weak forces of adhesion.

Flocculator. A device for aggregating fine particles.

Floor load. (1) The weight that may be safely placed on a floor without danger of structural collapse. (2) The actual load (weight) placed on a floor.

Flora, microflora. Microorganisms present in a given situation (such as intestinal flora, soil flora).

Flotation. A method of ore concentration in which the mineral is caused to float due to chemical frothing agents while the impurities sink.

Flotation reagent. Chemical used in flotation separation of minerals. Added to a pulverized mixture of solids and water and oil, it caused preferential nonwetting by water of certain solid particles, making possible the flotation and separation of nonwet particles.

Flow coefficient. A correction factor used for figuring the volume flow rate of a fluid through an orifice. This factor includes the effects of contraction and turbulence loss (covered by the coefficient of discharge), plus the compressibility effect and the effect of an upstream velocity other than zero. Because the latter two effects are negligible in many instances, that flow coefficient is often equal to the coefficient of discharge. *See also* Coefficient of discharge.

Flowmeter. An instrument for measuring the rate of flow of a fluid or gas.

Flow, turbulent. Fluid flow in which the fluid moves transversely as well as in the direction of the tube or pipe axis, as opposed to streamline or viscous flow.

Fluid. A substance tending to flow or conform to the outline of its container. It may be liquid, vapor, gas, or solid (such as raw rubber).

Fluorescence. Emission of light from a crystal, after the absorption of energy.

Fluorescent screen. A screen coated with a fluorescent substance so that it emits light when irradiated with X-rays.

Fluoroscope. A fluorescent screen mounted in front of an X-ray tube so that internal organs may be examined through their shadow cast by X rays. It may also be used for inspection of inanimate objects.

Fluoroscopy. Examination with the use of an X-ray fluoroscope.

Flux. Usually refers to a substance used to clean surfaces and promote fusion in soldering. However, fluxes of varying chemical nature are used in the smelting of ores, in the ceramic industry, in assaying silver and gold ores, and in other endeavors. The most common fluxes are silica, various silicate, lime, sodium and potassium carbonate, and litharge and red lead in the ceramics industry. *See also* Solder, Galvanize, and Luminous flux.

Fly ash. Finely divided particles of ash entrained in flue gases arising from the combustion of fuel.

Focus (pl. foci). A center or site of a disease process.

Fog. The visible presence of small water droplets suspended in air.

Follicle. A small anatomical cavity or deep, narrow mouthed depression; a small lymph node.

Folliculitis. Infection of a hair follicle, often caused by obstruction by natural or industrial oils.

Fomites. Clothing or other substances that can absorb and transmit contaminants, as in the case of poison ivy.

Footcandle. A unit of illumination. The illumination to a point on a surface that is 1 foot from, and perpendicular to, a uniform point source of one candle.

Foot-pounds of torque. A measurement of the physiological stress exerted upon any joint during the performance of a task. The product of the force exerted and the distance from the point of application to the point of stress. Physiologically, torque that does not produce motion nonetheless causes work stress, the severity of which depends on the duration and magnitude of the torque. In lifting an object or holding it elevated, torque is exerted and applied to the lumbar vertebrae.

Force. That which changes the state of rest or motion of matter. The SI (International System) unit of measurement in the newton (N).

Fovea. A depression or pit in the center or the macula of the eye; it is the area of clearest vision.

Fractionation. Separation of a mixture into different portions or fractions, usually by distillation.

Free sound field (free field). A field in a homogeneous, isotropic medium free from boundaries. In practice, it is a field in which the effects of the boundaries are negligible over the region of interest. *See also* Far field.

Frequency (in Hz). Rate at which pressure oscillations are produced; 1 hertz is equivalent to one cycle per second. A subjective characteristic of sound related to frequency is pitch.

Friable. Readily crumbled or crumbling state. Commonly used to describe airborne asbestos.

Friction factor. A factor used in calculating loss of pressure due to friction of a fluid flowing through a pipe or duct.

Friction loss. The pressure loss caused by friction.

Fuller's earth. A hydrated silica–alumina compound associated with ferric oxide. Used as a filter medium and a catalyst and catalyst carrier and in cosmetics and insecticides.

Fume. Airborne particulate formed by the condensation of solid particles from the gaseous state. Usually, fumes are generated after initial volatilization from a combustion process, or from a melting process (such as metal fumes emitted during welding). Usually less than 1 μm in diameter.

Fume fever, metal. An acute condition caused by a brief, high exposure to the freshly generated fumes of metals, such as zinc or magnesium, or their oxides.

Functional anatomy. Study of the body and its component parts, taking into account structural features directly related to physiological function.

Fundamental frequency. The lowest component frequency of a periodic quantity.

Fundus. The interior surface of a hollow organ, such as the retina of the eye.

Fungus (pl. fungi). Any of a major group of lower plants that lack chlorophyll and live on dead or other living organisms. Fungi include molds, rusts, mildews, smuts, and mushrooms.

Fuse. A wire or strip of metal with known electrical resistance, usually placed in an electrical circuit as a safeguard. As the electrical current increases, the resistance of the metals to flow causes it to heat until it reaches the point where the metal melts, breaking the current at the rated amperage.

Fusion. (1) The joining of atomic nuclei to form a heavier nucleus, accomplished under conditions of extreme heat (millions of degrees). If two nuclei of light atoms fuse, the fusion is accompanied by the release of a great deal of energy. The energy of the sun is believed to be derived from the fusion of hydrogen atoms to form helium. (2) In welding, the melting together of filler metal and base metal (substrate), or of base metal only, which results in coalescence.

FVC. Forced vital capacity.

Galvanizing. An old but still used method of providing a protective coating for metals by dipping them in a bath of molten zinc.

Gamete. A mature germ cell. An unfertilized ovum or spermatozoon.

Gamma rays (gamma radiation). The most penetrating of all radiation. Gamma (γ) rays are very high-energy X rays.

Ganglion (pl. ganglia). A knot or knotlike mass; used as a general term to designate a group of nerve cell bodies located outside the central nervous system. The term is also applied to certain nuclear groups within the brain or spinal cord.

Gangue. In mining or quarrying, useless chipped rock.

Gas. A state of matter in which the material has very low density and viscosity, can expand and contract greatly in response to change in temperature and pressure, easily diffuses into other gases, and readily and uniformly distributes itself throughout any container. A gas can be changed to the liquid or solid state only by the combined effect of increased pressure and decreased temperature (below the critical temperature).

Gas chromatography. A gaseous detection technique that involves the separation of mixtures by passing them through a column that enables the components to be held up for varying periods of time before they are detected and recorded.

Gas metal arc-welding (GMAW). An arc-welding process that produces coalescence of metals by heating them with an arc between a continuous filler metal (consumable) electrode and the work; shielding is obtained entirely from an externally supplied gas or gas mixture. Some methods of this process are called MIG or CO_2 welding.

Gastr-, gastro- (prefix). Pertaining to the stomach.

Gas tungsten arc-welding (GTAW). An arc-welding process that produces coalescence of metals by heating them with an arc between a tungsten (nonconsumable) electrode and the work; shielding is obtained from a gas or gas mixture. Pressure may not be used. This process has sometimes been called TIG welding.

Gate. A groove in a mold to act as a passage for molten metal.

Gauge pressure. The difference between two absolute pressures, one of which is usually atmospheric pressure.

Geiger counter. A gas-filled electrical device that counts the presence of an atomic particle or ray by detecting the ions produced. Sometimes called a Geiger-Muller counter.

General ventilation. System of ventilation consisting of either natural or mechanically induced fresh air movements to mix with and dilute contaminants in the workroom air. This is not the recommended type of ventilation to control contaminants that are toxic.

Generic name. A nonproprietary name for a material or product.

Genes. The ultimate biological units of heredity.

Genetically significant dose (GSD). The dose that, if received by every member of the population, would be expected to produce the same total genetic injury to the population as the actual doses received by the various individuals.

Genetic effects. Mutations or other change produced by irradiation of the germ plasma.

Germ. A microorganism; a microbe usually thought of as a pathogenic organism.

Germicide. An agent capable of killing germs.

GFCI. *See* Ground fault circuit interrupter.

GI. Gastrointestinal.

GI tract. Gastrointestinal tract.

Gingival. Pertaining to the gingivae (gums), the mucous membrane, with the supporting fibrous tissue, that overlies the crowns of unerupted teeth and encircles the necks of those that have erupted.

Gingivitis. Inflammation of the gums.

Gland. Any body organ that manufactures some liquid product and secretes it from its cells.

Globe thermometer. A thermometer set in the center of a metal sphere that has been painted black to measure radiant heat.

Globulin. General name for a group of proteins that are soluble in saline solutions but not in pure water.

Glossa- (prefix). Pertaining to the tongue.

Glove box. A sealed enclosure in which items inside the box are handled with long, impervious glove sealed to ports in the walls of the enclosure.

Gob.   Lump. Gob pole is waste mineral material, such as from coal mines, that contains sufficient coal that gob fires may arise from spontaneous combustion.

Gonads.   The male (testes) and female (ovaries) sex glands.

Grab sample.   A sample taken within a very short time period to determine the constituents at a specific time.

Gram (g).   A metric unit of weight; 1 ounce equals 28.4 grams.

Grams per kilogram (g/kg).   Indication of the dose of a substance given to test animals in toxicity studies.

Granuloma.   A mass or nodule of chronically inflamed tissue with granulations; usually associated with an infective process.

Graticule.   *See* Reticle.

Gravimetric.   Pertaining to measurement by weight.

Gravimetric method.   A procedure dependent upon the formation or use of a precipitate of residue, which is weighed to determine the concentration of a specific contaminant in a previously collected sample.

Gravitation.   The universal attraction existing between all material bodies. The gravitational attraction of the earth's mass for bodies at or near its surface is called gravity.

Gravity, specific.   The ratio of the mass of a unit volume of a substance to the mass of the same volume of a standard substance at a standard temperature. Water at 39.2°F (4°C) is usually the standard substance. For gases, dry air, at the same temperature and pressure as the gas, is often taken as the standard substance.

Gravity, standard.   A gravitational force that produces an acceleration equal to 32.17 ft (9.8 m) per second. The actual force of gravity varies slightly with altitude and latitude. The standard was arbitrarily established as that at sea level and 45° latitude.

Gray (Gy).   Unit of absorbed radiation dose equal to 1 joule of absorbed energy per kilogram of matter; also equal to 100 rad.

Gray iron.   The same as cast iron; in general, any iron with high carbon content.

Grooving.   Designing a tool with grooves on the handle to accommodate the fingers of the user — a bad practice because of the great variation in the sizes of workers' hands. Grooving interferes with sensory feedback. Intense pain may be caused by the grooves to the arthritic hand.

Ground.   A contact with the ground (earth) that becomes part of the electrical circuit.

Ground fault circuit interrupter (GFCI).   A device that measures the amount of current flowing to and from an electrical source. When a difference between the two is sensed, indicating a leakage of current, the device very quickly breaks the circuit.

Grounding.   The procedure used to carry an electrical charge to ground through a conductive path, usually a rod or fixed wire.

Gyn- gyne- (prefix).   Woman, female.

Gynecology.   The medical specialty concerned with diseases of women.

Gyratory crusher.   A device for crushing rock by means of a heavy steel pestle rotating in a steel cone, with the rock fed in at the top and passing out at the bottom.

Half-life, radioactive.   For a single radioactive decay process, the time required for the activity to decrease to half its value by that process.

Half-thickness.   The thickness of a specified absorbing material that reduces the dose rate to one half its original value.

Half-value layer (HVL).   The thickness of a substance necessary to reduce the intensity of a beam of gamma or X rays to half its original value. Also known as half-thickness.

Halogenated hydrocarbon.   A chemical material that has carbon plus one or more of these elements: chlorine, fluorine, bromine, and iodine.

Hammer mill.   A machine for reducing the size of stone or other bulk material by means of hammers usually placed on a rotating axle inside a steel cylinder.

Hand protection.   Coverings worn over the hands to protect against physical, chemical, biological, thermal, and electrical hazards.

Hard hat.   A helmet so constructed as to help prevent head injuries from falling objects of limited size.

Hardness.   A relative term to describe the penetrating quality of radiation. The higher the energy of the radiation, the more penetrating (harder) is the radiation.

Hardness of water.   A degree of hardness is the equivalent of 1 grain of calcium carbonate, $CaCO_3$, in 1 gallon of water.

Hazard.   An unsafe condition or activity that, if left uncontrolled, can contribute to an accident.

Hazard analysis.   An analysis performed to identify and evaluate hazards for the purpose of their elimination or control.

Hazard control.   A program to recognize, evaluate, eliminate, or control the existence of and exposure to hazards.

Hazardous material.   Any substance or compound that has the capacity of producing adverse effects on the health and safety of humans.

Hazwoper.   Hazardous waste operations and emergency response — an OSHA standard intended to protect workers engaged in hazardous waste operations.

Heading.   In mining, a horizontal passage or drift of a tunnel; also the end of a drift or gallery. In tanning, a layer of ground bark over the tanning liquor.

Health.   Personal freedom from physical or mental defect, pain, injury, or disease.

Health physicist.   A professional person specially trained in radiation physics and concerned with problems of radiation damage and protection.

Hearing conservation.   The prevention or minimizing of noise-induced deafness through the use of hearing protection devices; the control of noise through engineering methods, annual audiometric tests, and employee training.

Hearing level.   The deviation in decibels of an individual's threshold from the zero reference of the audiometer.

Hearing loss level.   The decrease in decibels of an individual's hearing from the accepted zero reference of an audiometer. ("He has a hearing loss of 10 decibels at 4000 hertz").

Heat cramps.   Painful muscle spasms as a result of exposure to excess heat.

Heat exhaustion.   A condition usually caused by loss of body water because of exposure to excess heat. Symptoms include headache, tiredness, nausea, and sometimes fainting.

Heat, latent.   The quantity of heat absorbed or given off per unit weight of material during a change of state, such as ice to water or water to steam.

Heat of fusion.   The heat given off by a liquid freezing to a solid or gained by a solid melting to a liquid, without a change in temperature.

Heat of vaporization.   The heat given off by a vapor condensing to a liquid or gained by a liquid evaporating to a vapor, without a change in temperature.

Heat rash.   Itchy skin disorder caused by sweating and inadequate hygiene practices.

Heat, sensible.   Heat associated with a change in temperature; specific heat exchange with environment, in contrast to a heat interchange in which only a change of state (phase) occurs.

Heat stress.   Relative amount of thermal strain from the environment.

Heat stress index (HIS).   Also known as the Belding–Hatch heat stress index, this index combines the environmental heat and metabolic heat into an expression of stress in terms of requirement for evaporation of sweat.

Heatstroke.   A serious disorder resulting from exposure to excess heat. It results from sweat suppression and increased storage of body heat. Symptoms include hot dry skin, high temperature, mental confusion, convulsions, and coma. Heatstroke is fatal if not treated promptly.

Heat syncope.   A heat-related disorder characterized by symptoms of blurred vision and brief fainting spells; caused by pooling of blood in the legs or skin during prolonged static postures in a hot environment.

Heat treatment.   Any of several process of metal modification, such as annealing.

Heavy hydrogen.   Same as deuterium.

Heavy metals.   Metallic elements with high molecular weights.

Heavy water.   Water containing heavy hydrogen (deuterium) instead of ordinary hydrogen. It is widely used in reactors to slow neutrons.

Helmet.   A device that shields the eyes, face, neck, and other parts of the head.

Hem-, hemato- (prefix).   Pertaining to blood. Hematuria means blood in the urine. When the roots occur internally in a word, the *h* is often dropped for the sake of pronunciation, leaving *-em-* to denote blood, as in anoxemia (deficiency of oxygen in the blood).

Hematology.   Study of the blood and the blood-forming organs.

Hematuria.   Blood in the urine.

Hemi- (prefix).   Half. The prefix is straightforward enough in *hemiplegia,* "half paralysis," affecting one side of the body. It is not so evident in *migraine* (one-sided headache), a word that shows how language changes through the centuries. The original word was *hemicrania*, "half-head."

Hemoglobin. The red coloring matter of the blood that carries the oxygen.

Hemolysis. Breakdown of red blood cells with liberation of hemoglobin.

Hemoptysis. Bleeding from the lungs, spitting blood, or blood-stained sputum.

Hemorrhage. Bleeding; especially profuse bleeding as from a ruptured or cut blood vessel (artery or vein).

Hemorrhagic. Pertaining to or characterized by hemorrhage.

HEPA filter. High-efficiency particulate air filter. A disposable, extended-medium, dry-type filter with a particle removal efficiency of no less than 99.97% for 0.3-μm particles.

Hepatitis. Inflammation of the liver.

Hepatitis B. A virus causing hepatitis. The virus may also cause liver cancer in some individuals. The virus is bloodborne and, as such, is one of the agents targeted by the OSHA bloodborne pathogen standard.

Hepatotoxin. Chemicals that produce liver damage.

Herpes. An acute inflammation of the skin or mucous membranes, characterized by the development of groups of vesicles on an inflammatory base.

Hertz. The frequency measured in cycles per second (cps). 1 cps = 1 Hz.

High-frequency loss. Refers to a hearing deficit starting with 2000 Hz and beyond.

HIV. Human immunodeficiency virus. Held to be the initiating case of acquired immunodeficiency syndrome (AIDS).

Hold harmless. A written agreement in which a party absolves or is absolved by another for liability arising from a specific cause.

Homeotherm. Uniform body temperature, or a warm-blooded creature remaining so regardless of environment.

Homogenizer. A machine that forces liquids under high pressure through a perforated shield against a hard surface to blend or emulsify the mixture.

Homoiotherm. *See* Homeotherm.

Hood. (1) Enclosure, part of a local exhaust system. (2) A device that completely covers the head, neck, and portions of the shoulders.

Hood entry loss. The pressure loss from turbulence and friction as air enters the ventilation system.

Hood, slot. A hood consisting of a narrow slot leading into a plenum chamber under suction to distribute air velocity along the length of the slot.

Hood static pressure. The suction or static pressure in a duct near a hood. It represents the suction that is available to draw air into the hood.

Hormones. Chemical substances secreted by the endocrine glands, exerting influence over practically all body activities.

Horsepower. A unit of power, equivalent to 33,000 foot-pounds per minute (746 W). *See also* Brake horsepower.

Host. A plant or animal harboring another as a parasite or as an infectious agent.

Hot. In addition to meaning having a relatively high temperature, a colloquial term meaning highly radioactive.

HSI. *See* Heat stress index.

Human–equipment. Areas of physical or perceptual contact between person and equipment. The design characteristics of the human–equipment interface determine the quality of information. Poorly designed interfaces may lead to excessive fatigue or localized trauma, e.g., calluses.

Human factors. *See* Ergonomics.

Human factors engineering. *See* Human–equipment; Ergonomics.

Humerus. The bone of the upper arm that starts at the shoulder joint and ends at the elbow. Muscles that move the upper arm, forearm, and hand are attached to this bone.

Humidify. To add water vapor to the atmosphere; to add water vapor or moisture to any material.

Humidity. Wetness of the atmosphere. (1) Absolute humidity is the weight of water vapor per unit volume: pounds per cubic foot or grams per cubic centimeter. (2) Relative humidity is the ratio of the actual partial vapor pressure of the water vapor in the space to the saturation pressure of pure water at the same temperature.

Humidity, specfic. The weight of water vapor per unit weight of dry air.

HVAC system. Heating, ventilating, and air-conditioning system. Ventilation system designed primarily for temperature, humidity, odor control, and air quality.

Hyalinization. Conversion into a substance resembling glass.

Hydration. The process of converting raw material into pulp by prolonged beating in water; to combine with water or the elements of water.

Hydrocarbons. Organic compounds composed solely of carbon and hydrogen. Several hundred thousand molecular combinations of C and H are known to exist. Basic building blocks of all organic chemicals. Main chemical industry sources of hydrocarbons are petroleum, natural gas, and coal.

Hydrogenation. A reaction of molecular hydrogen with numerous organic compounds. Examples are the hydrogenation of olefins to parraffins, or of the aromatics to the naphthenes, or the reduction of aldehydes and ketones to alcohols.

Hydrolysis. The interaction of water with a material resulting in decomposition.

Hydrometallurgy. Science of metal recovery by a process involving treatment of ores in an aqueous medium, such as an acid or cyanide solution.

Hydrophobic. Repelled by water, or water-hating.

Hygroscopic. Readily absorbing or retaining moisture.

Hyper- (prefix). Over, above, increased. The usual implication is overactivity or excessive production, as in hyperthyroidism.

Hyperkeratosis. Hypertrophy of the horny layer of the skin.

Hypertension. Abnormally high tension; especially high blood pressure.

Hypertrophy. Increase in cell size causing an increase in the size of the organ or tissue.

Hypnotic. Anything that induces or that produces the effects ascribed to hypnotism.

Hypo- (prefix). Under, below; less decreased. The two different meanings of this common prefix can be confusing. Hypodermic might reasonably be interpreted to mean that an unfortunate patient has too little skin. The actual meaning is "under or beneath the skin," a proper site for an injection. The majority of hypo- words, however, denote an insufficiency, lessening, or reduction from the norm, as in hypoglycemia, meaning too little glucose in the blood.

Hypothermia. A systemic effect of cold stress; condition of reduced body temperature.

Hysteresis. A retardation of the effect of cold stress; condition of reduced body temperature.

IAQ. The study, evaluation, and control of indoor air quality related to temperature, humidity, and airborne contaminants.

IARC. International Agency for Research on Cancer.

Iatro- (prefix). Pertaining to a doctor. A related root, -iatrist, denotes a specialist, as in psychiatrist.

Iatrogenic. Caused by the doctor.

ICC. Interstate Commerce Commission.

ICRP. International Commission on Radiological Protection and Measurements.

ICTERUS. *See* Jaundicea.

Idio- (prefix). Peculiar to, private, or distinctive, as in idiosyncrasy.

Idiopathic. Disease that originates in or by itself.

Idiosyncrasy. A special susceptibility to a particular substance introduced into the body.

IDLH. Immediately Dangerous to Life and Health. Concentrations of hazardous materials in air where exposure may cause serious injury or death within a short time, or serious delayed effect.

IES. Illumination Engineering Society.

Ignitable. Capable of being set afire.

Iliac crest. The upper rounded border of the hip bone. No muscles cross the iliac crest, which lies immediately below the skin. It is an important anatomical reference point because it can be felt through the skin. Seat backrests should clear the iliac crest.

Illness, occupational. Also, occupational disease. An unhealthy condition of the mind or body as a result of chronic exposure to physical, chemical, ergonomic, or biological agents or hazards (e.g., black lung as a result of long-term exposure to coal dust).

Image. The fluorescent picture produced by X rays hitting a fluoroscopic screen.

Image receptor. Any device, such as a fluorescent screen or radiographic film, that transforms incident X-ray photons either into a visible image or into another form that can be made into a visible image by further transformations.

Imminent danger. An impending or threatening hazard that could be expected to cause death or serious injury to persons in the immediate future unless corrective measures are taken.

Immune. Resistant to disease.

Immunity. The power of the body to resist infection and the effects of toxins. This resistance results from the possession by the body of certain "fighting substances," or antibodies. To immunize is to confer immunity. Immunization is the process of acquiring or conferring immunity.

Impaction. The forcible contact of particles of matter; a term often used synonymously with impingement, but generally reserved for the case where particles are contacting a dry surface.

Impervious. Pertaining to a material that does not allow another substance to pass through or penetrate it.

Impingement. As used in air sampling, refers to a process for the collection of particulate matter in which a particle-containing gas is directed against a wetted glass plate and the particles are retained by the liquid.

Impinger. A device containing an absorbing liquid used in air sampling for the collection of gaseous particulate constituents of an airstream directed by the device through the liquid. The impinger draws air at high velocity through a glass nozzle or jet. A commonly used type is called the midget impinger.

Impulse noise. Repetitive impulse noise is such as produced by rivet guns, pneumatic hammers, and other devices that create an impact. Impulse rates greater that 200 per minute are considered continuous noise.

Inches of mercury column. A unit used in measuring pressures; 1 inch of mercury column equals a pressure of 1.491 lb/in^2 (1.66 kPa).

Inches of water column. A unit used in measuring pressures; 1 inch of water column equals a pressure of 0.036 lb/in^2 (0.25 kPa).

Incidence. An undesired event that may cause personal harm or other damage. In the United States OSHA specifies that incidence of a certain severity be recorded on the OSHA 200 log.

Incidence rate. The number of injuries and/or illnesses or lost workdays per 100 full-time employees per year or 200,000 hours of exposure.

Incompatible. A term applied to liquid and solid systems to indicate that one material cannot be mixed with another specified material.

Incubation. Holding cultures of microorganisms under conditions favorable to their growth.

Incubation time. The elapsed time between exposure to infection and the appearance of disease symptoms, or the time period during which microorganisms inoculated into a medium are allowed to grow.

Indirect cost. Losses ultimately measurable in a monetary sense resulting from an accident other than those costs that are insurable (overtime to cover the absent employee, retraining cost, production losses, reporting time, etc.).

Induration. Heat hardening that may involve little more than thermal dehydration.

Industrial hygiene. The art and science of recognizing, evaluating, and controlling occupational hazards.

Industrial ventilation. The equipment or operation associated with the supply or exhaust of air, by natural or mechanical means, to control occupational hazards in the industrial setting.

Inert (chemical). Not having active properties.

Inert gas. A gas that does not normally combine chemically with the base metal or filler metal.

Inert welding. An electric welding operation using an inert gas such as helium to flush the air away to prevent oxidation of the metal being welded.

Inertial moment. As related to biomechanics, that moment of force-time caused by sudden accelerations or decelerations. Whiplash of the neck is caused by an inertial moment. In an industrial setting, sidestepping causes application of a lateral inertial moment on the lumbosacral joint, which may cause trauma, pain, and in any case lowers performance efficiency. The inertial moment is one of the seven elements of a lifting task.

Infection. Entrance into the body or its tissues of disease-causing organisms with the effect of damage to the body as a whole or to tissues or organs. It also refers to the entrance into the body of parasites, such as certain worms. On the other hand, parasites such as mites and ticks that attack the surface of the body are said to infest, not infect.

Infectious. Capable of invading a susceptible host, replicating and causing an altered host reaction; commonly referred to as a disease.

Infestation. Invasion of the body surface by parasites. *See also* infection.

Infiltration. Air leakage into a space through cracks and interstices, and through ceilings, floors, and walls.

Inflammation. The reaction of body tissue to injury; whether by infection or trauma. The inflamed area is red, swollen, hot, and usually painful.

Infrared. Wavelengths of the electromagnetic spectrum longer than those of visible light and shorter than radio waves, 10^{-4} to 10^{-1} cm wavelength.

Infrared radiation. Electromagnetic energy with wavelengths from 770 to 12,000 nm.

Ingestion. (1) The process of taking substances into the stomach, such as food, drink, or medicine. (2) With regard to certain cells, the act of engulfing or taking up bacteria and other foreign matter.

Ingot. A block of iron or steel cast in a mold for ease in handling before processing.

Inguinal region. The abdominal area on each side of the body occurring as a depression between the abdomen and the thigh; the groin.

Inhalation. The breathing in of a substance in the form of a gas, vapor, fume, mist, or dust.

Inhalation valve. A device that allows respirable air to enter a face piece and prevents exhaled air from leaving a face piece through the intake opening.

Inhibition. Prevention of growth or multiplication of microorganisms.

Inhibitor. An agent that arrests or slows chemical action or a material used to prevent or retard rust or corrosion.

Injury. Damage or harm to the body, as a result of violence, infection, or anything else that produces a lesion.

Innocuous. Harmless.

Inoculation. The artificial introduction of microorganisms or antigens into a system.

Inorganic. Used to designate compounds that generally do not contain carbon, whose source is matter other than vegetable or animal. Examples are sulfuric acid and salt. Exceptions are carbon monoxide and carbon dioxide.

Inrunning nip (point). A rotating mechanism that can seize loose clothing, belts, hair, body parts, etc; when two or more shafts or rollers rotate parallel to one another in opposite directions.

Insoluble. Incapable of being dissolved.

Insomnia. Inability to sleep; abnormal wakefulness.

Inspection. Monitoring function conducted in an organization to locate and report existing and potential hazards having capacity to cause accidents in the workplace. Also used to identify program, policy, inspection, and compliance adherence.

Instantaneous radiation. The radiation emitted during the fission process; often called prompt gamma rays or prompt neutrons. Most fission products continue to emit radiation after the fission process.

Inter- (prefix). Between.

Interlock. A device that interacts with another device or mechanism to govern succeeding operations. For example, an interlocked machine guard will prevent the machine from operating unless the guard is in its proper place. An interlock on an elevator door will prevent the car from moving unless the door is properly closed.

Intermediate. A chemical formed as a middle step in a series of chemical reactions, especially in the manufacture of organic dyes and pigments. In many cases, it may be isolated and used to form a variety of desired products. In other cases, the intermediate may be unstable or may be used up at once.

Internal biomechanical environment. The muscles, bones, and tissues of the body, all of which are subject to the same Newtonian force as external objects in their interaction with other bodies and natural forces. When designing for the body, one must consider the forces that the internal biomechanical environment must withstand.

Interphalangeal joints. The finger or toe joints. The thumb has one interphalangeal joint; the fingers have two interphalangeal joints each.

Interstitial. (1) Pertaining to the small spaces between cells or structures. (2) Occupying the interstices of a tissue or organ. (3) Designating connective tissue occupying spaces between the functional units of an organ or a structure.

Intoxication. Either drunkenness or poisoning.

Intra- (prefix). Within.

Intraperitoneal. Inside the space formed by the membrane that lines the interior wall of the abdomen and covers the abdominal organs.

Intravenous. Into or inside the vein.

Intrinsically safe. Said of an instrument that is designed and certified to be operated safely in flammable or explosive atmospheres.

Inverse square law. The propagation of energy through space is inversely proportional to the square of the distance it must travel. An object 3 m away from an energy source receives one ninth as much energy as an object 1 m away.

Inversion. Phenomenon of a layer of cool air trapped by a layer of warmer air above it so that the bottom layer cannot rise. This is a special problem in polluted areas because the contaminating substances cannot be dispersed.

Investment casting. There are numerous types of investment casting, and the materials include fire clay, silicon dioxide, zirconium oxide, and others. The Mercast process uses mercury poured into a steel die. A ceramic shell mold is built around the pattern, and then the pattern is frozen. The mercury is subsequently recovered at room temperature. The potential harm from exposure to mercury is often unrecognized.

Ion. An electrically charged atom. An atom that has lost one or more of its electrons is left with a positive electrical charge. Those that have gained one or more extra electrons are left with a negative charge.

Ion-exchange resin. Synthetic resins containing active groups that give the resin the property of combining with or exchanging ions between the resin and the solution.

Ionization. The process whereby one or more electrons is removed from a neutral atom by the action of radiation. Specific ionization is the number of ion pairs per unit distance in matter, usually air.

Ionization chamber. A device roughly similar to a Geiger counter; used to measure radioactivity.

Ionizing radiation. (1) Electrically charged or neutral particles. (2) Electromagnetic radiation that interacts with gases, liquids, or solids to produce ions. There are five major types: alpha, beta, X (for X ray), gamma, and neutrons.

Ion pair. A positively charged atom (ion) and an electron formed by the action of radiation on a neutral atom.

Irradiation. Exposure of something to radiation.

Irritant. A substance that produces an irritating effect when it contact skin, eyes, nose, or respiratory system.

Ischemia. Loss of blood supply to a particular part of the body.

Ischial tuberosity. A rounded projection on the ischium. It is a point of attachment for several muscles involved in moving the femur and the knee. It can be affected by improper chair design and by situations involving trauma to the pelvic region. When seated, pressure is borne at the site of the ischial tuberosities. Chair design should provide support to the pressure projection of the ischial tuberosity through the skin of the buttocks.

Isometric work. Refers to state of muscular contraction without movement. Although no work in the "physics" sense is done, physiological work (energy use and heat production) occurs. In isometric exercise, muscles are tightened against immovable objects. In work measurements, isometric muscular contractions must be considered as a major factor of task severity.

Isotope.   One of two or more atomic species of an element differing in atomic weight but having the same atomic number. Each contains the same number of protons but a different number of neutrons. Uranium-238 contains 92 protons and 146 neutrons; the isotope uranium-235 contains 92 protons and 143 neutrons. Thus, the atomic weight (atomic mass) of uranium-238 is 3 higher than that of uranium-235. *See also* Radioisotope.

Isotropic.   Exhibiting properties with the same values when measured along axes in all directions.

-itis (suffix).   Inflammation.

Jaundice.   Icterus. A serious symptom of disease that causes the skin, the whites of the eyes, and even the mucous membranes to turn yellow.

Jigs and fixtures.   Often used interchangeably; precisely, a jig holds work in position and guides the tools acting on the work, whereas a fixture holds but does not guide.

Job safety analysis.   A method for studying a job to (1) identify hazards or potential accidents associated with each step or task and (2) develop solutions that will eliminate, nullify, or prevent such hazards or accidents. Sometimes called a Job Hazard Analysis.

Joint.   Articulation between two bones that may permit motion in one or more planes. They may become the sites for work-induced trauma (such as tennis elbow or arthritis) or other disorders.

Joule.   Unit of energy used in describing a single pulsed output of a laser. It is equal to 1 watt-second or 0.239 calories. It equals 1×10^7 ergs.

Joule/cm² (J/cm²).   Unit of energy density used in measuring the amount of energy per area of absorbing surface or per area of a laser beam. It is a unit for predicting the damage potential of a laser beam.

Kaolin.   A type of clay composed of mixed silicates used for refractories, ceramics, tile, and stoneware. In some deposits, free silica may be present as an impurity.

Kaolinosis.   A disorder induced by inhalation of the dust released in the grinding and handling of kaolin (china clay).

Kelvin scale.   The fundamental temperature scale, also called the absolute or thermodynamic scale, in which the temperature measure is based on the average kinetic energy per molecule of a perfect gas. The zero of the Kelvin scale is $-273.18°C$.

Keratin.   Sulfur-containing proteins that form the chemical basis for epidermal tissues; found in nails, hair, and feathers.

Keratinocyte.   An epidermal cell that produces keratin.

Keratitis.   Inflammation of the cornea.

KeV.   A unit of energy equal to 1000 electron volts.

Kilocurie.   1000 curies. A unit of radioactivity.

Kilogram (kg).   A unit of weight in the metric system equal to 2.2 pounds.

Kinesiology.   The study of human movement in terms of functional anatomy.

Kinetic energy.   Energy due to motion. *See* work.

Kyphosis.   Abnormal curvature of the spine of the upper back in the anteroposterior plane.

l.   Liter.

Laboratory-acquired infection.   Any infection resulting from exposure to biohazardous materials in a laboratory environment. Exposure may be the result of a specific accident or inadequate biohazard control procedure or equipment.

Lacquer.   A colloidal dispersion or solution of nitrocellulose or similar film-forming compounds, resins, and plasticizers in solvents and diluents used as a protective and decorative coating for various surfaces.

Laminar airflow.   Streamlined airflow in which the entire body of air within a designated space moves with uniform velocity in one direction along parallel flow lines.

LAN.   Local area network. A network of computers linked electronically and by software. Located geographically locally, usually in one office or office building.

Lapping.   The operation of polishing or sanding surfaces such as metal or glass to a precise dimension.

Laryngitis.   Inflammation of the larynx.

Larynx.   The organ by which the voice is produced. It is situated at the upper part of the trachea.

Laser.   The acronym for Light Amplification by Stimulated Emission of Radiation.

Laser light.   A portion of the electromagnetic spectrum that includes ultraviolet, visible, and infrared light.

Laser system.   An assembly of electrical, mechanical, and optical components that includes a laser.

Latent heat.   The amount of heat energy absorbed or emitted by one unit of mass of a substance during a change of state (e.g., vaporization, freezing).

Latent period.   The time that elapses between exposure and the first manifestation of damage.

Latex.   Originally, a milky extract from the rubber tree, containing about 35% rubber hydrocarbon, with the remainder water, proteins, and sugars. Also applied to water emulsions of synthetic rubbers or resins. In emulsion paints, the film-forming resins are in the form of latex.

Lathe.   A machine tool used to perform cutting operations on wood or metal by the rotation of the workpiece.

Latissimus dorsi.   A large, flat muscle of the back humerus near the armpit. It adducts the upper arm, and when the elbow is abducted, it rotates the arm medially. It is actively used in operating equipment such as the drill press, where a downward pull by the arm is required.

LC.   Lethal concentration; a concentration of a substance being tested that will kill a test animal.

LC$_{50}$.   Lethal concentration that kills 50% of the test animals within a specified time. *See also* LD$_{50}$.

LD. Lethal dose; an amount of a substance being tested that will kill a test animal.

LD$_{50}$. Lead Poisoning. Poisoning that results when lead compounds are swallowed or inhaled. Inorganic lead compounds commonly cause symptoms of lead colic and lead anemia. Organic lead compounds can attack the nervous system.

Leakage radiation. Radiation emanating from the diagnostic source assembly, except for the useful beam and radiation, produced when the exposure switch or timer is not activated.

LEL. Lower explosive limit. *See* Flammable limits.

Lens, crystalline. Lens of the eye; a transparent biconvex body situated between the anterior chamber (aqueous) and the posterior chamber (vitreous) through which the light rays are further focused on the retina. The cornea provides most of the refractive power of the eye.

Lesion. Injury, damage, or abnormal change in a tissue or organ.

Lethal. Capable of causing death.

Leuk, leuko- (prefix). White.

Leukemia. A group of malignant blood diseases distinguished by overproduction of white blood cells.

Leukemogenic. Having the ability to cause leukemia.

Leukocyte. White blood cell.

Leukocytosis. An abnormal increase in the number of white blood cells.

Leukopenia. A serious reduction in the number of white blood cells.

LFL. Lower flammable limit. *See* Flammable limits.

Liability. The state of being bound or obligated in law to do, pay, or make good on something. Regarding the law of torts, usually based on the law of negligence.

Liability, strict. The imposition of liability for damages resulting from any and all defects and hazardous products without requiring proof of negligence. Disclaimers are not valid; traditional warranty concepts, privity, and notice of injury are eliminated.

Lifestyle. The way one lives and behaves.

Lig- (prefix). Binding. A ligament ties two or more bones together.

Linear accelerator. A machine for speeding up charged particles such as protons. It differs from other accelerators in that the particles move in a straight line at all times instead of in circles or spirals.

Line-voltage regulation. The difference between the no-load and the load-line potentials expressed as a percent of the load-line potential.

Lipo-(prefix). Fat, fatty.

Liquefied petroleum (LP) gas. A compressed or liquefied gas usually composed of propane, some butane, and lesser quantities of other light hydrocarbons and impurities; obtained as a by-product in petroleum refining. Used chiefly as a fuel and in chemical synthesis.

Liquid. A state of matter in which the substance is a formless fluid that flows in accord with the law of gravity.

Liter (l).   A measure of capacity; 1 quart equals 0.9 l.

Liver.   The largest gland or organ in the body, situated on the right side of the upper part of the abdomen. It has many important functions, including regulating the amino acids in the blood; storing iron and copper for the body; forming and secreting bile, which aids in absorption and digestion of fats; transforming glucose into glycogen; and detoxifying exogenous substances.

Live room.   A reverberant room that is characterized by an unusually small amount of sound absorption.

Load limit.   The upper weight limit capable of safe support by a vehicle, floor, or roof structure.

Local exhaust ventilation.   A ventilation system that captures and removes the contaminants at the point at which they are being produced before they escape into the workroom air.

Localized.   Restricted to one spot or area in the body, and not spread throughout it; contrasted to systemic.

Lockout/tagout.   A basic safety concept and OSHA standard requiring implementation of practices and procedures to prevent the release of potentially hazardous energy from machines or parts of machines and equipment while maintenance, servicing, or alteration activity is performed. The energy in question may be electrical, mechanical, chemical, or any other form. Also called lockout/ tagout/ blockout.

Long-term sample.   Sample taken over a sufficiently long period of time that the variation in exposure cycles are averaged.

Lordosis.   The curvature of the lower back in the anteroposterior plane.

Loss.   Usually refers to the conversion of static pressure to heat in components of the ventilation system, i.e., "the hood entry loss."

Loss control.   A program designed to minimize accident-related financial losses. The concept of total loss control is based on detailed analysis of both indirect and direct accident costs. Property damages, as well as injurious and potentially injurious accidents, are included in the analysis.

Loss prevention.   A before-the-loss program designed to identify and correct hazards before they result in incidents that produce actual financial loss and injury.

Loss workday.   The number of workdays (consecutive or not), beyond the day of injury or onset of illness, that an employee was away from work or limited to restricted work activity because of an occupational injury or illness.

Loudness.   The intensive attribute of an auditory sensation, in terms of which sounds may be ordered on a scale extending from soft to loud. Loudness depends primarily upon the sound pressure of the stimulus, but it also depends upon the frequency and waveform of the stimulus.

Louver.   A slanted panel.

Lower confidence limit (LCL).   In analyzing sampling data, a statistical procedure used to estimate the likelihood that the true value of the sampled quantity is lower than that obtained.

Lower explosive limit (LEL). The lower limit of flammability of a gas or vapor at ordinary ambient temperatures expressed by a percentage of the gas or vapor in air by volume. This limit is assumed constant for temperatures up to 250°F (120°C); above this, it should be decreased by a factor of 0.7, because explosibility increases with higher temperatures.

Low-pressure tank. A storage tank designed to operate at pressures between 0.5 and 15 psig (3.5 to 103 kPa).

LP gas. *See* Liquefied petroleum gas.

Lumbar spine. The section of the lower spinal column or vertebral column immediately above the sacrum. Located in the small of the back and consisting of five large lumbar vertebrae, it is a highly stressed area in work situations and in supporting the body structure.

Lumbosacral joint. The joint between the fifth lumbar vertebrae and the sacrum. Often the site of spinal trauma from lifting tasks.

Lumen. The luminous flux on 1 square foot of a sphere 1 foot in radius with a light source of one candle at the center that radiates uniformly in all directions.

Luminous flux. The rate of light flow measured in lumens.

Luminous lux. A unit of illumination equal to 10 footcandles.

Lyme disease. A disease transmitted to humans by the deer tick.

Lymph. A pale, coagulable fluid consisting of a liquid portion resembling blood plasma and containing white blood cells (lymphocytes).

Lymph node. Small oval bodies with a glandlike structure scattered throughout the body in the course of the lymph vessels. Also known as lymphatic nodes, lymph glands, and lymphatic glands.

Lymphoid. Resembling lymph.

Lyophilized. Freeze-dried, as in freeze-dried bacterial cultures.

Lysis. The distribution or breaking up of cells by internal or external means.

MAC. Maximum allowable concentration.

Maceration. Softening of the skin by action of a liquid.

Macrophage. Immune system cell whose normal function is to engulf and remove foreign matter from the body's tissues.

Macroscopic. Visible without the aid of a microscope.

Macula. An oval area in the center of the retina devoid of blood vessels; the area most responsible for color vision.

Magnification. The number of times the apparent size of an object has been increased by the lens system of a microscope.

Makeup air. Clean, tempered outdoor air supplied to a work space to replace air removed by exhaust ventilation or by some industrial process.

Malaise. A vague feeling of bodily discomfort.

Malignant. As applied to a tumor, cancerous and capable of undergoing metastasis (invasion of surrounding tissue).

Manometer. Instrument for measuring pressure; essentially a U-tube partially filled with a liquid (usually water, mercury, or a light oil) and constructed in such a way that the amount of displacement of the liquid indicates the pressure being exerted on the instrument.

Maser. Microwave amplification by stimulated emission of radiation. When used in the term optical maser, it is often interpreted as molecular amplification by stimulated emission of radiation.

Masking. The stimulation of a person's ear with controlled noise to prevent that person from hearing with one ear the tone or signal given to the other ear. This procedure is used when there is at least a 15 to 20 dBA difference in the hearing level between ears.

Mass. Quantity of matter; measured in grams or pounds.

Material safety data sheet (MSDS). As part of hazard communication standards (right-to-know laws), federal and state OSHA programs require manufacturers and importers of chemicals to prepare compendia of information on their products. Categories of information that must be provided on MSDSs include physical properties, recommended exposure limits, personal protective equipment, spill-handling procedures, first aid, health effects, and toxicological data.

Matter. Anything that has mass or occupies space.

Maximum evaporative capacity. The maximum amount of evaporating sweat from a person that an environment can accept.

Maximum line current. The root mean square current in the supply line of an X-ray machine operating at its maximum rating.

Maximum permissible concentration (MCP). Concentration set by the National Committee on Radiation Protection (NCRP); recommended maximum average concentrations of radionuclides to which a worker may be exposed assuming that the worker works 8 hours a day, 5 days a week, and 50 weeks a year.

Maximum permissible power. The intensity of laser radiation not expected to cause detectable bodily injury to a person or energy density at any time during the person's life.

Maximum use concentration (MUC). The product of the protection factor of the respiratory protection equipment and the permissible exposure limit (PEL).

Mechanical efficiency curve. A graphical representation of the relative efficiency of a fan in moving air at different airflow rates and static pressures.

Mechanical filter respirator. A respirator used to protect against airborne particulate matter like dusts, mist, metal fumes, and smoke. Mechanical filter respirators do not provide protection against gases, vapors, or oxygen-deficient atmospheres.

Mechanical ventilation. A powered device, such as a motor-driven fan or vacuum hose attachment, for exhausting contaminants from a workplace, vessel, or enclosure.

Mechanotactic stress. Stress caused by contact with a mechanical environment.

Mechanotaxis. Contact with a mechanical environment consisting of forces (pressure, moment, vibration, and so on); one of the ecological stress vectors. Improper design of he mechanotactic interface may lead to instantaneous trauma, cumulative pathogenesis, or death.

Median nerve. A major nerve controlling the flexor muscles of the wrist and hand. Tool handles and other grasped objects should make solid contact with the sensory feedback area of this nerve, located in the palmar surface of the thumb, index finger, middle finger, and part of the ring finger.

Medium. *See* Culture medium.

Medulla. The part of the brain that controls breathing.

Mega. One million. For example, 1 megacurie = 1 million curies.

Mega-, megalo- (perfix). Large, huge. The prefix macro- has the same meaning.

Meiosis. The process whereby chromosome pairs undergo nuclear division as the germ cell matures.

Melanocyte. An epidermal cell containing dark pigments.

Melanoderma. Abnormal darkening of the skin.

Melt. In the glass industry, the total batch of ingredients that may be introduced into pots or furnaces.

Melting point. The transition point between the solid and liquid states. Expressed as the temperature at which this change occurs.

Membrane. A thin, pliable layer of animal tissue that covers a surface, lines the interior of a cavity or organ, or divides a space.

Membrane filter. A filter medium made from various polymeric material such as cellulose, polyethylene, and tetrapolyethylene. Usually exhibit narrow ranges of effective pore diameters and are therefore useful in collecting and sizing microscopic and submicroscopic particles and in sterilizing liquids.

Men-, meno- (prefix). Pertaining to menstruation; from the Greek for month.

Méniér's disease. A combination of deafness, tinnitus, and vertigo.

Meson. A particle that weighs more than an electron but generally less than a proton. Mesons can be produced artificially or by cosmic radiation (natural radiation from outer space). Mesons are not stable and disintegrate in a fraction of a second.

Mesothelioma. Cancer of the membranes that line the chest and abdomen.

Metabolism. The flow of energy and the associated physical and chemical changes constantly taking place in the billions of cells that make up the body.

Metal fume fever. A flulike condition caused by inhaling heated metal fumes.

Metallizing. Melting wire in a special device that sprays the atomized metal onto a surface. The metal can be steel, lead, or another metal or alloy.

Metastasis. Transfer of the causal agent (cell or microorganism) of a disease from a primary focus to a distant one through the blood or lymphatic vessels. Also, spread of malignancy from a site of primary cancer to secondary sites.

Meter (m). A unit of length in the metric system; 1 meter is about 39.37 inches.

Methemoglobinemia. The presence of methemoglobin in the blood. (Methemoglobin is a compound formed when the iron moiety of hemoglobin is oxidized from the ferrous to the ferric state.) This protein inactivates hemoglobin as an oxygen carrier.

Mev.   Million electron volts.

mg.   Milligram. A metric unit of weight. There are 1000 milligrams in 1 gram (g) of a substance; 1 gram is equivalent to almost 4/100 of an ounce.

mg/kg.   Milligram per kilogram.

mg/m³.   Milligram per cubic meter.

Mica.   A large group of silicates of varying composition that are similar in physical properties. All have excellent cleavage and can be split into very thin sheets. Used in electrical insulation.

Microbar.   A unit of pressure commonly used in acoustics; equals 1 syne/cm². A reference point for the decibel, which is accepted as 0.0002 dyne/cm².

Microbe.   A microscopic organism.

Microcurie (μc).   One millionth of a currie. A still smaller unit is the micromicrocurie [μμc].

Micron (μm).   A unit of length equal to 10^{-4} cm, approximately 1/25,000 in. Also called micrometer.

Microorganism.   A minute organism; microbes, bacteria, cocci, viruses, and molds, among others.

Microphone.   An electroacoustic transducer that responds to sound waves and delivers essentially equivalent electric waves.

Midsagittal plane.   A reference plane formed by bisecting the human anatomy into a right and left aspect. Human motor function can be described in terms of movement relative to the midsagittal plane.

Miliary.   Characterized or accompanied by seedlike blisters or inflamed raised portions of tissue.

Milliampere.   1/1000 of an ampere.

Milligram (mg).   A unit of weight in the metric system; 1 thousand milligrams equal 1 gram.

Milligrams per cubic meter (mg/m³).   Unit used to measure air concentrations of dusts, gases, mists, and fumes.

Milliliter (ml).   A metric unit used to measure volume; 1 milliliter equals 1 cubic centimeter.

Millimeter of mercury (mmHg).   The unit of pressure equal to the pressure exerted by column of liquid mercury 1 millimeter high at standard temperature.

Milliroentgen.   One one-thousandth of a roentgen.

Millwright.   A mechanic engaged in the erection and maintenance of machinery.

Mineral pitch.   Tar from petroleum or coal as opposed to wood tar.

Mineral spirits.   A petroleum fraction with a boiling range between 300 and 400°F (149 and 240°C).

Minimum transport velocity (MTV).   The minimum velocity that will transport dry particles in a duct with little settling; the MTV depends upon article size, material density, particulate loading, and other factors; scrubbing velocity.

Miosis.   Excessive smallness or contraction of the pupil of the eye.

Mists.   Suspended liquid droplets generated by condensation from the gaseous to the liquid state or by breaking up a liquid into a dispersed state, such as by splashing, foaming, or atomizing. Formed when a finely divided liquid is suspended in air.

Mitosis.   Nuclear cell division in which resulting nuclei have the same number and kinds of chromosomes as the original cell.

Mixture.   A combination of two or more substances that may be separated by mechanical means. The components may not be uniformly dispersed. *See also* Solution.

ml.   *See* milliliter.

mmHg.   Millimeters (mm) of mercury (Hg).

Moderator.   A material used to slow neutrons in a reactor. These slow neutrons are particularly effective in causing fission. Neutrons are slowed when they collide with atoms of light elements such as hydrogen, deuterium, and carbon — three common moderators.

Mold.   (1) A growth of fungi forming a furry patch, as on stale bread or cheese. *See also* Spore. (2) A hollow form or matrix into which molten material is poured to produce a cast.

Molecule.   A chemical unit composed of one or more atoms.

Moment.   Magnitude of force times distance of application.

Moment concept.   A concept based on theoretical and experimental foundations that lifting stress depends on the bending moment exerted at susceptible points of the vertebral column rather than depending on weight alone.

Monaural hearing.   Hearing with one ear only.

Monitoring.   Testing to determine if the parameters being measured are within acceptable limits. This includes environmental and medical (biological) monitoring in the workplace.

Monochromatic.   Single fixed wavelength.

Monomer.   A compound of relatively low molecular weight that, under certain conditions, either alone or with another monomer, forms various types and lengths of molecular chains called polymers or copolymers of high molecular weight. Styrene, for example, is a monomer that polymerizes readily to form polystyrene. *See also* Polymer.

Morphology.   The branch of biological science that deals with the study of the structure and form of living organisms.

MORT.   Management Oversight and Risk Tree.

Motile.   Capable of spontaneous movement.

MPC.   *See* Maximum permissible concentration.

MPD.   *See* Dose equivalent, maximum permissible.

MPE.   Maximum permissible exposure.

MPL.   Maximum permissible level, limit, or dose; refers to the tolerable dose rate for humans exposed to nuclear radiation.

Mppcf.   Million particles per cubic foot.

mr.   millirem.

mR.   milliroentgen.

MSDS.   *See* Material safety data sheet.

MSHA.   The Mine Safety and Health Administration; a federal agency that regulates safety and health in the mining industry.

MUC. *See* Maximum use concentration.

Mucous membranes Lining of the hollow organs of the body, notably the nose, mouth, stomach, intestines, bronchial tubes, and urinary tract.

Muff. A covering on the outside ear to reduce noise exposure.

Musculoskeletal system The combined system of muscles and bones that comprise the internal biomechanical environment.

Mutagen. Anything that can cause a change (mutation) in the genetic material in the living cell.

Mutation. A transformation of the gene that may result in the alteration of characteristics of offspring.

MWD. Megawatt days, usually per ton. The amount of energy obtained from 1 megawatt power in 1 day, normally a measure of the extent of nuclear fuel burn-up. 10,000 MWD per ton is about 1% burn-up.

My-, myo- (prefix). Pertaining to muscle. Myocardium is the heart muscle.

Myelo- (prefix). Pertaining to marrow.

NAAQS. EPA National Ambient Air Quality Standard. Nine compounds have established standards (e.g., lead at 1.5 $\mu g/m^3$, carbon monoxide at 9 ppm, annual average.).

Nanometer. A unit of length equal 10^{-7} cm.

Naphthas. Hydrocarbons of the petroleum type that contain substantial portions of paraffins and naphthalenes.

Narcosis. Stupor or unconsciousness produced by chemical substances.

Narcotics. Chemical agents that completely or partially induce sleep.

Narrow band. Applies to a narrow band of transmitted waves, with neither the critical or cutoff frequencies of the filter being zero or infinite.

Nasal septum. Narrow partition that divides the nose into right and left nasal cavities.

Nascent. Just forming, as from a chemical or biological reaction.

Nasopharynx. Upper extension of the throat.

National Electronic Injury Surveillance System (NEISS). Collects data from 119 representative hospital emergency rooms on product-related injuries receiving emergency room treatment. A part of the U.S. Consumer Product Safety Commission.

Natural gas. A combustible gas composed largely of methane and other hydrocarbons with variable amounts of nitrogen and noncombustible gases; obtained from natural earth fissures or from driven wells. Used as a fuel in the manufacture of carbon black and in chemical synthesis of many products. Major source of hydrogen for the manufacture of ammonia.

Natural radioactivity. The radioactive background or, more properly the radioactivity that is associated with the naturally occurring heavy elements.

Natural uranium. Purified from the naturally occurring ore, as opposed to uranium enriched in fissionable content by processing at separation facilities.

Natural ventilation. The movement of outdoor air into space through intentionally provided openings, such as windows, doors, or other nonpowered ventilators, or by infiltrations.

Nature of injury.   The type of injury inflicted, such as sprain, burn contusion, laceration, etc.

Nausea.   An unpleasant sensation, vaguely referred to the epigastrium and abdomen. Often proceeds vomiting.

NC.   Noise criteria. NC curves have been developed for in-room use (e.g., offices, laboratories, conference rooms, lunchrooms, and other nonindustrial sites). Based on a series of octave band curves, the NC curves show the maximum allowable sound level at each octave.

NCRP.   National Committee on Radiation Protection; an advisory group of scientists and professionals that makes recommendations for radiation protection in the United States.

Near field.   In noise measurement, refers to a field in the immediate vicinity of the noise source where the sound-pressure level does not follow the inverse square law.

Necro- (prefix).   Dead.

Necrosis.   Death of body tissue.

Negligence.   The lack of required, expected, or reasonable conduct or care that a prudent person would ordinarily exhibit. There need not be a legal duty.

NEISS.   *See* National Electronic Injury Surveillance System.

Neoplasm.   A cellular outgrowth characterized by rapid cell multiplication; may be benign (semicontrolled and restricted) or malignant.

Nephr-, nephro- (prefix).   From the Greek for kidney.

Nephritis.   Inflammation of the kidneys.

Nephrotoxins.   Chemicals that produce kidney damage.

Neur- neuro- (prefix).   Pertaining to the nerves.

Neural loss.   Hearing loss. *See also* Sensorineural.

Neuritis.   Inflammation of a nerve.

Neurological (neurology).   The branch of medical science dealing with the nervous system.

Neurotoxin.   Chemicals that produce their primary effect on the nervous system.

Neutral wire.   Wire carrying electrical current back to its source, thus completing the circuit.

Neutrino.   A particle, resulting from nuclear reactions, that carries energy away from the system but has no mass or charge and is absorbed only with extreme difficulty.

Neutron.   A constituent of the atomic nucleus. A neutron weighs about as much as a proton, and has no electric charge. Neutrons make effective atomic projectiles for the bombardment of nuclei.

NFPA.   National Fire Protection Association; a voluntary membership organization whose aim is to promote and improve fire protection and prevention. NFPA publishes the National Fire Codes.

NIOSH.   The National Institute for Occupational Safety and Health; a federal agency that conducts research on health and safety concerns, tests and certifies respirators, and trains occupational health and safety professionals.

Nip point. The point of intersection or contact between two or more surfaces when one or more are moving.

Nitrogen fixation. Chemical combination or fixation of atmospheric nitrogen with hydrogen, as in the synthesis of ammonia. Bacteria fixate nitrogen in soil. Provides an industrial and agricultural source of nitrogen.

Node. (1) A point, line, or surface in a standing wave where some characteristic of the wave field has essentially zero amplitude. (2) A small, round, or oval mass of tissue; a collection of cells. (3) One of several constrictions occurring at regular intervals in a structure.

Nodule. A small mass of rounded or irregularly shaped cells or tissue; a small node.

Nodulizing. Simultaneous sintering and drum balling, usually in a rotary kiln.

NOEL. *See* No observable effect level.

Noise. Any unwanted sound.

Noise-induced hearing loss. Slowly progressive inner ear hearing loss resulting from exposure to continuous noise over a long period of time, as contrasted to acoustic trauma or physical injury to the ear.

Nonauditory effects of noise. Refers to stress, fatigue, health, work efficiency, and performance effects of loud, continuous noise.

Nonferrous metal. Metal such as nickel, brass, or bronze that does not include any appreciable amount of iron.

Nonflammable. Not easily ignited, or if ignited, not burning with a flame (smolders).

Nonionizing radiation. Electromagnetic radiation that does not cause ionization. Includes ultraviolet, laser, infrared, microwave, and radio frequency radiation.

Nonpolar solvents. The aromatic and petroleum hydrocarbon groups characterized by low dielectric constants.

Nonvolatile matter. The portion of a material that does not evaporate at ordinary temperature.

No Observable Effect Level (NOEL). In toxicology the concentration of a substance at (and below) which exposure produces no evidence of injury or impairment.

Normal pulse (conventional pulse). Heartbeat; also, a single output event whose pulse duration is between 200 μs and 1 ms.

Nosocomial. Pertaining to (1) a hospital; (2) disease caused or aggravated by hospital life.

NRC. Nuclear Regulatory Commission of the U.S. Department of Energy.

NRR. Noise Reduction Rating of hearing protection. EPA requires manufacturers to print attenuation potential (NRR) on the packaging of all hearing protection devices.

NTP. National Toxicology Program.

Nuclear battery. A device in which the energy emitted by decay of a radioisotope is first converted to heat and then directly to electricity.

Nuclear bombardment. The shooting of atomic projectiles at nuclei, usually in an attempt to split the atom or to form a new element.

Nuclear energy. The energy released in a nuclear reaction such as fission or fusion. Nuclear energy is popularly, although mistakenly, called atomic energy.

Nuclear explosion. The rapid fissioning of a large amount of fissionable material; creates intense heat, a light flash, a heavy blast, and a large amount of radioactive fission products. These may be attached to dust and debris, forming fallout. Nuclear explosions also result from nuclear fusion, which does not produce radioactive fission products.

Nuclear reaction. Result of the bombardment of a nucleus with atomic or subatomic particles or very high-energy radiation. Possible reactions are emission of other particles, fission, fusion, and the decay of radioactive material.

Nuclear reactor. A machine for producing a controlled chain reaction in fissionable material. It is the heart of nuclear power facilities, where it serves as a heat source. *See* Reactor.

Nucleonics. The application of nuclear science and techniques in physics, chemistry, astronomy, biology, industry, and other fields.

Nucleus. The inner core of the atom; consists of neutrons and protons tightly locked together..   .

Nuclide. A type of atom characterized by its mass number, atomic number, and energy state of the nucleus, provided that the mean life in that state is long enough to be observable.

Nuisance dust. Dust with a long history of little adverse effect on the lungs; does not produce significant organic disease or toxic effect when exposures are kept at reasonable levels.

Null point. The distance from a contaminant source at which the initial energy or velocity of the contaminants is dissipated, allowing the material to be captured by a hood.

N-unit (or n-unit). A measure of radiation dose caused by fast neutrons.

Nutrient. A substance that can be used for food.

Occupational Health Nursing (OHN). Specialized nursing practice providing health-care service to workers and worker populations.

Occupational Safety and Health Review Commission (OSHRC). An independent body established to review actions of federal OSHA that are contested by employers, employees, or their representatives.

Occupied zone. The region within an occupied space between the floor and 72 inches above the floor and more than 2 feet from the walls for fixed air-conditioning equipment (from ASHRAE Standard 55-1992).

Octave. The interval between two sounds having a basic frequency ratio of two.

Octave band. An arbitrary spread of frequencics. The top frequency in an octave band is always twice the bottom frequency. The octave band may be referred to by a center frequency.

Ocul-, oculo-, ophthalmo- (prefix). The eye; ophth- refers more often to eye diseases.

Odor. That property of a substance that affects the sense of smell.

Odorant. Any substance with an odor.

Odor control. Absorption (e.g., activated charcoal), oxidation, liquid scrubbing, odor masking.

Odor threshold. The minimum concentration of a substance at which a majority of test subjects can detect and identify the characteristic odor of a substance.

Ohm (Ω). The unit of electrical resistance.

Ohm's law. Voltage in a circuit is equal to the current times the resistance.

Oil dermatitis. Blackheads and acne caused by oils and waxes that plug the hair follicles and sweat ducts.

Olecranon fossa. A depression in the back of the lower end of the humerus in which the ulna rests when the arm is straight.

olf. From "olfactory." A perceived air quality term that attempts to quantify a given pollution load. One person creates 1 olf of bioeffluents. If there are 10 m³ of floor space per person, then people create 0.1 olf per square meter. Other sources are compared and quantified by olfs. For example, if 40% of the people smoke, this adds 0.2 olf/m² to the load.

Olfactory. Pertaining to the sense of smell.

Olfactory fatigue. The tendency for the odor threshold to shift following extended exposure to an odor.

Olfactory thresholds. Upper and lower concentration at which people detect odors; varies with individuals.

Olefins. A class of unsaturated hydrocarbons characterized by relatively great chemical activity. Obtained from petroleum and natural gas. Examples are butene, ethylene, and propylene. Generalized formula: C_nH_2n.

Olig-, oligo- (prefix). Scanty, few, little. Oliguria means scanty urination.

Oncogenic. Tumor generating.

Oncology. Study of causes, development, characteristics, and treatment of tumors.

Opacity. The condition of being nontransparent; a cataract.

Ophthalmologist. A physician who specializes in the structure, function, and diseases of the eye.

Optical density (OD). A logarithmic expression of the attenuation afforded by a filter.

Optically pumped laser. A type of laser that derives its energy from a noncoherent light source, such as a xenon flash lamp; usually pulsed and commonly called a solid state laser.

Organ. An organized collection of tissues that have a special and recognized function.

Organic. Chemicals that contain carbon. To date, nearly 1 million organic compounds have been synthesized or isolated. *See also* Inorganic.

Organic disease. Disease in which some charge in the structure of body tissue could either be visualized or positively inferred from indirect evidence.

Organic matter. Compounds containing carbon.

Organism. A living thing, such as a human being, animal, germ, plant, and so on, especially one consisting of several parts, each specializing in a particular function.

Organ of Corti. The heart of the hearing mechanism; an aggregation of nerve cells in the ear lying on the basilar membrane that picks up vibrations and converts them to electrical energy, which is sent to the brain and interpreted as sound.

Orifice. (1) The opening that serves as an entrance and/or outlet of a body cavity or organ, especially the opening of a canal or a passage. (2) A small hole in a tube or duct. A critical or limiting orifice is used to control rate of flow of a gas in rotometers and other air-sampling equipment.

Orifice meter. A flowmeter, employing as the measure of a flow rate the difference between pressures measured on the upstream and downstream sides of a restriction within a pipe or duct.

Ortho- (prefix). Straight, correct, normal. Orthopsychiatry is the specialty concerned with "straightening out" behavioral disorders.

Orthoaxis. The true anatomical axis about which a limb rotates, as opposed to the assumed axis. The assumed axis is usually the most obvious or geometric one; the orthoaxis is less evident and can only be found by the use of anatomical landmarks.

Os-, oste-, osteo- (prefix). Pertaining to bone. The Latin *os-* is most often associated with anatomical structures, whereas the Greek *osteo-* usually refers to conditions involving bone. Osteogenesis means formation of bone.

Oscillation. The variation, usually with time, of the magnitude of a quantity with respect to a specified reference when the magnitude is alternatively greater and smaller than the reference.

OSHA. U.S. Occupational Safety and Health Administration.

OSHA 200 Log. Record keeping of employee injuries and illnesses is required by OSHA standard; OSHA 200 Log is a format that contains the necessary required details. It may be used by employers and is available from OSHA.

Osmosis. The passage of fluid through a semipermeable membrane as a result of osmotic pressure.

Osseous. Pertaining to bone.

Ossicle. One of three small bones in the ear that connects the eardrum with the oval window of the inner ear.

Ot-, oto- (prefix). Pertaining to the ear. Otorrhea means ear discharge.

Otitis media. An inflammation and infection of the middle ear.

Otologist. A physician specializing in surgery and diseases of the ear.

Otosclerosis. A condition of the ear caused by growth of body tissue about the foot plate of the stapes and oval window of the inner ear; results in a gradual loss of hearing.

Outdoor air (OA). "Fresh" air mixed with return air (RA) to dilute contaminants in the supply air (SA).

Output power and output energy. Power is used primarily to rate CW lasers because the energy delivered per unit time remains relatively constant (output measured in watts). In contrast, pulsed lasers deliver their energy output in pulses and their effects may be best categorized by energy output per pulse. The output power of CW lasers is usually expressed in milliwatts or watts, pulsed lasers in kilowatts, and q-switched pulsed lasers in megawatts or gigawatts. Pulsed energy output is usually expressed as joules per pulse.

Overexposure.   Exposure beyond the specified limits.

Oxidation.   Process of combining oxygen with some other substance; technically, a chemical change in which an atom loses one or more electrons. Opposite of reduction.

Oxygen deficiency.   A point at which the concentration (or partial pressure) of oxygen in air is inadequate. Various concentrations may be "deficient"; two are 18% and 16% by volume in air. Another definition of deficient is oxygen partial pressure less than 120 mmHg.

PAH.   Polynuclear aromatic hydrocarbons.

Pair production.   The conversion of a gamma ray into a pair of particles: an electron and a positron. That is an example of direct conversion of energy into matter according to Einstein's famous formula, $E = mc^2$: energy = mass × velocity of light squared.

Palmar arch.   Blood vessels in the palm of the hand from which the arteries supplying blood to the fingers are branched. Pressure against the palmar arch by poorly designed tool handles may cause ischemia of the fingers and loss of tactile sensation and precision of movement.

Palpitation.   Rapid heartbeat of which a person is acutely aware.

Papilloma.   A small growth or tumor of the skin or mucous membrane; warts and polyps, for example.

Papule.   A small, solid, usually conical elevation of the skin.

Papulovesicular.   Characterized by the presence of papules and vesicles.

Para- (prefix).   Alongside, near, abnormal; as in paraproctitus, inflammation of tissues near the rectum. A Latin suffix with the same spelling, *-para,* denotes bearing or giving birth, as in multipara, a woman who has given birth to two or more children.

Paraffins, paraffin series.   From *parm affinis* — small affinity. Straight or branched-chain hydrocarbon components of crude oil and natural gas whose molecules are saturated (that is, carbon atoms attached to each other by single bonds) and therefore very stable. Examples are methane and ethane. Generalized formula: C_nH_2n+2.

Parasite.   An organism that derives its nourishment from a living plant of animal host. Does not necessarily cause a disease.

Parenchyma.   The distinguished or specific (working) tissue of a bodily gland or organ, contained in and supported by the connected tissue framework, or stroma.

Parent.   Precursor; the name given to a radioactive product or daughter.

Partial barrier.   An enclosure constructed so that sound transmission between its interior and its surroundings is minimized.

Particle.   Also, aerosol. A small, discrete solid or liquid matter in air. Could include smokes, fogs, mists, dusts, fumes, or sprays.

Particle concentration.   Concentration expressed in terms of number of particles per unit volume of air or other gas. When expressing particle concentrations, the method of determining the concentration should be stated.

Particle size.   The measured dimension of liquid or solid particles, usually in microns.

Particle size distribution.   The statistical distribution of the sizes or ranges of size of a population of particles.

Particulate.   A particle of solid or liquid.

Particulate matter.   A suspension of fine solid or liquid particles in air, such as dust, fog, fume, mist, smoke, or sprays. Particulate matter suspended in air is commonly known as an aerosol.

Path-, patho-, (prefix), -pathy (suffix).   Feeling, suffering, disease. Pathogenic means producing disease; enteropathy means disease of the intestines; pathology is the medical specialty concerned with all aspects of disease. The root appears in the everyday word sympathy (feeling with).

Pathogen.   Any microorganism capable of causing disease.

Pathogenesis.   Describes how a disease takes hold on the body and spreads.

Pathogenic.   Producing or capable of producing disease.

Pathognomonic.   Distinctive or characteristic of a specific disease or pathological condition; a sign or symptom from which a diagnosis can be made.

Pathological.   Abnormal or diseased.

Pathology.   The study of disease processes.

PEL.   Permissible exposure limit. OSHA version of acceptable or upper-limit exposure in the workplace.

Pelleting.   In various industries, powdered material may be made into pellets or briquettes for convenience. The pellet is a distinctly small briquette. *See* Pelletizing.

Pelletizing.   Refers primarily to extrusions and to some balled products. Generally regarded as being larger than grains and smaller than briquettes.

Percent impairment of hearing.   Or percent hearing loss. An estimate of a person's ability to hear correctly; usually determined by the pure tone audiogram. The specific rule for calculating this quantity varies from state to state according to law.

Percutaneous.   Performed through the unbroken skin, as by absorption of an ointment through the skin.

Peri- (prefix).   Around, about, surrounding. Periodontium is tissue that surrounds and supports the teeth.

Periodic table.   Systematic classification of the elements according to atomic numbers (nearly the same order as by atomic weights) and by physical and chemical properties.

Peripheral neuropathy.   Deterioration of peripheral nerve function; affects the hands, arms, feet, and legs. Certain hydrocarbon solvents are known to cause peripheral neuropathies in overexposed individuals.

Permanent disability or permanent impairment.   The partial or complete loss or impairment of any part or function of the body.

Permeation.   Process by which a chemical moves through a protective clothing material on a molecular level.

Permissible dose.   *See* MPC, MPL.

Permissible exposure limit (PEL).   An exposure limit published and enforced by OSHA as legal standard.

Personal protective equipment.   Devices worn by the worker to protect against hazards in the environment. Respirators, gloves, and hearing protectors are examples.

Person–machine Interface.   Points of perceptual or physical contact between people and the equipment they work with; also "man/machine" interface.

Pesticides.   General term for chemical used to kill such pests as rats, insects, fungi, bacteria, weeds, and so on, that prey on humans or agricultural products. Among these are insecticides, herbicides, fungicides, rodenticides, miticides, fumigants, and repellents.

Petrochemical.   A term applied to chemical substances produced from petroleum products and natural gas.

PF.   Protection factor. In respirator usage, PF refers to the ratio of the concentration outside the mask to that inside (e.g., if the concentration outside is 100 ppm and the concentration inside is 5 ppm, then PF = 100/5 = 20).

PH.   The degree of acidity or alkalinity of a solution, with neutrality indicated as 7.

Phagocyte.   A cell in the body that engulfs foreign material and consumes debris and foreign bodies.

Phalanx (pl. phalanges).   Any of the bones of the fingers or toes. Often used as anatomical reference points in ergonomic work analysis.

Pharmaceuticals.   Drugs and related chemicals reaching the public primarily through drug suppliers. In government reports, this category includes not only such medicinals as aspirin and antibiotics but also such nutrients as vitamins and amino acids for both human and animal use.

Pharyngeal.   Pertaining to the pharynx (the musculo-membranous sac between the mouth, nares, and esophagus).

Phenol. C_6H_5OH. Popularly known as carbolic acid; important chemical intermediate and base for plastics, pharmaceuticals, explosives, antiseptics, and many other end products.

Phenolic resins.   A class of resins produced as the condensation product of phenol or substituted phenol and formaldehydes.

Phosphors.   Fluorescent or luminescent materials.

Photochemical process.   Chemical changes brought about by radiant energy acting upon various chemical substances. *See* Photosynthesis.

Photoelectric effect.   Occurs when an electron is thrown out of an atom by a light ray or gamma ray. This effect is used in an "electric eye," where light falls on a sensitive surface throwing out electrons that can then be detected.

Photoionization detector (PID).   A direct-reading monitoring instrument that operates by detecting and distinguishing between ions of vapors and gases following ionization by the ultraviolet light source of the instrument.

Photomultiplier tube. A vacuum tube that multiplies electron input.

Photon. A bundle (quantum) of radiation. Constitutes, for example, X rays, gamma rays, and light.

Photophobia. Abnormal sensitivity to light.

Photopic vision. Vision attributed to cones of the eye; the ability to discriminate small detail and color; usually associated with vision in daylight.

Photosynthesis. The process by which plants produce carbohydrates and oxygen from carbon dioxide and water.

Physical hazards of chemicals. A chemical for which there is scientifically valid evidence that it is a combustible liquid, a compressed gas, explosive, flammable, an organic peroxide, an oxidizer, pyrochloric, unstable (reactive), or water reactive.

Physiology. The study of the functions in disease or modified by a disease.

Physiopathology. The science of functions in disease or modified by a disease.

Pig. (1) A container (usually lead) used to ship or store radioactive materials. The thick walls protect workers from radiation. (2) In metal refining, a small ingot from the casting of blast furnace metal.

Pigment. A finely divided insoluble substance that imparts color to a material.

Pilot facility. Small-scale operation preliminary to major enterprises. Common in the chemical industry.

Pinch point. Any point at which it is possible to be caught between the moving parts, stationary parts, or the material being processed. *See also* Nip point.

Pink noise. Noise that has been weighted, especially at the low end of the spectrum, so that the energy per band (usually octave band) is approximately constant over the spectrum.

Pinna. Ear flap; the part of the ear that projects from the head. Also known as the auricle.

Pitch. The attribute of auditory sensation in terms of which sounds may be ordered on a scale extending from low to high. Pitch depends primarily on the frequency of the sound stimulus, but also on the sound pressure and waveform of the stimulus.

Pitot tube. A device consisting of two concentric tubes: one serving to measure the total or impact pressure existing in the airstream; the other to measure the static pressure only. When the annular space between the tubes and the interior of the center tube are connected across a pressure-measuring device, the pressure difference automatically nullifies the static pressure, and the velocity pressure alone is registered.

Plasma. (1) The fluid part of the blood in which the blood cells are suspended. Also called protoplasm. (2) Gas that has been heated to a partially or completely ionized condition, enabling it to conduct an electric current.

Plasma arc welding (PAW). A process that produces coalescence of metals by heating them with a constricted arc between an electrode and the workpiece (transferred arc) or between the electrode and the constricting nozzle (nontransferred arc). Shielding is obtained by the hot, ionized gas issuing from the orifice, which may be supplemented by an auxiliary source of shielding gas. Shielding gas can be an inert gas or a mixture of gases.

Pressure may or may not be used and filler metal may or may not be supplied.

Plastics. Any one of a large group of materials that contains as an essential ingredient an organic substance of large molecular weight. Two basic types are thermosetting (irreversibly rigid) and thermoplastic (reversibly rigid). Before compounding and processing, plastics often are referred to as (synthetic) resins. Final form may be a film, sheet, solid, or foam (flexible or rigid).

Plasticizers. Organic chemicals used in modifying plastics, synthetic rubber, and processing, and to impart flexibility to the end product.

Plenum. Pressure equalizing chamber.

Plenum chamber. An air compartment connected to one or more ducts or connected to a slot in a hood; used for air distribution.

Pleura. The thin membrane investing the lungs and lining the thoracic cavity, completely enclosing a potential space known as the pleural cavity. There are two pleurae, right and left, entirely distinct from each other. The pleura is moistened with a secretion that facilitates the movements of the lungs and the chest.

Pleurisy. Condition caused when the outer lung lining (visceral pleura) and the chest cavity's inner lining (parietal pleura) lose their lubricating properties; the resultant friction causes irritation and pain.

Plumbism. A name for lead intoxication.

Plume trap. An exhaust ventilation hood designed to remove the plume given off the target on impact of a laser beam.

Plutonium. A heavy element that undergoes fission under the impact of neutrons. It is a useful fuel in nuclear reactors. Plutonium cannot be found in nature, but can be produced and "burned" in reactors.

Pneumo- (Greek), pulmo- (Latin) (prefix). Pertaining to the lungs.

Pneumoconiosis. Dusty lungs; a result of the continued inhalation of various kinds of dust or other particles.

Pneumoconiosis-producing dust. Dust that, when inhaled, deposited, and retained in the lungs, may produce signs, symptoms, and findings of pulmonary disease.

Pneumonitis. Inflammation of the lungs.

Poison. (1) A material introduced into the reactor core to absorb neutrons. (2) Any substance that, when taken into the body, is injurious to health.

Poison, Class A. A U.S. DOT hazard class for extremely dangerous poisons, that is, poisonous gases or liquids of such nature that a very small amount of the gas, or vapor of the liquid, mixed with air is dangerous to life. Some examples: phosgene, cyanogen, hydrocyanic acid, and nitrogen peroxide.

Poison, Class B. A U.S. DOT hazard class for liquid, solid, paste, or semisolid substances (other than Class A poisons or irritating materials) that are known (or presumed on the basis of animal tests) to be so toxic to humans as to afford a hazard to health during transportation. Some examples: arsenic, beryllium chloride, cyanide, mercuric oxide.

Polarography. A physical analysis method for determining certain atmospheric pollutants that are electroreducible or electro-oxidizable and are in true solution and stable for the duration of the measurement.

Polar solvents. Solvents (such as alcohols and keytones) that contain oxygen and that have high dielectric constants.

Pollution. Synthetic contamination of soil, water, or atmosphere beyond that which is natural.

Poly- (prefix). Many.

Polycythemia. A condition marked by an excess in the number of red corpuscles in the blood.

Polymer. A high-molecular-weight material formed by the joining together of many simple molecules (monomers). There may be hundreds or even thousands of the original molecules linked end to end and often cross-linked. Rubber and cellulose are naturally occurring polymers. Most resins are chemically produced polymers.

Polymerization. A chemical reaction in which two or more small molecules combine to form larger molecules (polymers) that contain repeating structural units of the original molecules. A hazardous polymerization is one with an uncontrolled release of energy.

Polystyrene resins. Synthetic resins formed by polymerization of styrene.

Popliteal clearance. Distance between the front of the seating surface and the popliteal crease. This should be about 5 in. in good seat design to prevent pressure on the popliteal artery.

Popliteal crease (or line). The crease in the hollow of the knee when the lower leg is flexed. Important anatomical reference point.

Popliteal height of chair. The height of the highest part of the seating surface above the floor.

Popliteal height of individual. The distance between the crease in the hollow of the knee and the floor.

Porphyrin. One of a group of complex chemical substances that forms the basis of he respiratory pigments of animals and plants; hemoglobin and chlorophyll are other examples.

Portal. Place of entrance.

Portland cement. *See* Cement, Portland.

Positive displacement pump. Any type of air mover pump in which leakage is negligible, so that the pump delivers a constant volume of fluid, building up to any pressure necessary to deliver that volume.

Positron. A particle that has the same weight and charge as an electron but is electrically positive rather than negative. The existence of the positron was predicted in theory years before it was actually detected. It is not stable in matter because it reacts readily with an electron to yield two gamma rays.

Potential energy. Energy due to position of one body with respect to another or to the relative parts of the same body.

Pound-mole. The molecular weight equivalent of a material in pounds, e.g., one pound-mole of water is 18 lb (h = 1 + 1, O =16).

Power.   Rate at which work is done; measured in watts (1 joule per second) and horsepower (33,000 foot-pounds per minute); 1 horsepower equals 746 watts.

Power density.   The intensity of electromagnetic radiation per unit area, expressed as watts/cm.

Power level.   Ten times the logarithm to the base 10 of the ratio of a given power to a reference power; measured in decibels.

ppb.   Parts per billion.

PPE.   *See* Personal protective equipment.

ppm.   Parts per million parts of air by volume of a vapor or gas or other contaminant.

Precision.   The degree of agreement (expressed in terms of distribution of test results about the mean result) of repeated measurements of the same property, obtained by repetitive testing of a homogeneous sample under specified conditions. The precision of a method is expressed quantitatively as the standard deviation, computed from the results of a series of controlled determinations.

Presby- (prefix).   Old. As in presbyopia, eye changes associated with aging.

Presbycusis.   Hearing loss due to aging. Usually occurs in nerve cells in cochlea.

Presence-sensing device.   A device designed to detect an intrusion into a defined danger zone and to cause the potential harmful action to cease.

Pressure.   Force applied to or distributed over a surface; measured as force per unit area. *See also* Atmospheric pressure; Gauge pressure; Standard temperature and pressure; Static pressure; Pressure total; Vapor pressure; and Velocity pressure.

Pressure drop.   The difference in static pressure measured at two locations in a ventilation system; caused by friction or turbulence.

Pressure loss.   Energy lost from a pipe or duct system through friction or turbulence.

Pressure, static.   The potential pressure exerted in all directions by a fluid at rest. It is the tendency either to burst or to collapse the pipe. Usually expressed in "inches of water gauge" when dealing with air. *See also* Static pressure.

Pressure, total.   In the theory of the flow of fluids, the sum of the static pressure and the velocity pressure at the point of measurement. Also called dynamic pressure.

Pressure, vapor.   *See* Vapor pressure.

Pressure vessel.   A storage tank or vessel designed to operate at pressures greater than 15 psig (103 kPa).

PRF laser.   A pulsed recurrence frequency laser, which is a pulsed-type laser with properties similar to a CW laser when the frequency is very high.

Probe.   A tube used for sampling or for measuring pressures at a distance from the actual collection or measuring apparatus; commonly used for reaching inside stacks or ducts.

Process safety management (PSM).   Encompassing safety concept for the chemical processing industry that is mandated and regulated in the OSHA process safety management standard. In PSM, potential hazards are systematically analyzed for each step of a chemical process.

Product liability. The liability a merchant or a manufacturer may incur as the result of some defect in the product sold or manufactured, or the liability a contractor might incur after job completion from improperly performed work.

Prokaryote. Single-celled organism lacking mitochondria and a defined nucleus. Usually has a cell wall. Describes primarily bacterial organisms.

Proliferation. The reproduction or multiplication of similar forms, especially of cells and morbid cysts.

Pronation. Rotation of the forearm in a direction to face the palm downward when the forearm is horizontal, and backward when the forearm is vertical.

Propagation of flame. The spread of flame through the entire volume of a flammable vapor–air mixture from a single source of ignition. A vapor–air mixture below the lower flammable limit may burn at the point of ignition without propagating from the ignition source.

Prophylactic. Preventive treatment for protection against disease.

Protection factor (PF). In respiratory protective equipment, the ratio of the ambient airborne concentration of the contaminant to the concentration inside the face piece.

Protective atmosphere. A gas envelope surrounding an element to be brazed, welded, or thermal-sprayed with the gas composition controlled with respect to chemical composition, dew point, pressure, flow rate, and so on.

Protective coating. A thin layer of metal or organic material, applied as paint to a surface to protect it from oxidation, weathering, and corrosion.

Proteins. Large molecules found in the cells of all animal and vegetable matter containing carbon, hydrogen, nitrogen, and oxygen, and sometimes sulfur and phosphorus. The fundamental structural units of proteins are amino acids.

Proteolytic. Capable of splitting or digesting proteins into simpler compounds.

Proton. A fundamental unit of matter having a positive charge and a mass number of one.

Protoplasm. The basic material from which all living tissue is made. Physically, it is a viscous, translucent, semifluid colloid, composed mainly of proteins, carbohydrates, fats, salts, and water.

Protozoa. Single-celled microorganisms belonging to the animal kingdom.

Proximal. The part of a limb that is closest to the point of attachment. The elbow is proximal to the wrist, which is proximal to the fingers.

psi. Pounds per square inch. For technical accuracy, pressure must be expressed as psig (pounds per square inch gauge) or psia (pounds per square inch absolute); that is, gauge pressure plus sea level atmospheric pressure, of psig plus about 14.7 pounds per square inch).

psig. Pounds per square inch gauge. *See* psi.

Psittacosis. Parrot fever. An infectious disease of birds to which poultry handlers and other workers exposed to dried bird feces are at risk. Caused by *Chlamydia psittaci.* The most noted symptom of the disease among humans is fever.

Psych-, psycho- (prefix). Pertaining to the mind, from the Greek for soul.

Psychogenic deafness. Loss originating in or produced by the mental reaction of individuals to their physical or social environment. It is sometimes called functional deafness or feigned deafness.

Psychogenic disease. Real or perceived health effects that originate in mental or emotional reactions to physical or social environments.

Psychrometer. An instrument consisting of wet- and dry-bulb thermometers for measuring relative humidity.

Psychrometric chart. A graphical representation of the thermodynamic properties of moist air.

Pterygium. A growth of the conjunctiva caused by a degenerative process brought on by long, continued irritation (as from exposure to wind, dust, and possibly ultraviolet radiation).

Pulmonary. Pertaining to the lungs.

Pulsed laser. A class of laser characterized by operation in a pulsed mode; that is, emission occurs in one or more flashes of short duration (pulse length).

Pulse length. Duration of a pulsed laser flash; may be measured in milliseconds, microseconds, or nanoseconds.

Pulvation. The act of particles being emitted or induced to become airborne; term coined by Hemeon.

Pumice. A natural silicate from volcanic ash or lava. Used as an abrasive.

Pupil. The variable aperture in the iris through which light travels toward the interior regions of the eye. The pupil size varies from 2 to 8 mm.

Pur-, pus- (Latin), Pyo- (Greek) (prefix). Indicates pus, as in purulent, supportive, pustulant, and pyoderma.

Purefaction. Decomposition of proteins by microorganisms, producing disagreeable odors.

Pure tone. A sound wave characterized by its singleness of frequency.

Purpura. Extensive hemorrhage into the skin or mucous membrane.

Push–pull hood. A local exhaust hood that has an air jet, opposite the hood intake, which pushes a stream of air across the control surface toward the exhaust opening.

Pyel-, pyelo- (prefix). Pertaining to the urine-collecting chamber of the kidney.

Pyloric stenosis. Obstruction of the pyloric opening of the stomach caused by hypertrophy of the pyloric sphincter.

Pylorus. The orifice of the stomach leading to the small intestine.

Pyr-, pyret- (prefix). Fever.

Pyrethrum. A pesticide obtained from the dried, powdered flowers of the plant of the same name; mixed with petroleum distillates, it is used as an insecticide.

Pyrolysis. The breaking apart of complex molecules into simpler units by the use of heat, as in the pyrolysis of heavy oil into gasoline.

QF. *See* Quality factor.

Q fever.   Disease caused by a rickettsial organism that infects meat and livestock handlers; similar but not identical to tick fever.

Q-switched laser.   Also known as Q-spoiled. A pulsed laser capable of extremely high peak powers for very short durations (pulse length of several nanoseconds).

Qualitative fit testing.   A method of assessing the effectiveness of a particular size and brand of respirator based on an individual's subjective response to a test atmosphere. The most common test agents are isoamyl acetate (banana oil), irritant smoke, and sodium saccharin. Proper respirator fit is indicated by the individual reporting no indication of the test agent inside the face piece during the performance of a full range of facial movements.

Quality.   A term used to describe the penetrating power of X rays or gamma rays.

Quality assurance (Quality control).   A management function to ensure that the products or goods are produced as intended.

Quality factor.   A linear energy transfer–dependent factor by which absorbed radiation doses are to be multiplied to obtain the dose equivalent.

Quantitative fit testing.   A method of assessing the effectiveness of a particular size and brand of respirator on an individual. Instrumentation is used to measure both the test atmosphere (a gas, vapor, or aerosol, such as DOP) and the concentration of the test contaminant inside the face piece of the respirator. The quantitative fit factor thus obtained is used to determine if a suitable fit has been obtained by referring to a table or to the software of the instrumentation. Quantitative fit factors obtained in this way do not correlate well with Assigned Protection Factors, which are based on actual measurements of levels of contaminant inside the face piece during actual work.

Quantum.   "Bundle of energy"; discrete particle of radiation; pl. quanta.

Quartz.   Vitreous, hard, chemically resistant, free silica, the most common form in nature. The main constituent in sandstone, igneous rocks, and common sands.

Quenching.   A heat-treating operation in which metal raised to the desired temperature is quickly cooled by immersion in an oil bath.

Rabbit.   A capsule that carries samples in and out of an atomic reactor through a pneumatic tube to permit study of the effect of intense radiation on various materials.

rad.   Roentgen absorbed dose or radiation absorbed dose; a standard unit of absorbed ionizing radiation dose equal to 100 erg absorbed per gram.

Radar (Radio Detection and Ranging).   A radio detecting instrument able to measure distance to an object, among other characteristics.

Radial deviation.   Flexion of the hand that decreases the angle between its longitudinal axis and radius. Tool design should minimize radial deviation. Strength of grasp is diminished in radial deviation.

Radian.   An arc of a circle equal in length to the radius.

Radiant temperature. The temperature resulting from a body absorbing radiant energy.

Radiation (nuclear). The emission of atomic particles or electromagnetic radiation from the nucleus of an atom.

Radiation protection guide (RPG). The radiation dose that should not be exceeded without careful consideration of the reasons for doing so; every effort should be made to encourage the maintenance of radiation doses as far below this guide as practicable.

Radiation (radioactivity). *See* Ionizing radiation.

Radiation source. An apparatus or material emitting or capable of emitting ionizing radiation.

Radiation (thermal). The transmission of energy by means of electromagnetic waves longer than visible light. Radiant energy of any wavelength may, when absorbed, become thermal energy and result in an increase in the temperature of the absorbing body.

Radiator. That which is capable of emitting energy in waveform.

Radioactive. The property of an isotope or element that is characterized by spontaneous decay to emit radiation.

Radioactivity. Emission of energy in the form of alpha, beta, or gamma radiation from the nucleus of an atom. Always involves change of one kind of atom into a different kind. A few elements, such as radium, are naturally radioactive. Other radioactive forms are induced. *See also* Radioisotope.

Radioactivity concentration guide (RCG). The concentration of radioactivity in the environment that is determined to result in organ doses equal to the radiation protection guide (RPG).

Radiochemical. Any compound or mixture containing a sufficient portion of radioactive elements to be detected by a Geiger counter.

Radiochemistry. The branch of chemistry concerned with the properties and behavior of radioactive materials.

Radiodiagnosis. A method of diagnosis that involves X-ray examination.

Radiohumeral joint. Part of the elbow. Not truly a joint, but a thrust bearing.

Radioisotope. A radioactive isotope of an element. A radioisotope can be produced by placing material in a nuclear reactor and bombarding it with neutrons. Many of the fission products are radioisotopes. Sometimes used as tracers, as energy sources for chemical processing or food pasteurization, or as heat sources for nuclear batteries. Radioisotopes are at present the most widely used outgrowth of atomic research and are one of the most important peacetime contributions of nuclear energy.

Radionuclide. A radioactive nuclide; one that have the capability of spontaneously emitting radiation.

Radioresistant. Relatively invulnerable to the effect of radiation.

Radiosensitive. Tissues that are more easily damaged by radiation.

Radiotherapy. Treatment of human ailments with the application of relatively high roentgen dosages.

Radium. One of the earliest-known naturally radioactive elements. It is far more radioactive than uranium and is found in the same ores.

Radius.   The long bone of the forearm in line with the thumb; the active element in the forearm during pronation (inward rotation) and supination (outward notation). Also provides the forearm connection to the wrist joint.

Rale.   Any abnormal sound or noise in the lungs.

Random noise.   A sound or electrical wave whose instantaneous amplitudes occur as a function of time, according to a normal (Gaussian) distribution curve. Random noise is an oscillation whose instantaneous magnitude is not specified for any given instant of time. The instantaneous magnitudes of a random noise are specified only by probability functions giving the fraction of the total time that the magnitude, or some sequence of the magnitudes, lies within a specified range.

Rare earths.   Originally, the elements in the periodic table with atom numbers 57 through 71. Often included are numbers 39 and, less often, 21 and 90. Emerging uses include the manufacture of special steels and glasses.

Rash.   Abnormal reddish coloring or blotch on some part of the skin.

Rated-line voltage.   The range of potentials, in volts, of the supply line specified by the manufacturer at which an X-ray machine is designed to operate.

Rated output current.   The maximum allowable lead current of an X-ray high-voltage generator.

Rated output voltage.   The allowable peak potential, in volts, at the output terminals of an X-ray high-voltage generator.

Raynaud's syndrome phenomenon.   Abnormal constriction of the blood vessels of the fingers on exposure to cold temperature.

RBE.   Relative biological effectiveness; the relative effectiveness of the same absorbed dose of two ionizing radiations in producing a measurable biological response.

RCG.   *See* Radioactivity concentration guide.

Reaction.   A chemical transformation or change; the interaction of two or more substances to form new substances.

Reactivity (chemical).   The susceptibility of a substance to undergo a chemical reaction or change that may result in dangerous side effects, such as an explosion, burning, and corrosive or toxic emissions.

Reactor.   An atomic "furnace" or nuclear reactor. In a reactor, nuclei of the fuel undergo controlled fission under the influence of neutrons. The fission produces new neutrons in a chain reaction that releases large amounts of energy. This energy is removed as heat that can be used to make steam. The moderator for the first reactor was piled-up blocks of graphite. Thus, a nuclear reactor was formerly referred to as a pile. Reactors are usually classified now as research, test, process heat, and power, depending on their principal function. No workable design for a controlled fusion reactor has yet been devised.

Reagent.   Any substance used in a chemical reaction to produce, measure, examine, or detect another substance.

Receiving hood.   A one- or two-sided overhead hood that receives rising hot air or gas.

Recommended exposure limit (REL). An exposure limit, generally a time-weighted average, to a substance; developed by NIOSH based on toxicological and industrial hygiene data.

Recoil energy. The energy emitted and shared by the reaction products when a nucleus undergoes a nuclear reaction such as fission or radioactive decay.

Reduction. Addition of one or more electrons to an atom through chemical change.

Reflectance. The ratio of light reflected from a surface to the light arriving at the surface.

Refractories. Materials exceptionally resistant to the action of heat and hence used for lining furnaces; examples are fire clay, magnesite, graphite, and silica.

Regenerative process. Replacement of damaged cells by new cells.

Regimen. A regulation of the mode of living, diet, sleep, exercise, and so on for a hygienic or therapeutic purpose; sometimes mistakenly called regime.

Reid method. A method of determining the vapor pressure of a volatile hydrocarbon by the Standard Method of Test for Vapor Pressure of Petroleum Products, ASTM D232.

Relative biological effectiveness. *See* RCG.

Relative humidity. The ratio of the quantity of water vapor present in the air to the quantity that would saturate it at any specific temperature.

Reliability. The degree to which an instrument, component, or system retains its performance characteristics over a period of time.

rem. Roentgen equivalent man; a radiation dose unit that equals the dose in rads multiplied by the appropriate value of relative biological effect or quality factor for the particular radiation.

Renal. Having to do with the kidneys.

Replacement air. Also, compensating air, makeup air. Air supplied to a space to replace exhausted air.

Replication. A fold or folding back; the act or process of duplicating or reproducing something.

Resin. A solid or semisolid amorphous (noncrystalline) organic compound or mixture of such compounds with no definite melting point and no tendency to crystallize. May be of vegetable (gum arabic), animal (shellac), or synthetic (celluloid) origin. Some resins may be molded, cast, or extruded. Others are used as adhesives, in the treatment of textiles and paper, or as protective coatings.

Resistance. (1) Opposition to the flow of air, as through a canister, cartridge, particulate filter, or orifice. (2) A property of conductors, depending on their dimensions, material, and temperature, that determines the current produced by a given difference in electrical potential.

Resonance. Each object or volume of air resonates or strengthens a sound at one or more particular frequencies. The frequency depends on the size and construction of the object or air volume.

Resonate. An object that resonates, strengthening the sound at a particular frequency.

Respirable-size particulates. Paticulates in a size range that permits them to penetrate deep into the lungs upon inhalation.

Respirator. A device that protects the wearer from inhalation of harmful contaminants.

Respiratory protection. Devices that protect the wearer's respiratory system from overexposure by inhalation to airborne contaminants.

Respiratory system. System consisting of the nose, mouth, nasal passages, nasal pharynx, pharynx, larynx, trachea, bronchi, bronchioles, air sacs (alveoli) of the lungs, and muscles of respiration.

Reticle. A scale or grid or other pattern located in the focus of the eyepiece of a microscope.

Retina. The light-sensitive inner surface of the eye that receives and transmits images formed by the lens.

Retro- (prefix). Backward or behind.

Return air. Air that is returned from the primary space to the fan for recirculation.

Reverberatory furnace. A furnace in which heat is supplied by burning fuel in a space between the charge and the low roof.

Reynolds number (Re). A unitless measure that describes the turbulence of a material (e.g., water, air) as it flows; Re takes into consideration the size of space (e.g., the diameter of a duct), and the velocity, density, and viscosity of the material.

Rheumatoid. Resembling rheumatism, a disease marked by inflammation of the connective tissue structures of the body, especially the membranous linings of the joints, and by pain in these parts; eventually the joints become stiff and deformed.

Rhin-, rhino- (prefix). Pertaining to the nose.

Rhinitis. Inflammation of the mucous membrane lining in the nasal passages.

Rickettsia. Rod-shaped microorganisms that grow within the cells of animals. These human pathogens are often carried by arthropods.

Riser. In metal casting, a channel in a mold to permit escape of gases.

Risk. (1) An insurance term for insured value and another name for the insured or prospective insured. (2) A term applied to the individual or combined assessments of "probability of loss" and potential amount of loss.

rms average. Root mean square average. Root mean square is obtained by squaring each entry in a timed set of numbers, adding all the squares, dividing by the total time, and then taking the square root of that number.

Roasting of ores. A refining operation in which ore is heated to a high temperature, sometimes with catalytic agents, to drive off certain impurities; an example is the roasting of copper ore to remove sulfur.

Roentgen (R). A unit of radioactive or exposure. *See* rad.

Roentgenogram. A film produced by exposing X-ray film to X rays.

Roentgenography. Photography by means of roentgen rays. Special techniques for roentgenography of different areas of the body have been given specific names.

Rosin. Specifically applies to the resin of the pine tree and chiefly derives from the manufacture of turpentine. Widely used in the manufacture of soap and flux.

Rotometer. A flowmeter consisting of a precision-bored, tapered, transparent tube with a solid float inside.

Rotary kiln. Any of several types of ovens used to heat material, as in the Portland cement industry.

Rouge. A finely powdered form of iron oxide used as a polishing agent.

Route of entry. A path by which chemicals can enter the body. There are three main routes of entry: inhalation, ingestion, and skin absorption.

RTECS. Registry of Toxic Effects of Chemical Substances.

SAE. Sampling and analytical error. The reason a particular sampling result may vary from the true value. Quantitative estimates of SAE are often used to develop a clear picture of the potential range of a given exposure.

Safe. A condition of relative freedom from danger.

Safeguarding. The term used to encompass all methods of protection against injury or illness.

Safety. The control of recognized hazards to attain an acceptable level of risk.

Safety belt. A life belt worn by linesmen, window washers, etc., attached to a secure object (window sill, etc.) to prevent falling. A seat or torso belt securing a passenger in an automobile or airplane to provide body protection during a collision, sudden stop, air turbulence, etc.

Safety can. An approved container of not more than 5-gal (19-l) capacity having a spring-loaded lid and spout cover and designed to relieve internal pressure safely when exposed to fire.

Safety factor. *See* Factor of safety.

Safety program. Activities designed to assist employees in the recognition, understanding, and control of hazards in the workplace.

Safety shoes. A term commonly used to describe protective footwear meeting ANSI Z-41 requirements.

Sagittal plane. A plane from back to front vertically dividing the body into the right and left portions. Important in anthropometric definitions. Midsagittal plane is the sagittal plane symmetrically dividing the body.

Salamander. A small furnace, usually cylindrical in shape, without grates.

Salivation. An excessive discharge of saliva; ptyalism.

Salmonella. A genus of Gram-negative, rod-shaped pathogenic bacteria.

Salt. A product of the reaction between an acid and a base. Table salt, for example, is a compound of sodium and chlorine. It can be made by reacting sodium hydroxide with hydrochloric acid.

Sampling. The withdrawal or isolation of a fractional part of a whole. In air analysis, the separation of a portion of an ambient atmosphere with subsequent analysis to determine concentration.

Sandblasting.   A process for cleaning metal castings and other surfaces with sand projected by a high-pressure airstream.

Sandhog.   Any worker performing tunneling work requiring atmospheric pressure control.

Sanitize.   To reduce the microbial flora in or on articles such as eating utensils to levels judged safe by public health authorities.

Saprophyte.   An organism living on dead organic matter.

Saprophytic.   Obtaining nourishment from nonliving organic materials; related to the growth of microorganisms in HVAC systems.

SARA.   Superfund Amendments and Reauthorization Act.

Sarcoma.   Malignant tumors that arise in connective tissue.

Scattered radiation.   Radiation that is scattered by interaction with object or within tissue.

SCBA.   Self-contained breathing apparatus.

scfm.   Standard cubic feet per minute. Airflow rate at standard conditions; dry air at 29.92 in. Hg gauge, 70°F.

Scintillation counter.   A device for counting atomic particles by means of the tiny flashes of light (scintillations) that particles produce when they strike certain crystals or liquids.

Scler- (prefix).   Hard, tough.

Sclera.   The tough white outer coat of the eyeball.

Scleroderma.   Hardening of the skin.

Scotoma.   A blind or partially blind area in the visual field.

Scotopic vision.   Vision attributed to rods of the eye; the inability to discriminate small detail and color; usually associated with vision at night and for the detection of movement and low-light-intensity vision.

Sealed source.   A radioactive source sealed in a container or having a bonded cover, in which the container or cover has sufficient mechanical strength to prevent contact with and dispersion of the radioactive material.

Sebaceous.   Of or related to fatty material.

Seborrhea.   An oily skin condition caused by an excess output of sebum from the sebaceous glands of the skin.

Self-ignition.   *See* Autoignition temperature.

Self-insurance.   A term used to describe the assumption of one's own financial risk.

Semicircular canals.   The special organs of balance closely associated with the hearing mechanism and the eighth cranial nerve.

Semiconductor or junction laser.   A class of laser that normally produces relatively low CW power outputs; can be tuned in wavelength and has the greatest efficiency.

Sensation.   The translation into consciousness of the effects of a stimulus exciting a sense organ.

Sensible.   Capable of being perceived by the sense organs.

Sensible heat.   Heat manifested to body senses; the heat that fixes the temperature of air; heat that, when added to air, changes its temperature.

Sensitivity. The minimum amount of contaminate that can repeatedly be detected by an instrument.

Sensitization. The process of rendering an individual sensitive to the action of a chemical.

Sensitizer. A material that can cause an allergic reaction of the skin or respiratory system.

Sensorineural. Type of hearing loss that affects millions of people. If the inner ear is damaged, the hearing loss is sensory; if the fibers of the eighth nerve are affected, it is a neural hearing loss. Because the pattern of hearing loss is the same in either case, the term sensorineural is used... .

Sensory end organs. Receptor organs of the sensory nerves located in the skin. Each end organ can sense only a specific type of stimulus. Primary stimuli are heat, cold, or pressure, each requiring different end organs.

Sensory feedback. Use of external signals perceived by sense organs to indicate quality or level of performance of an event triggered by voluntary action. On the basis of sensory feedback information, decisions may be made; for example, permitting or not permitting an event to run its course or enhancing or decreasing activity levels.

Septicemia. Blood poisoning; growth of infectious organisms in the blood.

Septum. A dividing wall or partition; used as a general term in anatomical nomenclature.

Sequestrants. Chelates used to deactivate undesirable properties of metal ions without removing these ions from solution. Sequestrants have many uses, including application as antigumming agents in gasoline, antioxidants in rubber, and rancidity retardants in edible fats and oils.

Serious violation. Any violation in which there is a substantial probability that death or serious physical harm could result from violating the conditions of the OSHA Act.

Serum. (1) The clear fluid that separates from the blood during clotting. (2) Blood serum-containing antibodies.

Shakeout. In the foundry industry, the separation of the solid, but still not cold, casting from its molding sand.

Shakes. Common term for metal fume fever.

Shale. Many meanings in industry, but in geology, a common fossil rock formed from clay, mud, or silt; somewhat stratified but without characteristic cleavage.

Shale oil. Tarry oil distilled from bituminous shale.

Shaver's disease. Bauxite pneumoconiosis.

Shell. Spheres centered on the nucleus of an atom. The electrons around the nucleus of an atom are arranged in shells. The innermost shell is called the K-shell, the next is called the L-shell, and so on to the Q-shell. The nucleus itself may also have a shell-type structure.

Shielded-metal arc welding (SMAW). An arc-welding process that produces coalescence of metal by heating them with an arc between a covered metal electrode and the work. Shielding is obtained from decomposition of the electrode covering. Pressure is not used and filler metal is obtained from the electrode.

Shield, shielding. Interposed material (such as a wall) that protect workers from harmful radiations released by radioactive materials.

Shock. Primarily, the rapid fall in blood pressure following injury, a surgical operation, or the administration of anesthesia.

Short-term exposure limit (STEL). ACGIH-recommended exposure limit. Maximum concentration to which workers can be exposed for a short period of time (15 min) only four times throughout the day with at least 1 hr between exposures. *See* TLV.

Shotblasting. A process for cleaning metal castings or other surfaces by small steel shot in a high-pressure airstream; a substitute for sandblasting to avoid silicosis.

SI. Systeme International d'Unités (International System of Units); the metric system that is being adopted throughout the world. It is a modern version of the MKSA (meter, kilogram, second, ampere) system, whose details are published and controlled by an international treaty organization financed by member states of the Metre Convention, including the United States.

SIC. *See* Standard Industrial Classification Code.

Sick building syndrome (SBS). Usually refers to a class of complaints or symptoms (e.g., discomfort, headache, urinary tract irritation) seen in IAQ episodes.

Siderosis. The deposition of iron pigments in the lung; can be associated with disease.

Sievert. Unit of absorbed radiation dose in gray times the quality factor of the radiation in comparison with gamma radiation; 1 sievert = 100 rem.

Silica gel. Compounds of silicon, oxygen, and one or more metals with or without hydrogen. These dusts cause nonspecific dust reactions, but generally do not interfere with pulmonary function or result in disability.

Silicates. Compounds consisting of oxygen, silicon, and one or more metals.

Silicon. A nonmetallic element that, next to oxygen, is the chief elementary constituent of the earth's crust.

Silicones. Unique group of compounds made by molecular combination of silicon (or certain silicon compounds) with organic chemicals. Produced in a variety of forms, including silicone fluids, resins, and rubber. Silicones have special properties, such as water repellency, wide temperature resistance, high durability, and great dielectric strength.

Silicosis. A disease of the lungs caused by the inhalation of silica dust.

Silver solder. A solder of varying components but usually containing an appreciable amount of cadmium.

Simple tone (pure tone). (1) A sound wave whose instantaneous sound pressure is a simple sinusoidal function of time. (2) A sound sensation characterized by its singularity of pitch.

Sintering. Process of making coherent powder of earthy substances by heating without melting.

Skin dose.   A special instance of tissue dose referring to the dose immediately on the surface of the skin.

Skin toxicity.   *See* Dermal toxicity.

Slag.   The dross of flux and impurities that rise to the surface of molten metal during melting and refining.

Slot velocity.   Linear flow rate through the opening in a slot-type hood (planting, degreasing operations, and so on).

Sludge.   Any muddy or slushy mass. Specifically, mud from a drill hole in boring, muddy sediment in a steam boiler, or precipitated solid matter arising from sewage treatment processes.

Slurry.   A thick, creamy liquid resulting from the mixing and grinding of limestone, clay, water, and other raw materials.. .

SMACNA.   Sheet Metal and Air Conditioning National Association.

Smelting.   One step in the procurement of metals from ore; hence, to reduce, to refine, to flux, or to scorify.

Smog.   Irritating haze resulting from the sun's effect on certain pollutants in the air, notably automobile and industrial exhaust.

Smoke.   An air suspension (aerosol) of particles originating from combustion or sublimation; generally contains droplets as well as dry particles. Tobacco, for example, produces a wet smoke composed of minute tarry droplets.

Soap.   Ordinarily a metal salt of a fatty acid, usually sodium stearate, sodium oleate, sodium palmitate, or some combination of these.

Soapstone.   Complex silicate of varied composition, similar to some talcs, with wide industrial application, including rubber manufacture.

Solder.   A material used for joining metal surfaces together by filling a joint or covering a junction. The most commonly used solder contains lead and tin; silver solder may contain cadmium. Zinc chloride and fluorides are commonly used as fluxes to clean the soldered surfaces.

Solid-state laser.   A type of laser that uses a solid crystal such as ruby or glass; commonly used in pulsed lasers.

Solution.   Mixture in which the components lose their individual properties and are uniformly dispersed. All solutions are composed of a solvent (water or other fluid) and a solute (the dissolved substance). A true solution is homogeneous, as salt in water.

Solvent.   A substance that dissolves another substance. Usually refers to organic solvents.

Soma.   Body, as distinct from psyche (mind).

Somatic.   Pertaining to all tissue other than reproductive cells.

Somnolence.   Sleepiness; also unnatural drowsiness.

Soot.   Agglomerations of carbon particles impregnated with tar; formed in the incomplete combustion of carbonaceous material.

Sorbent.   (1) A material that removes toxic gases and vapors from air inhaled through a canister or cartridge. (2) Material used to collect gases and vapors during air sampling.

Sound. An oscillation in pressure, stress, particle displacement, particle velocity, and so on, propagated in an elastic material, in a medium with internal forces (elastic or viscous, for example); or the superposition of such propagated oscillations. Also the sensation produced through the organs of hearing usually by vibrations transmitted in a material medium, commonly air.

Sound absorber. A material (e.g., air, fiberglass) capable of changing sound energy into another form of energy, usually heat. Sound absorbers reduce the sound by "absorbing" its energy.

Sound absorption. The change of sound energy into some other form, usually heat, on passing through a medium or on striking a surface. Also, the property possessed by materials and objects, including air, of absorbing sound energy.

Sound absorption coefficient. The ratio of the sound energy absorbed by the surface of a medium (or material) exposed to a sound field (or to sound radiation) to the sound energy incident on that surface.

Sound analyzer. A device for measuring the band pressure level or pressure-spectrum level of a sound as a function of frequency.

Sound level. A weighted sound-pressure level obtained by use of metering characteristics and the weighting A, B, or C, as specified in ANSI S1.4.

Sound-level meter and octave-band analyzer. Instruments for measuring sound-pressure levels in decibels referenced to 0.0002 microbars. Readings can also be made in specific octave bands, usually beginning at 75 Hz and continuing through 10,000 Hz.

Sound-pressure level (SPL). The level, in decibels, of a sound is 20 times the logarithm to the base 10 of the ratio of the pressure of this sound to the reference pressure, which must be explicitly stated.

Sound pressure units. 1 dyne\cm^2 = 1 μbar = 0.1 N/m^2 = 0.1 Pa.

Sound transmission. The word *sound* usually means sound waves traveling in air. However, sound waves also travel in solids and liquids. These sound waves may be transmitted to air to make sound that can be heard.

Sound transmission loss. The ability of a barrier to block transmission; measured in decibels.

Sound wave. A pattern of fluctuations on air pressure over distance and/or time. The fluctuations in pressure can be modeled as waves.

Sour gas. Slang for either natural gas or a gasoline contaminated with odor-causing sulfur compounds. In natural gas, the contaminant is usually hydrogen sulfide; in gasoline, usually mercaptans.

Source. Any substance that emits radiation. Usually refers to a piece of radioactive material conveniently packaged for scientific or industrial use.

Spasm. Tightening or contraction of any set of muscles.

Spray coat painting. The result of the application of a spray in painting as a substitute for brush painting or dipping. Typically creating mist that presents possible respiratory hazards.

Specific absorption (SA). Quantity of radiofrequency energy in joules per kilogram.

Specific absorption rate (SAR).   Radiofrequency dosage term expressed as watts of power per kilogram of tissue.

Specific gravity.   The ratio of the mass of a unit volume of a substance to the mass of the same volume of a standard substance at a standard temperature. Water at 39.2°F (4°C) is usually the standard for liquids; for gases, dry air (at the same temperature and pressure as the gas) is often taken as the standard substance. *See* Density.

Specificity.   The degree to which an instrument or detection method is capable of accurately detecting or measuring the concentration of a single contaminant in the presence of other contaminants.

Specific ionization.   *See* Ionization.

Specific volume.   The volume occupied by a unit mass of a substance under specified conditions of temperature and pressure.

Specific weight.   The weight per unit volume of a substance; same as density.

Spectrography, spectral emission.   An instrument method for detecting trace contaminants using a spectrum formed by exciting the subject contaminants by various means, causing characteristic radiation to be formed, which is dispersed by a grating or prism and photographed.

Spectrophotometer.   A direct-reading instrument used for comparing the relative intensities of corresponding electromagnetic wavelengths produced by absorption of ultraviolet, visible, or infrared radiation from a vapor or gas.

Spectroscopy.   Observation of the wavelength and intensity of light or other electromagnetic waves absorbed or emitted by various materials. When excited by an arc or spark, each element emits light of certain well-defined wavelengths.

Spectrum.   The frequency distribution of the magnitudes (and sometimes phases) of the components of the wave. Also used to signify a continuous range of frequencies. Usually wide in extent, within which waves have some specified common characteristics. Also, the pattern of red-to-blue light observed when a beam of sunlight passes through a prism and then projects upon a surface.

Speech interference level (SIL).   The average, in decibels, of the sound-pressure levels of a noise in the three octave bands of frequency: 600, 1200 to 2400, and 2400 to 4800 Hz.

Speech perception test.   A measurement of hearing acuity by the administration of a carefully controlled list of words. The identification of correct responses is evaluated in terms of norms established by the average performance of normal listeners.

Speech reading.   Lip reading or visual hearing.

Sphincter.   A muscle that surrounds an orifice and functions to close it.

Sphygmomanometer.   Apparatus for measuring blood pressure.

SPL formula.   Sound-pressure levels are defined by a formula that compares the measured pressure, P, to a base or reference pressure, P_0 (usually 0.0002 dynes/cm^2 = 0.00002 Pa, the weakest sound that a healthy ear can hear under ideal listening conditions; also known as the threshold of hearing.) The formula is: SPL (in decibels, dB) = 20 log P/P_0.

Spontaneously combustible. A material that ignites as a result of retained heat from processing, or that will oxidize to generate heat and ignite, or that absorbs moisture to generate heat and ignite.

Spore. A resistant body formed by certain microorganisms; resistant resting cells. Mold spores: unicellular reproductive bodies.

Spot size. Cross-sectional area of laser beam at the target.

Spot welding. One form of electrical resistance welding in which the current and pressure are restricted to the spots of metal surfaces directly in contact.

Spray-coating painting. The result of the application of spray in painting as a substitute for brush painting or dipping.

Squamous. Covered with or consisting of scales.

Stack. The device on the end of a ventilation system that disperses exhaust contaminates for dilution by the atmosphere.

Stain. A dye used to color microorganisms as an aid to visual inspection.

Stamping. A term with many different usages in industry; a common one is the crushing of ores by pulverizing.

Standard. A written guide that may or may not be a legal requirement.

Standard air. Air at standard temperature and pressure. The most common values are 70°F (21.1°C) and 29.92 in. Hg (103.3 kPa). Also, air with a density of 0.075 lb/ft³ (1.2 kg/m³) is substantially equivalent to dry air at 70°F and 29.92 in. Hg.

Standard air-density. The density of air, 0.075 lb/ft³ (1.2 kg/m³), at standard conditions.

Standard conditions. In industrial ventilation, 70°F (21.1°C), 50% relative humidity, and 29.92 in. Hg (101.3 kPa) atmospheric pressure.

Standard gravity. Standard accepted value for the force of gravity. It equals the force that produces an acceleration of 32.17 ft/s (9.8 m/s).

Standard Industrial Classification Code (SIC). Classification system for places of employment according to major type of activity.

Standard man. A theoretical physically fit man of standard (average) height, weight, dimensions, and other parameters (blood composition, percent water, mass of salivary glands, to name a few).

Standard temperature and pressure. *See* Standard air.

Standing wave. A periodic wave having a fixed distribution in space that is the result of interference of progressive waves of the same frequency and kind. Such waves are characterized by the existence of nodes or partial nodes and antinodes that are fixed in space.

Stannosis. A form of pneumoconiosis caused by the inhalation of tin-bearing dusts.

Static pressure. The potential pressure exerted in all directions by a fluid at rest. For a fluid in motion, it is measured in a direction normally (at right angles) to the direction of flow; thus, it shows the tendency to burst or collapse the pipe. When added to velocity pressure, it gives total pressure.

Static pressure curve. A graphical representation of the volumetric output and fan static pressure relationship for a fan operating at a specific rotating speed.

Static pressure regain. The increase in static pressure in a system as air velocity decreases and velocity pressure is converted into static pressure according to Bernoulli's theorem.

STEL. *See* Short-term exposure limit.

Sterile. Free of living microorganisms.

Sterility. Inability to reproduce.

Sterilization. The process of making sterile; the killing of all forms of life.

Sterilize. To perform any act that results in the absence of all life on or in an object.

Sternomastoid muscles. A pair of muscles connecting the breastbone and lower skull behind the ears, which flex or rotate the head.

Stink damp. In mining, hydrogen sulfide.

Stoke's law. A law that describes the relationships between particle shape, size, density, and settling velocity. (Spheres tend to settle most easily. Particle settling tends to increase as the square of particle diameter.).

Stp flow rate. The rate of flow of fluid, by volume, corrected to standard temperature and pressure.

Stp volume. The volume that a quantity of gas or air would occupy at standard temperature and pressure.

Stress. A physical, chemical, or emotional factor that causes bodily or mental tension and may be a factor in disease causation or fatigue.

Stressor. Any agent or thing causing a condition of stress.

Strict liability. *See* Liability, strict.

Strip mine. A mine in which coal or ore is extracted from the earth's surface after removal of overlayers of soil, clay, and rock.

STS. Standard threshold shift in hearing acuity. OSHA defines STS as a 10-decibel shift in the 2000, 3000, or 4000 Hz bands.

Stupor. Partial unconsciousness or nearly complete unconsciousness.

Sublimation. A process in which a material passes directly from a solid to a gaseous state or condenses from a gaseous state to form solid crystals, without liquefying.

Suction pressure. Archaic; refers to static pressure on upstream side of fan. *See* Static pressure.

Sulcus (pl. sulci). A groove, trench, or furrow; used in anatomical nomenclature as a general term to designate a depression, especially on the surface of the brain, separating the gyri; also, a linear depression in the surface of a tooth, the sloping sides of which meet at an angle.

Superfund. *See* CERCLA.

Supination. Rotation of the forearm about its own longitudinal axis. Supination turns the palm upward when the forearm is horizontal, and forward when the body is in anatomical position. Supination is an important element of available motions inventory for industrial applications, particularly where tools such as screwdrivers are used. Efficiency in supination depends on arm position. Workplace design should provide for elbow flexion at 90°.

Supplied-air suit.   A one- or two-piece suit that is impermeable to most particulate and gaseous contaminants and is provided with an adequate supply of respirable air.

Supra- (prefix).   Above, on.

Surface-active.   Any of a group of compounds added to a liquid to modify an agent, surfactant, or interfacial tension. In synthetic detergents, which is the best known use of surface-active agents, reduction of interfacial tension provides cleansing action.

Surface coating.   Paint, lacquer varnish, or other chemical composition used for protecting and/or decorating surfaces. *See also* Protective coating.

Suspect carcinogen.   A material believed to be capable of causing cancer, based on limited scientific evidence.

Sweating.   (1) Visible perspiration. (2) The process of uniting metal parts by heating solder so that it runs between the parts.

Swing grinder.   A large, power-driven grinding wheel mounted on a counterbalanced, swivel-supported arm guided by two handles.

Symptom.   Evidence from a patient indicating illness; the subjective feelings of the patient.

Syncope.   Fainting spell.

Syndrome.   A collection, constellation, or concurrence of signs and symptoms, usually of disease.

Synergism.   Cooperative action of substances whose total effect is greater than the sum of their separate effects.

Synergistic.   Pertaining to an action of two or more substances, organs, or organisms to achieve an effect greater than the additive effects of the separate elements.

Synonym.   Another name by which a chemical may be known.

Synthesis.   The reaction or series of reactions by which a complex compound is obtained from simpler compounds or elements.

Synthetic.   From the Greek *synthetikos,* "that which is put together." "Manmade 'synthetic' should not be thought of as a substitute for the natural," according to the *Encyclopedia of the Chemical Process Industries*, which adds, "synthetic chemicals are frequently more pure and uniform than those obtained naturally." A classic example is synthetic indigo.

Synthetic detergents.   Chemically tailored cleaning agents soluble in water or other solvents. Originally developed as soap substitutes; because they do not form insoluble precipitates, they are especially valuable in hard water. They may be composed of surface-active agents alone, but generally are combinations of surface-active agents and other substances, such as complex phosphates, to enhance detergency.

Synthetic rubber.   Artificial polymer with rubberlike properties. Types have varying composition and properties. Major types are designated as S-types, butyl, neoprene (chloroprene polymers), and N-type. Several synthetics duplicate the chemical structure of natural rubber.

System curve. The actual or design curve of pressure vs. flow rate for a specific ventilation system. System design determines the optimum operating point on the system curve.

Systemic. Spread throughout the body; affecting all body systems and organs, not localized in one spot or area.

Systemic toxicity. Adverse effects caused by a substance that affects the body in a general rather than local manner.

Tachy- (prefix). Indicates fast or speedy, as in tachycardia, abnormally rapid heartbeat.

Tailings. In mining or metal recovery processes, the gangue rock residue after all or most of the metal has been extracted.

Talc. A hydrous magnesium silicate used in ceramics, cosmetics, paint, and pharmaceuticals, and as a filler in soap, putty, and plaster.

Tall oil. Derived from the Swedish *tallolja*; a material first investigated in Sweden, not synonymous with U.S. pine oil. Natural mixture of rosin acids, fatty acids, sterols, high-molecular-weight alcohols, and other materials, derived primarily from waste liquors of sulfate wood pulp manufacture. Dark brown, viscous, oily liquid often called liquid rosin.

Tar. A loose term embracing wood, coal, or petroleum exudations. In general represents complex mixture of chemicals of top fractional distillation systems.

Tar crude. Organic raw material derived from distillation of coal tar and used for chemicals.

Tare. A deduction of weight, made in allowance for the weight of a container or medium. The initial weight of a filter, for example.

Tare weight. The initial weight of a filter; the deduction of the weight of a container or holder.

Target. The material into which the laser beam is fired or at which electrons are fired in an X-ray tube.

Temper. To relieve the internal stresses in metal or glass and to increase ductility by heating the material to a point below its critical temperature and cooling slowly. *See also* Anneal.

Temperature. The condition of a body that determines the transfer of heat to or from other bodies. Specifically, it is a manifestation of the average translational kinetic energy of the molecules of a substance caused by heat agitation. *See* Celsius; Kelvin scale.

Temperature, dry-bulb. The temperature of a gas or mixture of gases indicated by an accurate thermometer after correction for radiation.

Temperature, effective. An arbitrary index that combines into a single value the effect of temperature, humidity, and air movement on the sensation of warmth or cold felt by the human body. The numerical value is the temperature of still, saturated air that would induce an identical sensation.

Temperature, mean radiant (MRT). The temperature of a uniform black enclosure in which a solid body or occupant would exchange the same amount of radiant heat as in the existing nonuniform environment.

Temperature, wet-bulb. The temperature indicated by a wet-bulb psychrometer. Thermodynamic wet-bulb temperature is the temperature at which liquid or solid water, by evaporating into air, can bring the air to saturation adiabatically at the same temperature.

Tempering. The process of heating or cooling makeup air to the proper temperature.

Temporary threshold shift (TTS). The hearing loss suffered as the result of noise exposure, all or part of which is recovered during an arbitrary period of time when one is removed from the noise. It accounts for the necessity of checking hearing acuity at least 16 hours after a noise exposure.

Temporary total disability. An injury or illness that does not result in death or permanent disability, but that renders the injured person unable to perform regular duties or activities on one or more calendar days after the day of injury. (This is a definition established by U.S. OSHA.)

Tendon. Fibrous component of a muscle. It often attaches to bone at the area of application of tensile force. When its cross section is small, stresses in the tendon are high, particularly because the total force of many muscle fibers is applied at the single terminal tendon. *See also* Tenosynovitis.

Tennis elbow. Sometimes called lateral epicondylitis, an inflammatory reaction of tissues in the lateral elbow region.

Tenosynovitis. Inflammation of the connective tissue sheath of a tendon.

Teratogen. An agent or substance that may cause physical defects in the developing embryo or fetus when a pregnant female is exposed to the substance.

Terminal velocity. The terminal rate of fall of a particle through a fluid as induced by gravity or other external force; the rate at which frictional drag balances the accelerating force (or the external force).

Tetanus. A disease of sudden onset caused by the toxin of the bacterium called *Clostridium tetani*. It is characterized by muscle spasms. Also called lockjaw.

Therm. A quantity of heat equivalent to 100,000 Btu.

Thermal burns. Result of the application of too much heat to the skin. First-degree burns show redness of the unbroken skin; second-degree burns, skin blisters and some breaking of the skin; third-degree burns, skin blisters and destruction of the skin and underlying tissues, which can include charring and blackening.

Thermal pollution. Discharge of heat into bodies of water to the point that the increased warmth activates all sewage, depletes the oxygen the water must have to cleanse itself, and eventually destroys some of the fish and other organisms in the water. Eventually, thermal pollution causes the water to smell and taste bad.

Thermonuclear reaction. A fusion reaction, that is, a reaction in which two light nuclei combine to form a heavier atom, releasing a large amount of energy. This is believed to be the sun's source of energy. It is called thermonuclear because it occurs only at a very high temperature.

Thermoplastic. Capable of being repeatedly softened by heat.

Thermoplastic plastics. Plastics that can repeatedly melt or that soften with heat and harden on cooling. Examples: vinyls, acrylics, and polyethylene.

Thermosetting. Capable of undergoing a chemical change from a soft to a hardened substance when heated.

Thermosetting plastics. Plastics that are heat-set in their final processing to a permanently hard state. Examples: phenolics, ureas, and melamines.

Thermostable. Resistant to changes by heat.

Thinner. A liquid used to increase the fluidity of paints, varnishes, and shellac.

Threshold. The level where the first effects occur; also, the point at which a person begins to notice that a tone is audible.

Threshold limit value (TLV). Chemical or physical exposure limit established by the ACGIH.

Thromb- (prefix). Pertaining to a blood clot.

Throw distance. The distance contaminants, primarily particles, are dispersed by the initial emitting velocity before being slowed by air fiction to the settling velocity.

Tight building syndrome. Commonly related to problems associated with buildings designed and operated at minimum outdoor air supply, or poor distribution.

Timbre. The quality given to a sound by its overtones; the tone distinctive of a singing voice or a musical instrument. Pronounced "TAMbra" or "TIM-ber."

Time-weighted average Concentration (TWA). Refers to concentrations of airborne toxic materials weighted for a certain time duration, usually 8 hours.

Tinning. Any work with tin such as tin roofing; in particular, in soldering, the primary coating with solder of the two surfaces to be united.

Tinnitus. A perception of sound arising in the head. Most often perceived as a ringing or hissing sound in the ears. Can be the result of high-frequency hearing loss.

Tissue. A large group of similar cells bound together to form a structural component. An organ is composed of several kinds of tissue, and in this respect it differs from a tissue as a machine differs from its parts.

TLV. Threshold limit value. A time-weighted average concentration under which most people can work consistently for 8 hours a day, day after day, with no harmful effects. A table of these values and accompanying precautions is published annually by the American Conference of Governmental Industrial Hygienist (ACGIH).

TLV-C. The ceiling limit; the concentration that should not be exceeded even for an instant.

TLV-STEL. The short term exposure limit, or maximum concentration for a continuous 15-min exposure period (maximum of four such periods per day, with at least 60 min between exposure periods, provided that the daily TLV-TWA is not exceeded).

TLV-TWA. The allowable time-weighted average concentration for a normal 8-hour workday, 40-hour work week.

Tolerance.   (1) The ability of the living organism to resist the usually anticipated stress. (2) The limits of permissible inaccuracy in the fabrication of an article above and below its design specification.

Tolerance dose.   *See* Maximum permissible concentration; MPL.

Toluene, C₆H₅CH₃.   Hydrocarbon derived mainly from petroleum but also from coal. Source of TNT, lacquers, saccharin, and many other chemicals.

Tone deafness.   The inability to discriminate between fundamental tones close together in pitch.

Topography.   Configuration of a surface, including its relief and the position of its natural and anthropogenic features.

Tort.   A civil wrong, other than breach of contract, for which the law allows compensation by payment of damages.

Total pressure.   *See* Pressure, total.

Total suspended.   The mass of particles suspended in a unit volume of air particulate matter (typically $1 m^3$) when collected by a high-volume sampler.

Toxemia.   Poisoning by the way of the bloodstream.

Toxicant.   A poison or poisonous agent.

Toxicity.   The property of a material or agent that is capable of causing a harmful effect; the relative amount of material required to induce a specific harmful biological outcome.

Toxic substance.   Any substance that can cause acute or chronic injury to the human body, or that is suspected to be able to cause disease or injury under some conditions.

Toxin.   A poisonous substance derived from an organism.

Tracer.   A radioisotope mixed with a stable material. The radioisotope enables scientists to trace the material as it undergoes chemical and physical changes. Tracers are used widely in science, industry, and agriculture today. When radioactive phosphorus, for example, is mixed with a chemical fertilizer, the radioactive substance can be traced through the plant as it grows.

Trachea.   The windpipe, or tube that conducts air to and from the lungs. It extends between the larynx above and the point where it divides into two bronchi below.

Trade name.   The commercial name or trademark by which a chemical is known. One chemical may have a variety of trade names depending on the manufacturers or distributors involved.

Trade secret.   Any confidential formula, pattern, process, device, information, or compilation of information (including chemical name or other unique chemical modifier) that is used in an employer's business and that gives the employer an opportunity to obtain an advantage over competitors who do not know or use it.

Transducer.   Any device or element that converts an input signal into an output signal of a different form; examples include the microphone, phonograph pickup, loudspeaker, barometer, photoelectric cell, automobile horn, doorbell, and underwater sound transducer.

Transmission loss. The ratio, expressed in decibels, of the sound energy incident on a structure to the sound energy that is transmitted. The term is applied both to building structures (wall, floors, etc.) and to air passages (muffler, ducts, etc.).

Transmutation. Any nuclear process that involves a change in energy or identity of the nucleus.

Transport (conveying) velocity. Minimum air velocity required to move suspended particulates in the airstream.

Trauma. An injury or wound brought about by an outside force.

Tremor. Involuntary shaking, trembling, or quivering.

Triceps. The large muscle at the back of the upper arm that extends the forearm when contracted.

Tridymite. Vitreous, colorless form of free silica formed when quartz is heated to 1598°F (870°C).

Trigger finger. Also known as snapping finger, a condition of partial obstruction in flexion or extension of a finger. Once past the point of obstruction, movement is eased. Caused by constriction of the tendon sheath.

Tripoli. Rottenstone. A porous, siliceous rock resulting from the decomposition of chert or siliceous limestone. Used as a base in soap and scouring powders, in metal polishing, as a filtering agent, and in wood and paint fillers. A cryptocrystalline form of free silica.

Tritium. Often called hydrogen-3, extraheavy hydrogen whose nucleus contains two neutrons and one proton. It is three times as heavy as ordinary hydrogen and is radioactive.

TSCA. Toxic Substance Control Act; U.S. environmental legislation, administrated by the EPA, for regulating the manufacture, handling, and use of materials classified as "toxic substances."

Tuberculosis. A contagious disease caused by infection with the bacterium *Mycobacterium tuberculosis*. It usually affects the lung, but bone, lymph glands, and other tissues may be affected.

Tularemia. A bacterial infection of wild rodents, such as rabbits. It may be generalized or localized in the eyes, skin, lymph nodes, or respiratory tract. It can be transmitted to humans.

Tumbling. An industrial process, as in founding, in which small castings are cleaned by friction in a revolving drum (tumbling barrel), which may contain sand, sawdust, stone, etc.

Turbid. Cloudy.

Turbidity. Cloudiness; disturbances of solids (sediment) in a solution, so that it is not clear.

Turbinates. A series of scroll-like bones in the nasal cavity that serve to increase the amount of tissue surface exposed in the nose, permitting incoming air to be moistened and warmed prior to reaching the lungs.

Turbulence loss. The pressure or energy lost from a ventilation system through air turbulence.

Turbulent flow. Airflow characterized by transverse velocity components as well as velocity in the primary direction of flow in a duct; mixing velocities.

Turning vanes. Curved pieces added to elbows or fan inlet boxes to direct air and so reduce turbulence losses.

TWA. Time-weighted average.

Tympanic cavity. Another name for the chamber of the middle ear.

UCL. *See* Upper confidence limit.

UEL. *See* Upper explosive limit.

Ulcer. The destruction of an area of skin or mucous membrane.

Ulceration. The formation or development of an ulcer.

Ulna. One of the two bones of the forearm. It forms the hinge joint at the elbow and does not rotate about its longitudinal axis. It terminates at the wrist on the same side as the little finger. Task design should not impose thrust loads through the ulna.

Ulnar deviation. A position of the hand in which the angle on the little finger side of the hand with the corresponding side of the forearm is decreased. Ulnar deviation is a poor working position for the hand and may cause nerve and tendon damage.

Ultrasonics. The technology of sound at frequencies above the audible range.

Ultraviolet. Wavelengths of the electromagnetic spectrum that are shorter than those of visible light and longer than X rays, 10^{-5} to 10^{-6} cm wavelength.

Unstable. Refers to all radioactive elements, because they emit particles and decay to form other elements.

Unstable (reactive) liquid. A liquid that in the pure state or as commercially produced or transported, vigorously polymerizes, decomposes, condenses, or becomes self-reactive under conditions of shocks, pressure, or temperature.

Upper confidence limit (UCL). In sampling analysis, a statistical procedure used to estimate the likelihood that a particular value is above the obtained value.

Upper explosive limit (UEL). The highest concentration (expressed as the percentage of vapor or gas in the air by volume) of a substance that will burn or explode when an ignition source is present.

Uranium. A heavy metal. The two principal isotopes of natural uranium are uranium-235 and uranium-238. Uranium-235 has the only readily fissionable nucleus that occurs in appreciable quantities in nature, hence its importance as a nuclear fuel. Only 1 part of 140 of natural uranium is uranium-235. Highly toxic and a radiation hazard that requires special consideration.

Urethr-, urethro- (prefix). Relating to the urethra, the canal leading from the bladder for discharge of urine.

URT. Upper respiratory tract.

Urticaria. Hives.

USC. United States Code. The official compilation of federal statutes. New editions are issued approximately every 6 years. Cumulative supplements are issued annually.

USDA. U.S. Department of Agriculture.

Vaccine. A suspension of disease-producing microorganisms modified by killing or attenuation so that it does not cause disease and can facilitate the formation of antibodies upon inoculation into humans or animals.

Valence. A number indicating the capacity of an atom and certain groups of atoms to hold others in combination. The term also is used in more complex senses.

Valve (air oxygen). A device that controls the direction of air or fluid flow or the rate and pressure at which air or fluid is delivered, or both.

Vapor. The monomolecular "gaseous" form of a material, which is normally a liquid at room temperature (e.g., when water evaporates, it forms water vapor).

Vapor pressure. Pressure (measured in pounds per square inch absolute, psia) exerted by a vapor. If a vapor is kept in confinement over its liquid so that the vapor can accumulate above the liquid (with the temperature held constant), the vapor pressure approaches a fixed limit called the maximum (or saturated) vapor pressure, dependent only on the temperature and the liquid.

Vapors. The gaseous form of substances that are normally in the solid or liquid state (at room temperature and pressure). The vapor can be changed back to the solid or liquid state either by increasing or decreasing the temperature alone. Vapors also diffuse. Evaporation is the process by which a liquid is changed to the vapor state and mixed with the surrounding air. Solvents with low boiling points volatilize readily.

Vapor volume. The number of cubic feet of pure solvent vapor formed by the evaporation of 1 gallon of liquid at 75°F (24°C).

van der Waall's forces. The forces of attraction and retention between the molecules of a gas or vapor coming in contact with a solid sorbent. (Adsorption increases directly with the number or carbon atoms and inversely with the number of hydrogen atoms in a molecule.).

Variable air volume (VAV). Refers to HVAC systems in which the air volume is varied by dampers or fan speed controls to maintain the temperature; primarily used for energy conservation.

Vasoconstriction. Decrease in the cross-sectional area of blood vessels. This may result from contraction of a muscle layer within the walls of the vessels or may be the result of mechanical pressure. Reduction in blood flow results.

Vat dyes. Water-insoluble, complex coal tar dyes that can be chemically reduced in a heated solution to a soluble form that can impregnate fibers. Subsequent oxidation then produces insoluble color dyestuffs that are remarkably fast to washing, light, and chemicals.

Vector. (1) Term applied to an insect or any living carrier that transports a pathogenic microorganism from the sick to the well, inoculating the latter;

the organism may or may not pass through any developmental cycle. (2) Any quantity (for example, velocity, mechanical force, electromotive force) having magnitude and direction that can be represented by a straight line of appropriate length and direction.

Velocity.   A vector that specifies the time rate of change of displacement with respect to a reference.

Velocity, capture.   The air velocity required to draw contaminants into a hood.

Velocity, face.   The inward air velocity in the plane of openings into an enclosure.

Velocity pressure.   The kinetic pressure in the direction of flow necessary to cause a fluid at rest to flow at a given velocity. When added to static pressure, it gives total pressure.

Velometer.   A device for measuring air velocity.

Vena contracta.   The reduction in the diameter of a flowing airstream at hood entries and other locations.

Veni-, veno- (prefix).   Relating to the veins.

Ventilation.   One of the principal methods to control health hazards; may be defined as causing fresh air to circulate to replace foul air simultaneously removed.

Ventilation, dilution.   Airflow designed to dilute contaminants to acceptable levels. Also called general ventilation.

Venilation, local exhaust.   Ventilation near the point of generation of a contaminant.

Ventilation, mechanical.   Air movement caused by a fan or other air-moving device.

Ventilation, natural.   Air movement caused by wind, temperature difference, or other nonmechanical factors.

Vermiculite.   An expanded mica (hydrated magnesium–aluminum–iron silicate) used in lightweight aggregates, insulation, fertilizer, and soil conditioners; as a filler in rubber and paints; and as a catalyst carrier.

Vertigo.   Dizziness; more exactly, the sensation that the environment is revolving around one.

Vesicant.   Anything that produces blisters on the skin.

Vesicle.   A small blister on the skin.

Vestibular.   Relating to the cavity at the entrance to the semicircular canals of the inner ears.

Viable.   Any living organism.

Vibration.   An oscillation motion about an equilibrium position produced by a disturbing force.

Vinyl.   A general term applied to a class of resins such as polyvinyl chloride, acetate, butyryl, etc.

Virulence.   The capacity of a microorganism to produce disease.

Virulent.   Extremely poisonous or venomous; capable of overcoming bodily defense mechanisms.

Viruses.   A group of pathogens consisting mostly of nucleic acids and lacking cellular structure.

Viscera. Internal organs of the abdomen.

Viscose. Term applied to viscous liquid composed of cellulose xanthate.

Viscose rayon. The type of rayon produced from the reaction of carbon disulfide with cellulose and the hardening of the resulting viscous fluid by passing it through dilute sulfuric acid; this final operation causes the evolution of hydrogen sulfide gas.

Viscosity. The property of a fluid that resists internal flow by releasing counteracting forces.

Viscosity, absolute. A measure of the tendency of a fluid to resist flow without regard to density. The product of the kinematic viscosity of a fluid times its density, expressed in dyne-seconds per centimeter or poises (or pascal-seconds).

Viscosity, kinematic. The relative tendency of a fluid to resist flow. The value of the kinematic viscosity is equal to the absolute viscosity of the fluid divided by the fluid density and is expressed in units of stoke (or square meters per second).

Visible light. Wavelengths of the electromagnetic spectrum visible to the human eye (e.g., 10^{-4} to 10^{-5} cm in length).

Visible radiation. The wavelengths of the electromagnetic spectrum between 10^{-4} cm and 10^{-5} cm.

Vision, photopic. Vision attributed to cone function characterized by the ability to discriminate colors and small details; daylight vision.

Vision, scotopic. Vision attributed to rod function characterized by the lack of ability to discriminate colors and small details and effective primarily in the detection of movement and low luminous intensities; might vision.

Visual Acuity. Ability of the eye to perceive sharply the shapes of objects in the direct line of vision.

Volatile. Percent volatile by volume; the percentage of a liquid or solid (by volume) that will evaporate at an ambient temperature or 70°F (unless some other temperature is stated). Examples: butane, gasoline, and paint thinner (mineral spirits) are 100% volatile; their individual evaporation rates vary, but over a period of time each will evaporate completely.

Volatility. The tendency or ability of a liquid to vaporize. Such liquids as alcohol and gasoline, because of their well-known tendency to evaporate rapidly, are called volatile liquids.

Volt. A practical unit of electric force or difference in potential between two points in an electrical field.

Volume flow rate. The quantity (measured in units of volume) of a fluid flowing per unit of time, such as cubic feet per minute, gallons per hour, or cubic meters per second.

Volume, specific. The volume occupied by 1 pound of a substance under specified conditions of temperature and pressure.

Volumetric analysis. A statement of the various components of a substance (usually applied to gases only), expressed in percentages by volume.

Vortex tube. A tube that uses compressed air to provide personal cooling.

Vulcanization. The process of combining rubber (natural, synthetic, or latex) with sulfur and accelerators in the presence of zinc oxide under heat and usually pressure to change the material permanently from a thermoplastic to a thermosetting composition, or from a plastic to an elastic condition. Strength, elasticity, and abrasion resistance also are improved.

Vulcanizer. A machine in which raw rubber that has been mixed with chemicals is cured by heat and pressure to render it less plastic and more durable.

WAN. Wide-area network of linked computers or LANs, whose elements are usually geographically distant.

Warranty. A promise that a proposition of fact is true and, if not true, a consideration is available.

Wart. A characteristic growth on the skin, appearing most often on the fingers; generally as a result of a virus infection. Synonym: verruca.

Water column. A unit used in measuring pressure. *See also* Inches of water column.

Water curtain or waterfall booth. A term with many different meanings in industry; but in spray painting, a stream of water running down a wall into which the excess paint spray is drawn or blown by fans, and which carries the paint downward to a collecting point.

Waterproofing gents. Usually formulations of three distinct material: a coating material, a solvent, and a plasticizer. Among the materials used in water-proofing are cellulose esters and ether, polyvinyl chloride resins or ace-tates, and variations of vinyl chloride–vinylidine chloride polymers.

Watt (W). A unit of power equal to 1 joule per second. *See* erg.

Watts/cm². A unit of power density used in measuring the amount of power per area of absorbing surface, or per area of a CW laser beam.

Wavelength. The distance in the line of advance of a wave from any point to a like point on the next wave. It is usually measured in angstroms, microns, or nanometers.

WBGT index. Wet-bulb globe temperature index. An empirical index of the effects of heat on humans (e.g., heat stress).

Weight. The force with which a body is attracted toward the earth. Although the weight of the body varies with its location, the weights of various standards of mass are often used as units of force. *See* Force.

Weighting network (sound). Electrical networks (A, B, C) associated with sound-level meters. The C network provides a flat response over the frequency range 20 to 10,000 Hz; the B and A networks selectively discriminate against low (less than 1 kHz) frequencies.

Weld. A localized coalescence of metals or nonmetals produced either by heating the materials to suitable temperatures, with or without the application of pressure, or by the application of pressure alone, with or without the use of filler material.

Welding. Fusing the several types of welding are electric arc-welding, spot welding, and inert or shielded gas welding using helium or argon. The hazards involved in welding stem from the fumes from the weld metal such as lead or cadmium metal, the gases created by the process, and the fumes or gases arising from the flux.

Welding rod.   A rod or heavy wire that is melted and fused to metals in arc-welding.

Wellness.   The practice of a healthy lifestyle.

Wet-bulb globe temperature (WBGT).   A temperature obtained with a WBGT thermometer, which takes into account the effects of humidity and infrared radiation.

Wet-bulb globe temperature index.   An index of the heat stress in humans when work is performed in a hot environment.

Wet-bulb temperature (WB).   The temperature of air as influenced by humidity; the WB-emperature is obtained by covering a dry-bulb thermometer bulb with a wet wick and moving air past the wick; the WB-temperature is almost always less than the dry-bulb temperature because of the cooling effects of the evaporation of water from the wick.

Wet-bulb thermometer.   A thermometer having the bulb covered with a cloth saturated with water.

Wheatstone bridge.   A type of electrical circuit used in one type of combustible gas monitor. Combustion in small quantities of the ambient gas and changes in electrical resistivity by this circuitry are detected.

White damp.   In mining, carbon monoxide.

White noise.   A noise whose spectrum density (of spectrum level) is substantially independent of frequency over a specified range.

Wideband.   Applied to broad band of transmitted waves, with neither of the critical or cutoff frequencies of the filter being zero or infinite.

Wideband noise.   Noise associated with a broad band of frequencies.

Wind load.   The pressure exerted on a building or structure from moving air.

Work.   When a force acts against resistance to produce motion in a body, the force is said to do work. Work is measured by the product of the force acting and the distance moved against the resistance. The units of measurement are the erg (a joule is 1×10^7 ergs) and the foot-pound.

Work hardening.   The property of metal to become harder and more brittle on being worked (bent repeatedly or drawn).

Work hours.   The total numbers of hours worked by all employees.

Work injuries.   Injuries (including occupational illnesses) that arise out of or in the course of gainful employment regardless of where the accident occurs. Excluded are work injuries to private household workers and injuries occurring in connection with farm chores, which are classified as home injuries.

Work strain.   The natural physiological response of the body to the application of work stress. The focus of the reaction may be remote from the point of application of work stress. Work strain is not necessarily traumatic but may appear as trauma when excessive, either directly or cumulatively, and must be considered by the industrial engineer in equipment and task design.

Work stress.   Biomechanically, any external force acting on the body during the performance of a task. It always produces work strain. Application of work stress to the human body is the inevitable consequence of performance of any task, and is therefore synonymous with stressful work conditions only when excessive. Work stress analysis is an integral part of task design.

Worker's Compensation.   An insurance system under law, financed by employers, that provides payment to injured and diseased employees or relatives for job-related injuries and illnesses.

Working level (WL).   Any combination of radon daughters in 1 liter of air that result in the ultimate emission of 1.3×10^5 MeV of alpha energy.

Xanth- (prefix).   Yellow.

Xero-(prefix).   Dryness, as in xerostomia, dryness of the mouth.

Xeroderma.   Dry skin; may be rough as well as dry.

X rays.   Highly penetrating radiation similar to gamma rays. Unlike gamma rays, X rays do not come from the nucleus of the atom but from the surrounding electrons. They are produced by electron bombardment. When these rays pass through an object, they leave shadow picture of the denser portions.

X-ray diffraction.   Because all crystals act as three-dimensional gratings for X rays, the pattern of diffracted rays is a characteristic for each crystalline material. The method is of particular value in determining the presence or absence of a crystalline silica in a industrial dust.

X-ray tube.   Any electron tube designed for the conversion of electrical energy into X-ray energy.

Z.   Symbol for atomic number. The atomic number of an element is the same as the number of protons found in one of its nuclei. All isotopes of a given element have the same Z number.

Zero energy state.   Zero mechanical energy.

Zero mechanical energy (ZME).   An old term, now called energy isolation, that indicates a piece of equipment without any source of power that could harm someone.

Zinc protoporphyrin (ZPP).   Hematopoietic enzyme used as a measure of exposure.

Zoonoses.   Diseases biologically adapted to and normally found in lower animals, but that under some conditions also infect humans.

Zygote.   Cell produced by the joining of two gametes (sex or germ cells).

Index

A

ABIH, *see* American Board Industrial Hygiene
Acceptable entry conditions, 81
Accident investigations, 223–227
Accident reports, 227
ACGIH, *see* American Conference on
        Government Industrial Hygienists
Acids, 149
Aerosols
    chemical profile, 139–143
    flammable, 119
Affected employee, 103, 210
AFFT, *see* Aqueous film forming foam
After-flame, 70
Air-purifying respirator, 21, 27, 31
Alcoholic beverages, 116, 117
American Board of Industrial Hygiene (ABIH),
      229, 230
American Conference on Government Industrial
      Hygienists (ACGIH), 6
American Industrial Health Association, 5
American Society of Safety Engineers (ASSE), 5
Approved, definition, 70–71
Aqueous film forming foam (AFFT), 70
Article, definition, 118
ASSE, *see* American Society of Safety Engineers
Atmosphere, accident investigations, 225
Atmosphere-supplying respirator, 22, 27
Attendant, definition, 81
Authorized employee, 103, *see also* Employees
Authorized entrant, 82
Automatic fire detection device, 71

B

Bases, 149
BCSP, *see* Board of Certified Safety Professional
Benchmarking, 4–7
Binders, 161, *see also* Paint
Birdcage, 246, 247
Bitrex solution aerosol protocol, 49–51
Blanking, 82
Blinding, *see* Blanking
Board of Certified Safety Professional (BCSP),
      231

Breathing, difficulties, 44, *see also* Respirators
Breathing air quality, 37

C

Canisters, 22, 27
Capable of being locked out, 103, 210
Carbon dioxide, 71
Carcinogenicity, 158
Cardiopulmonary resuscitation (CPR)
    confined space entry, 99
    insulation, 160
    paint, 163
    pesticides, 168
    refractories, 174
    toluene/xylene, 181
    waste, 185
Categorizations
    accident investigations, 223, 224
    chemicals, 138–139
Caution labels, 165
Certification, 71, 95, 108
Certified Industrial Hygienist (CIH), 12, 229–231
Certified Safety Professional (CSP), 231–232
CFCs, *see* Chlorofluorocarbons
Check valves, 146
Checklists
    bad habits and short cuts, 189
    computer workstations, 190
    confined space entry, 191
    emergency preparedness, 192
    fire protection, 193
    hazardous atmospheres, 194
    hazardous communications, 195
    ladders, 196
    lockout/tagout policy and safety standard
        operating manual, 221
    machine guarding, 197
    new hire employees, 198
    noise and hearing protection, 199
    personal protective equipment, 200
    powered platforms, 201
    scaffolds, 202
    twenty questions, 188
    walking surfaces, 203
Chemical explosions, 3

Milton Keynes UK
Ingram Content Group UK Ltd.
UKHW021817071024
449327UK00021B/1334